Vibrations of Shells and Plates

MECHANICAL ENGINEERING

A Series of Textbooks and Reference Books

1. Spring Designer's Handbook, *by Harold Carlson*
2. Computer-Aided Graphics and Design, *by Daniel L. Ryan*
3. Lubrication Fundamentals, *by J. George Wills*
4. Solar Engineering for Domestic Buildings, *by William A. Himmelman*
5. Applied Engineering Mechanics: Statics and Dynamics, *by G. Boothroyd and C. Poli*
6. Centrifugal Pump Clinic, *by Igor J. Karassik*
7. Computer-Aided Kinetics for Machine Design, *by Daniel L. Ryan*
8. Plastics Products Design Handbook, Part A: Materials and Components, *edited by Edward Miller*
9. Turbomachinery: Basic Theory and Applications, *by Earl Logan, Jr.*
10. Vibrations of Shells and Plates, *by Werner Soedel*

OTHER VOLUMES IN PREPARATION

Vibrations of Shells and Plates

WERNER SOEDEL
Professor of Mechanical Engineering
School of Mechanical Engineering
Purdue University
West Lafayette, Indiana

MARCEL DEKKER, INC. New York and Basel

Library of Congress Cataloging in Publication Data

Soedel, Werner [date]
Vibrations of shells and plates.

(Mechanical engineering ; 10)
Includes index.
1. Shells (Engineering)--Vibration. 2. Plates (Engineering)--Vibration. I. Title. II. Series.
TA660.S5S65 624.1'776 81-9892
ISBN 0-8247-1193-9 AACR2

MARCEL DEKKER, INC.
270 Madison Avenue, New York, New York 10016

Current printing (last digit):
10 9 8 7 6 5 4 3 2 1

PRINTED IN THE UNITED STATES OF AMERICA

To my Family

PREFACE

This book attempts to give engineering graduate students and practicing engineers an introduction to the vibration behavior of shells and plates. It is also hoped that it will prove to be a useful reference to the vibration specialist. It fills a need in the present literature on this subject, since it is the current practice to either discuss shell vibrations in a few chapters at the end of texts on shell statics that may be well written but are too limited in the selection of material, or to ignore shells entirely in favor of plates and membranes, as in some of the better known vibration books. There are a few excellent monographs on very specialized topics, for instance, on natural frequencies and modes of cylindrical and conical shells. But a unified presentation of shell and plate vibration, both free and forced, and with complicating effects as they are encountered in engineering practice, is still missing. This collection attempts to fill the gap.

The state of the art of modern engineering demands that engineers have a good knowledge of the vibration behavior of structures beyond the usual beam and rod vibration examples. Vibrating shell and plate structures are not only encountered by the civil, aeronautical, and astronautical engineer, but also by the mechanical, nuclear, chemical, and industrial engineer. Parts or devices such as engine liners, compressor shells, tanks, heat exchangers, life support ducts, boilers, automotive tires, vehicle bodies, valve reed plates, and saw disks, are all composed of structural elements that cannot be approximated as vibrating beams. Shells especially exhibit certain effects that are not present in beams or even plates and cannot be interpreted by engineers who are only familiar with beam-type vibration theory. Therefore, this book stresses the understanding of basic phenomena in shell and plate vibrations and it is hoped that the material covered will be useful in explaining experimental measurements or the results of the ever-increasing number of finite element programs. While it is the goal of every engineering manager that these programs will eventually be used as black boxes, with input provided and output obtained by relatively untrained technicians, reality shows that the interpretation of results of these programs requires a good background in finite element theory and, in the case of shell and plate vibrations, in vibration theory of greater depth and breadth than usually provided in standard texts.

It is hoped that the book will be of interest to both the stress analyst whose task it is to prevent failure and to the acoustician whose task it is to control noise. The treatment is fairly complete as far as the needs of the stress analysts go. For acousticians, this collection stresses those applications in which boundary conditions cannot be ignored.

The note collection begins with a historical discussion of vibration analysis and culminates in the development of Love's equations of shells. These equations are derived in Chapter 2 in curvilinear coordinates. Curvilinear coordinates are used throughout as much as possible, because of the loss of generality that occurs when specific geometries are singled out. For instance, the effect of the second curvature cannot be recovered from a specialized treatment of cylindrical shells. Chapter 3 shows the derivation by reduction of the equations of some standard shell geometries that have a tendency to occur in standard engineering practice, like the circular cylindrical shell, the spherical shell, the conical shell, and so on. In Chapter 4 the equations of motion of plates, arches, rings, beams, and rods are obtained. Beams and rings are sometimes used as supplementary examples in order to tie in the knowledge of beams that the reader may have with the approaches and results of shell and plate analysis.

Chapter 5 discusses natural frequencies and modes. It starts with the transversely vibrating beam, followed by the ring and plate. Finally, the exact solution of the simply supported circular cylindrical shell is derived. The examples are chosen in such a way that the essential behavior of these structures is unfolded with the help of each previous example; the intent is not to exhaust the number of possible analytical solutions. For instance, in order to explain why there are three natural frequencies for any mode number combination of the cylindrical shell, the previously given case of the vibrating ring is used to illustrate modes in which either transverse or circumferential motions dominate.

In the same chapter, the important property of orthogonality of natural modes is derived and discussed. It is pointed out that when two or more different modes occur at the same natural frequency, a superposition mode may be created that may not be orthogonal, yet is measured by the experimenter as the governing mode shape. Ways of dealing with this phenomenon are also pointed out.

For some important applications, it is possible to simplify the equations of motion. Rayleigh's simplification, in which either the bending stiffness or the membrane stiffness is ignored, is presented. However, the main thrust of Chapter 6 is the derivation and use of the Donnell-Mushtari-Vlasov equations.

While the emphasis of Chapter 5 was on so-called exact solutions (series solutions are considered exact solutions), Chapter 7 presents some of the more common approximate techniques to obtain solutions for geometrical shapes and boundary condition combinations that do not lend themselves to exact analytical treatment. First, the variational techniques known as the Rayleigh-Ritz technique and Galerkin's method and variational method are presented. Next, the purely mathematical technique of finite differences is outlined, with examples. The finite element method follows. Southwell's and Dunkerley's principles conclude the chapter.

The forced behavior of shells and plates is presented in Chapters 8, 9, and 10. In Chapter 8, the modal analysis approach is used to arrive at the general solution for distributed dynamic loads in transverse and two orthogonal in-plane directions. The Dirac delta function is then used to obtain the solutions for point and line loads. Chapter 9 discusses the dynamic Green's function approach and applies it to traveling load problems. An interesting resonance condition that occurs when a load travels along the great circles of closed shells of revolution is shown. Chapter 10 extends the types of possible loading to the technically significant set of dynamic moment loading, and illustrates it by investigating the action of a rotating point moment as it may occur when rotating unbalanced machinery is acting on a shell structure.

The influence of large initial stress fields on the response of shells and plates is discussed in Chapter 11. First, Love's equations are extended to take this effect into account. It is then demonstrated that the equations of motion of pure membranes and strings are a subset of these extended equations. The effect of initial stress fields on the natural frequencies of structures is then illustrated by examples.

In the original derivation of Love's equations, transverse shear strains, and therefore shear deflections, were neglected. This becomes less and less permissible as the average distance between node lines associated with the highest frequency of interest approaches the thickness of the structure. In Chapter 12, the shear deformations are included in the shell equations. It is shown that these equations reduce in the case of a rectangular plate and the case of a uniform beam to equations that are well known in the vibration literature. Sample cases are solved to illustrate the effect shear deformation has on natural frequencies.

Rarely are practical engineering structures simple geometric shapes. In most cases the shapes are so complicated that finite element or difference methods have to be used for accurate numerical results. However, there is a category of cases in which the engineering structures can be interpreted as being assembled of two or more classic shapes or parts. In Chapter 13, the method of receptance is presented and used to obtain, for instance, very general design rules for stiffening panels by ring- or beam-type stiffeners. It is also shown that the receptance method gives elegant and easily interpretable results for cases in which springs or masses are added to the basic structure.

The formulation and use of equivalent viscous damping was advocated in the forced vibration chapters. For steady-state harmonic response problems a complex modulus is often used. In Chapter 14, this type of structural damping, also called hystereses damping, is presented and tied in with the viscous damping formulation.

Because of the increasing importance of composite material structures, the equations of motion of laminated shells are presented and discussed in Chapter 15, along with some simple examples.

This book evolved over a period of almost ten years from lecture notes on the vibration of shells and plates. To present the subject in a unified fashion made it necessary to do some original work in areas where the available literature did not provide complete information. Some of it was done with the help of graduate students attending my lectures, for instance R. G. Jacquot, U. R. Kristiansen, J. D. Wilken, M. Dhar, U. Bolleter, and D. P. Powder. Especially talented in detecting errors were M. G. Prasad, F. D. Wilken, M. Dhar, S. Azimi, and D. P. Egolf. Realizing that I have probably forgotten some significant contributions, I would like to single out in addition O. B. Dale, J. A. Adams, D. D. Reynolds, M. Moaveni, R. Shashaani, R. Singh, J. R. Friley, J. DeEskinazi, F. Laville, E. T. Buehlmann, N. Kaemmer, C. Hunckler, and J. Thompson, and extend my appreciation to all my former students.

I would also like to thank my colleagues on the Purdue University faculty for their direct or indirect advice.

If this book is used for an advanced course in structural vibrations of about forty-five lectures, it is recommended that Chapters 2 through 8 be treated in depth. If there is time remaining, highlights of the other chapters can be presented. Recommended prerequisites are a first course in mechanical vibrations and knowledge of boundary value problem mathematics.

Werner Soedel

CONTENTS

Vibrations of Shells and Plates

1

HISTORICAL DEVELOPMENT OF VIBRATION ANALYSIS OF CONTINUOUS STRUCTURAL ELEMENTS

Vibration analysis has its beginnings with Galileo Galilei (1564-1642), who solved by geometrical means the dependence of the natural frequency of a simple pendulum on the pendulum length [1.1]. He proceeded to make experimental observations on the vibration behavior of strings and plates, but could not offer any analytical treatment. He was partially anticipated in his observations of strings by his contemporary Marin Mersenne (1588-1648), a French priest. Mersenne recognized that the frequency of vibration is inversely proportional to the length of the string and directly proportional to the square root of the cross-sectional area [1.2]. This line of approach found its culmination in Joseph Sauveur (1653-1716), who coined the terminology "nodes" for zero displacement points on a string vibrating at its natural frequency and also actually calculated an approximate value for the fundamental frequency as a function of the measured sag at its center, similar to the way the natural frequency of a single degree of freedom spring-mass system can be calculated from its static deflection [1.3].

The foundation for a more precise treatment of the vibration of continuous systems was laid by Robert Hook (1635-1703) when he established the basic law of elasticity, by Newton (1642-1727) when he established that force was equal to mass times acceleration, and by Leibnitz (1646-1716) when he established differential calculus. An approach similar to differential calculus called *fluxions* was developed by Newton independently at the same time. In 1713 the English mathematician Brook Taylor (1685-1731) actually used the fluxion approach, together with Newton's second law applied to an element of the continuous string, to calculate the true value of the first natural frequency of a string [1.4]. The approach was based on an assumed first mode shape. This is where work in vibration analysis stagnated in England since the fluxion method and especially its notation proved to be too clumsy to allow anything but the attack of simple problems. Because of the controversy between followers of Newton and Leibnitz as to the origin of differential calculus, patriotic Englishmen refused to use anything but fluxions and left the fruitful use of the Leibnitz notation and approach to the investigators on the continent. There the mathematics of differential calculus prospered and paved the way for Jean Le Rond d'Alembert (1717-1783), who derived in 1747 the partial differential equation which today is referred to as the wave equation and who found the wave travel solution [1.5]. He was ably assisted in this by Daniel Bernoulli (1700-1782) and Leonhard Euler (1707-1783), both German speaking Swiss and friends, but did not give them due credit. It is still a controversial subject to decide who did actually what, especially since the participants were not too bashful to insult each other and claim credit right and left. However, it seems fairly clear that the principle of superposition of modes was first noted in 1747 by Daniel Bernoulli [1.6] and proven in 1753 by Euler [1.7]. These two must, therefore, be credited as being the fathers of the modal expansion technique or of eigenvalue expansion in general. The technique did not find immediate general acceptance. In 1822 Joseph Fourier (1768-1830) used it to solve certain problems in the theory of heat [1.8]. The resulting Fourier series can be viewed as a special case of the use of orthogonal functions and might as well carry the name of Bernoulli. However, it is almost a rule in the history of science that people that are credited with an achievement do not completely deserve it. Progress moves in small steps and it is often the individual who publishes at the right developmental step and at the right time who gets the public acclaim.

The longitudinal vibration of rods was investigated experimentally by Chladni [1.9] and Biot [1.10]. However, not until 1824 do we find the published analytical equation and solutions, done by Navier. This is interesting since the analogous problem of the longitudinal vibration of air columns was already done in 1727 by Euler [1.11].

The equation for the transverse vibration of flexible thin beams was derived in 1735 by Daniel Bernoulli [1.12] and the first solutions for simply supported ends, clamped ends, and free ends where found by Euler [1.13] and published in 1744.

The first torsional vibration solution, but not in a continuous sense, was given in 1784 by Coulomb [1.14]. But not until 1827 do we find an attempt to derive the continuous torsional equation [1.15]. This was done by Cauchy (1789-1857) in an approximate fashion. Poisson (1781-1840) is generally credited for having derived the one dimensional torsional wave equation in 1827 [1.16]. The credit for deriving the complete torsional wave equation and giving some rigorous results belongs to Saint-Venant (1797-1886), who published this in 1849 [1.17].

In membrane vibrations, Euler in 1766 published equations for a rectangular membrane that were incorrect for the general case but will reduce to the correct equation for the uniform tension case [1.18]. It is interesting to note that the first membrane vibration case investigated analytically was not the circular membrane, even while the latter, in form of the drum head, would have been the more obvious shape. The reason is that Euler was able to picture the rectangular membrane as a superposition of a number of crossing strings. In 1828 Poisson read a paper to the French Academy of Science on the special case of uniform tension and showed the circular membrane equation and solved it for the special case of axisymmetric vibration [1.16]. One year later, Pagani furnished the non axisymmetric solution [1.19]. In 1852 Lamé (1795-1870) published his lectures which gave a summary of the work on rectangular and circular membranes and contain an investigation of triangular membranes [1.20].

Work on plate vibration analysis went on in parallel. Influenced by Euler's success in deriving the membrane equation by considering the superposition of strings, James Bernoulli, a nephew of Daniel Bernoulli, attempted to derive the plate equation by considering the superposition of beams. The resulting equation was wrong. James, in his 1788 presentation to the St. Petersburg Academy [1.21], acknowledged that he was stimulated in his attempt by the German experimentalist Chladni [1.9], who demonstrated the beautiful node lines of vibrating plates at the courts of Europe. A presentation by Chladni before emperor Napoleon who was a trained military engineer and very interested in technology and science caused the latter to transfer money to the French Academy of Sciences for a prize to that person that would best explain the vibration behavior of plates. The prize was won, after several attempts, by a woman, Sophie Germain (1776-1831), in 1815. She gave an almost correct form of the plate equation [1.22]. The bending stiffness and the density constants were not defined. Neither were the boundary conditions stated correctly. These errors are the reason that her name is not associated today with the equation, despite the brilliance of her approach. Contributing to this was Todhunter [1.23], who compiled a fine history of the theory of elasticity which was published posthumously in 1886, in which he is unreasonably critical of her work, demanding a standard of perfection that he does not apply to the works of the Bernoullis, Euler, Lagrange, and others, where he is quite willing to accept partial results. Also, Lagrange (1736-1813) entered into the act by correcting errors that Germaine made when first competing for the prize in 1811. Thus, indeed we do find the equation first stated in its modern form by Lagrange in 1811 in response to Germaine's submittal of her first competition paper [1.24].

What is even more interesting is that Sophie Germaine published in 1821 a very simplified equation for the vibration of a cylindrical shell [1.22]. Unfortunately again it contained mistakes. This equation can be reduced to the current rectangular plate equation, but when it is reduced to the ring equation a sign mistake is passed on. But for the sign difference in one of its terms, the ring equation is identical to one given by Euler [1.25].

The correct bending stiffness was first identified in 1829 by Poisson [1.16]. Consistent boundary conditions were not developed until 1850 by Kirchoff (1824-1887) who also gave the correct solution for a circular plate example [1.26].

The problem of shell vibrations was first attacked by Sophie Germaine before 1821, as already pointed out. She assumed that the in plane deflection of the neutral surface of a cylindrical shell was negligible. Her result contained errors. In 1874, Aron derived a set of five equations [1.27], which he shows to reduce to the plate equation when curvatures are set to zero. The equations are complicated because of his reluctance to employ simplifications. They are in curvilinear coordinate form and apply in general. The simplifications that are logical extensions of the beam and plate equations both for transverse and in plane motion were introduced by Love (1863-1940) in 1888 [1.29]. In between Aron and Love, Lord Rayleigh (1842-1919) proposed in 1882 various simplifications that viewed the shell neutral surface as either extensional or inextensional [1.28]. His simplified solutions are special cases of Love's general theory. Love's equations brought the basic development of the theory of vibration of continuous structures that have a thickness that is much less than any length or surface dimensions to a satisfying end. Subsequent development was concerned with higher order or complicating effects and will be discussed in this book when appropriate.

REFERENCES

1.1 Galileo Galilei, *Dialogue Concerning Two New Sciences (1638)*, Northwestern University Press, Evanston, Ill., 1939.

1.2 M. Mersenne, *Harmonicorum Libri XII*, Paris, 1635.

1.3 J. Sauveur, *Systeme General des Intervalles des Sons*, L'Academie Royale des Sciences, Paris, 1701.

1.4 B. Taylor, *De Motu Nervi Tensi*, Phil. Trans. Roy. Soc. (London), vol. 28, 1713.

1.5 T. Le Rond d'Alembert, *Recherches sur le courbe que forme une Corde Tendue Mise en Vibration*, Royal Academy, Berlin, 1747.

1.6 D. Bernoulli, *Réflexions et Eclaircissements sur les Nouvelles Vibrations des Cordes*, Royal Academy, Berlin, 1755 (presented 1747).

1.7 L. Euler, *Remarques sur les Memoires Precedents de M. Bernoulli*, Royal Academy, Berlin, 1753.

1.8 J. B. J. Fourier, *La Theorie Analytique de la Chaleur*, Didot, Paris, 1822.

1.9 E. F. F. Chladni, *Entdeckungen über die Theorie des Klanges*, Weidmann und Reich, Leipzig, 1787.

1.10 J. B. Biot, *Traite de Physique Experimentale et Mathematique*, Deterville, Paris, 1816.

1.11 L. Euler, *Dissertatio Physica de Sono*, Basel, 1727.

1.12 D. Bernoulli, *Letters to Euler*, Basel, 1735.

1.13 L. Euler, "Methodus inveniendi lineas curvas maximi minimive proprietate gaudentes," Berlin, 1744.

1.14 C. A. Coulomb, "Recherches theoriques et experimentales sur la force torsion et sur l'elasticite des fils de metal," Memoirs of the Paris Academy, Paris, 1784.

1.15 A. Cauchy, *Exercices de mathematiques*, Paris, 1827.

1.16 S. D. Poisson, "Sur l'equilibre et le mouvement des corps elastiques," Memoirs of the Paris Academy, Paris, 1829.

1.17 B. de Saint-Venant, "Memoir sur les vibrations tournantes des verges elastiques", *Comp. Rend., vol. 28*, 1849.

1.18 L. Euler, "De motu vibratorio tympanorum," Novi Commentarii, St. Petersburg Academy, St. Petersburg, 1766.

1.19 M. Pagani, "Note sur le mouvement vibratoire d'une membrane elastique de forme circulaire," Royal Academy of Science at Brussels, Brussels, 1829.

1.20 G. Lamé, "Lecons sur la theorie mathematique de l'elasticite des corps solides," Bachelier, Paris, 1852.

1.21 James Bernoulli, "Essai theoretique sur les vibrations des plaques elastiques rectangulaires et libres," Nova Acta Academiae Scientiarum Petropolitanae, St. Petersburg, 1789.

1.22 S. Germaine, "Recherches sur la theorie des surfaces elastiques," Paris, 1821.

1.23 L. Todhunter, *A History of the Theory of Elasticity*, vol. I, Cambridge University Press, New York, 1886.

1.24 J. L. Lagrange, "Note communiquee aux Commissaires pour le prix de la surface elastique," Paris, 1811.

1.25 L. Euler, "Tentamen de sono campanarum," Novi Commentarii, St. Petersburg Academy, St. Petersburg, 1766.

1.26 G. R. Kirchhoff, "Über das Gleichgewicht und die Bewegung einer elastischen Scheibe," *J. Mathematik (Crelle)*, vol. 40, 1850.

1.27 H. Aron, "Das Gleichgewicht und die Bewegung einer unendlich dünnen, beliebig gekrümmten elastischen Schale," J. Mathematik (Crelle), vol. 78, 1874.

1.28 J. W. S. Lord Rayleigh, "On the infinitesimal bending of surfaces of revolution," London Math. Soc. Proc., vol. 13, 1882.

1.29 A. E. H. Love, "On the small free vibrations and deformations of thin elastic shells," *Phil. Trans. Royal Soc.* (London), *vol. 179A*, 1888.

2

DEEP SHELL EQUATIONS

The term *deep* is used to distinguish the set of equations used in this chapter from the so-called shallow shell equations that will be discussed later. The equations are based on the assumptions that the shells are thin with respect to their radii of curvature and that deflections are reasonably small. On these two basic assumptions secondary assumptions rest. They are discussed as the development warrants it.

The basic theoretical approach is due to Love [2.1], who published the equations in their essential form toward the end of the nineteenth century. He essentially extended Rayleigh's work on shell vibrations, who divided shells into two classes: one where the middle surface does not stretch and bending effects are the only important ones, and one where only the stretching of the middle surface is important and the bending stiffness can be neglected [2.2]. Love allowed the coexistance of these two classes. He used the principle of virtual work to derive his equations, following Kirchhoff [2.3], who had used it when deriving the plate equation. The derivation given here uses Hamilton's principle, following Reissner's [2.4, 2.5] derivation.

2.1 SHELL COORDINATES AND INFINITESIMAL DISTANCES IN SHELL LAYERS

We assume that thin, isotropic, and homogeneous shells of constant thickness have neutral surfaces like beams in transverse deflection have neutral fibers. That this is true will become evident later. Stresses in such a neutral surface can be of the membrane type but cannot be bending stresses. Locations on the neutral surface, placed into a three-dimensional cartesian coordinate system, can also be defined by two-dimensional curvilinear surface coordinates α_1, α_2. The location of a point P on the neutral surface (Fig.2.1.1) in cartesian coordinates is related to the location of the point in surface coordinates by

$$x_1 = f_1(\alpha_1,\alpha_2) \quad x_2 = f_2(\alpha_1,\alpha_2) \quad x_3 = f_3(\alpha_1,\alpha_2) \tag{2.1.1}$$

The location of P on the neutral surface can also be expressed by a vector, e.g.,

$$\bar{r}(\alpha_1,\alpha_2) = f_1(\alpha_1,\alpha_2)\bar{e}_1 + f_2(\alpha_1,\alpha_2)\bar{e}_2 + f_3(\alpha_1,\alpha_2)\bar{e}_3 \tag{2.1.2}$$

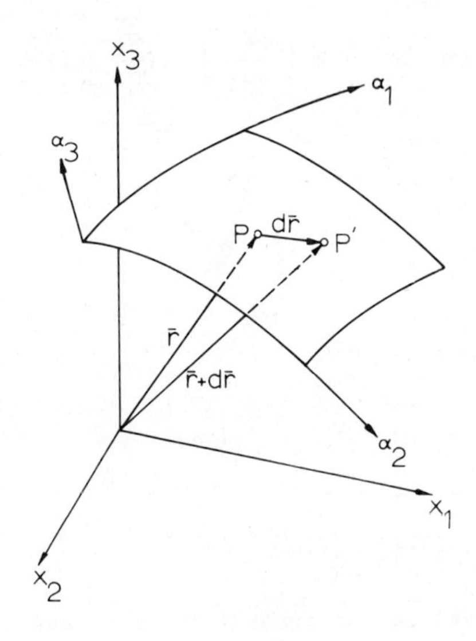

FIG. 2.1.1

Now let us define the infinitesimal distance between points P and P' on the neutral surface. The differential change $d\bar{r}$ of the vector $\bar{r}$ as we move from P to P' is

$$d\bar{r} = \frac{\partial \bar{r}}{\partial \alpha_1} d\alpha_1 + \frac{\partial \bar{r}}{\partial \alpha_2} d\alpha_2 \tag{2.1.3}$$

The magnitude ds of $d\bar{r}$ is obtained by

$$(ds)^2 = d\bar{r} \cdot d\bar{r} \tag{2.1.4}$$

or

$$(ds)^2 = \frac{\partial \bar{r}}{\partial \alpha_1} \cdot \frac{\partial \bar{r}}{\partial \alpha_1} (d\alpha_1)^2 + \frac{\partial \bar{r}}{\partial \alpha_2} \cdot \frac{\partial \bar{r}}{\partial \alpha_2} (d\alpha_2)^2 + 2\frac{\partial \bar{r}}{\partial \alpha_1} \cdot \frac{\partial \bar{r}}{\partial \alpha_2} d\alpha_1\, d\alpha_2 \tag{2.1.5}$$

In the following we will limit ourselves to orthogonal curvilinear coordinates which coincide with the lines of principal curvature of the neutral surface. The third term in Eq. (2.1.5) becomes, therefore,

$$2\frac{\partial \bar{r}}{\partial \alpha_1} \cdot \frac{\partial \bar{r}}{\partial \alpha_2} d\alpha_1\, d\alpha_2 = 2 \left|\frac{\partial \bar{r}}{\partial \alpha_1}\right| \left|\frac{\partial \bar{r}}{\partial \alpha_2}\right| \cos\frac{\pi}{2}\, d\alpha_1\, d\alpha_2 = 0 \tag{2.1.6}$$

Defining

$$\frac{\partial \bar{r}}{\partial \alpha_1} \cdot \frac{\partial \bar{r}}{\partial \alpha_1} = \left|\frac{\partial \bar{r}}{\partial \alpha_1}\right|^2 = A_1^2$$

$$\frac{\partial \bar{r}}{\partial \alpha_2} \cdot \frac{\partial \bar{r}}{\partial \alpha_2} = \left|\frac{\partial \bar{r}}{\partial \alpha_2}\right|^2 = A_2^2 \tag{2.1.7}$$

Eq. (2.1.5) becomes

$$(ds)^2 = A_1^2 (d\alpha_1)^2 + A_2^2 (d\alpha_2)^2 \tag{2.1.8}$$

This equation is called the *fundamental form* and A_1 and A_2 are the *fundamental form parameters* or *Lamé parameters*.

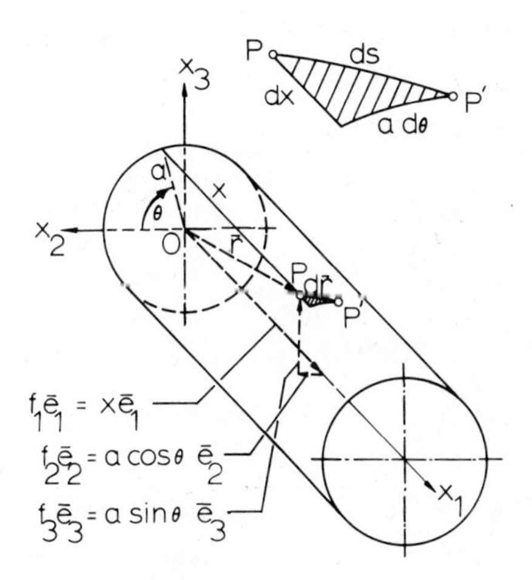

FIG. 2.1.2

As example, let us look at a circular cylindrical shell as shown in Fig. 2.1.2. The lines of principal curvature (for each shell surface point there exists a maximum and a minimum radius of curvature, the directions of which are at an angle of $\pi/2$) are in this case parallel to the axis of revolution, where the radius of curvature $R_x = \infty$ or the curvature $1/R_x = 0$, and along circles, where the radius of curvature $R_\theta = a$ or the curvature $1/R_\theta = 1/a$. We will proceed now to obtain the fundamental form parameters from definition (2.1.7). The curvilinear coordinates are

$$\alpha_1 = x \quad \alpha_2 = \theta \tag{2.1.9}$$

and Eq. (2.1.2) becomes

$$\bar{r} = x\bar{e}_1 + a\cos\theta\ \bar{e}_2 + a\sin\theta\ \bar{e}_3 \tag{2.1.10}$$

Thus

$$\frac{\partial\bar{r}}{\partial\alpha_1} = \frac{\partial\bar{r}}{\partial x} = \bar{e}_1 \tag{2.1.11}$$

or

$$\left|\frac{\partial\bar{r}}{\partial\alpha_1}\right| = A_1 = 1 \tag{2.1.12}$$

and $$\frac{\partial \bar{r}}{\partial \alpha_2} = \frac{\partial \bar{r}}{\partial \theta} = -a \sin \theta \, \bar{e}_2 + a \cos \theta \, \bar{e}_3 \qquad (2.1.13)$$

or $$\left| \frac{\partial \bar{r}}{\partial \theta} \right| = A_2 = a\sqrt{\sin^2\theta + \cos^2\theta} = a \qquad (2.1.14)$$

The fundamental form is therefore

$$(ds)^2 = (dx)^2 + a^2(d\theta)^2 \qquad (2.1.15)$$

Recognizing that the fundamental form can be interpreted as defining the hypotenuse ds of a right triangle whose sides are infinitesimal distances along the surface coordinates of the shell, we may obtain A_1 and A_2 in a simpler fashion by expressing ds directly using inspection:

$$(ds)^2 = (dx)^2 + a^2(d\theta)^2$$

end of example

By comparison with Eq. (2.1.8), we obtain $A_1 = 1$ and $A_2 = a$.

For the general case, let us now define the infinitesimal distance between a point P_1 which is normal to P and a point P_1' which is normal to P' (see Fig. 2.1.3). P_1 is located at a distance α_3 from the neutral surface (α_3 is defined to be along a normal straight line to the neutral surface). P_1' is located at a distance $\alpha_3 + d\alpha_3$ from the neutral surface. We may therefore express the location of P_1 as

$$\bar{R}(\alpha_1,\alpha_2,\alpha_3) = \bar{r}(\alpha_1,\alpha_2) + \alpha_3\bar{n}(\alpha_1,\alpha_2) \qquad (2.1.16)$$

where $\bar{n}$ is a unit vector normal to the neutral surface. The differential change $d\bar{R}$, as we move from P_1 to P_1', is

$$d\bar{R} = d\bar{r} + \alpha_3 \, d\bar{n} + \bar{n} \, d\alpha_3 \qquad (2.1.17)$$

where

$$d\bar{n} = \frac{\partial \bar{n}}{\partial \alpha_1} d\alpha_1 + \frac{\partial \bar{n}}{\partial \alpha_2} d\alpha_2 \qquad (2.1.18)$$

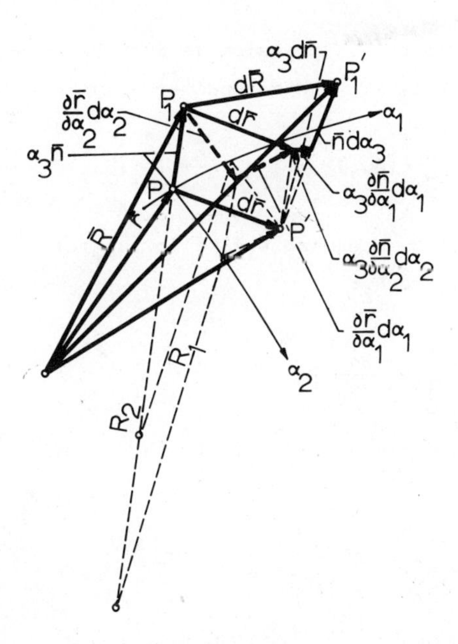

FIG. 2.1.3

The magnitude ds of $d\bar{R}$ is obtained by

$$(ds)^2 = d\bar{R} \cdot d\bar{R} \tag{2.1.19}$$

or

$$(ds)^2 = d\bar{r} \cdot d\bar{r} + \alpha_3^2 \, d\bar{n} \cdot d\bar{n} + \bar{n} \cdot \bar{n}(d\alpha_3)^2 + 2\alpha_3 \, d\bar{r} \cdot d\bar{n} + 2d\alpha_3 \, d\bar{r} \cdot \bar{n} + 2\alpha_3 d\alpha_3 \, d\bar{n} \cdot \bar{n} = d\bar{r} \cdot d\bar{r} + \alpha_3^2 \, d\bar{n} \cdot d\bar{n} + (d\alpha_3)^2 + 2\alpha_3 \, d\bar{r} \cdot d\bar{n} \tag{2.1.20}$$

We have already seen that

$$d\bar{r} \cdot d\bar{r} = A_1^2 (d\alpha_1)^2 + A_2^2 (d\alpha_2)^2 \tag{2.1.21}$$

Next

$$\alpha_3^2 \, d\bar{n} \cdot d\bar{n} = \alpha_3^2 \left[\frac{\partial \bar{n}}{\partial \alpha_1} \cdot \frac{\partial \bar{n}}{\partial \alpha_1} (d\alpha_1)^2 + \frac{\partial \bar{n}}{\partial \alpha_2} \cdot \frac{\partial \bar{n}}{\partial \alpha_2} (d\alpha_2)^2 + 2 \frac{\partial \bar{n}}{\partial \alpha_1} \cdot \frac{\partial \bar{n}}{\partial \alpha_2} \, d\alpha_1 \, d\alpha_2 \right] \tag{2.1.22}$$

The third term of this expression is zero because of orthogonality (see also Fig. 2.1.3). The second term may be written

$$\alpha_3^2 \frac{\partial\bar{n}}{\partial\alpha_2} \cdot \frac{\partial\bar{n}}{\partial\alpha_2} (d\alpha_2)^2 = \left| \alpha_3 \frac{\partial\bar{n}}{\partial\alpha_2} \right|^2 (d\alpha_2)^2 \tag{2.1.23}$$

From Fig. 2.1.3 we recognize the following relationship to the radius of curvature R_2:

$$\frac{\left| \dfrac{\partial\bar{r}}{\partial\alpha_2} \right|}{R_2} = \frac{\left| \alpha_3 \dfrac{\partial\bar{n}}{\partial\alpha_2} \right|}{\alpha_3} \tag{2.1.24}$$

Since

$$\left| \frac{\partial\bar{r}}{\partial\alpha_2} \right| = A_2 \tag{2.1.25}$$

we get

$$\left| \alpha_3 \frac{\partial\bar{n}}{\partial\alpha_2} \right| = \frac{\alpha_3 A_2}{R_2} \tag{2.1.26}$$

and therefore

$$\alpha_3^2 \frac{\partial\bar{n}}{\partial\alpha_2} \cdot \frac{\partial\bar{n}}{\partial\alpha_2} (d\alpha_2)^2 = \alpha_3^2 \frac{A_2^{\,2}}{R_2^{\,2}} (d\alpha_2)^2 \tag{2.1.27}$$

Similarly, the first term becomes

$$\frac{\partial\bar{n}}{\partial\alpha_1} \cdot \frac{\partial\bar{n}}{\partial\alpha_1} (d\alpha_1)^2 = \alpha_3^2 \frac{A_1^{\,2}}{R_1^{\,2}} (d\alpha_1)^2 \tag{2.1.28}$$

and expression (2.1.22) becomes

$$\alpha^3 \, d\bar{n} \cdot d\bar{n} = \alpha_3^2 \left[\frac{A_1^{\,2}}{R_1^{\,2}} (d\alpha_1)^2 + \frac{A_2^{\,2}}{R_2^{\,2}} (d\alpha_2)^2 \right] \tag{2.1.29}$$

Finally, the last expression of Eq. (2.1.20) becomes

$$2\alpha_3\, d\bar{r} \cdot d\bar{n} = 2\alpha_3 \left[\frac{\partial \bar{r}}{\partial \alpha_1} \cdot \frac{\partial \bar{n}}{\partial \alpha_1} (d\alpha_1)^2 + \frac{\partial \bar{r}}{\partial \alpha_2} \cdot \frac{\partial \bar{n}}{\partial \alpha_2} (d\alpha_2)^2 \right.$$

$$\left. + \frac{\partial \bar{r}}{\partial \alpha_1} \cdot \frac{\partial \bar{n}}{\partial \alpha_2} d\alpha_1 d\alpha_2 + \frac{\partial \bar{r}}{\partial \alpha_2} \cdot \frac{\partial \bar{n}}{\partial \alpha_1} d\alpha_1 d\alpha_2 \right] \tag{2.1.30}$$

The last two terms are zero because of orthogonality. The first term may be written

$$\frac{\partial \bar{r}}{\partial \alpha_1} \cdot \frac{\partial \bar{n}}{\partial \alpha_1} (d\alpha_1)^2 = \left| \frac{\partial \bar{r}}{\partial \alpha_1} \right| \left| \frac{\partial \bar{n}}{\partial \alpha_1} \right| (d\alpha_1)^2 = \frac{A_1^2}{R_1} (d\alpha_1)^2 \tag{2.1.31}$$

Similarly

$$\frac{\partial \bar{r}}{\partial \alpha_2} \cdot \frac{\partial \bar{n}}{\partial \alpha_2} (d\alpha_2)^2 = \frac{A_2^2}{R_2} (d\alpha_2)^2 \tag{2.1.32}$$

Expression (2.1.30) becomes, therefore,

$$2\alpha_3\, d\bar{r} \cdot d\bar{n} = 2\alpha_3 \left[\frac{A_1^2}{R_1} (d\alpha_1)^2 + \frac{A_2^2}{R_2} (d\alpha_2)^2 \right] \tag{2.1.33}$$

Substituting expressions (2.1.33), (2.1.29), and (2.1.21) in Eq. (2.1.20) gives

$$(ds)^2 = A_1^2 \left(1 + \frac{\alpha_3}{R_1}\right)^2 (d\alpha_1)^2 + A_2^2 \left(1 + \frac{\alpha_3}{R_2}\right)^2 (d\alpha_2)^2 + (d\alpha_3)^2 \tag{2.1.34}$$

2.2 STRESS-STRAIN RELATIONSHIPS

Having chosen the mutually perpendicular lines of principal curvature as coordinates, plus the normal to the neutral surface as the third coordinate, we have three mutually perpendicular planes of strain and three shear strains. Assuming that Hooke's law applies, we have for a three-dimensional element

$$\varepsilon_{11} = \frac{1}{E}\,[\sigma_{11} - \mu(\sigma_{22} + \sigma_{33})] \tag{2.2.1}$$

$$\varepsilon_{22} = \frac{1}{E}\,[\sigma_{22} - \mu(\sigma_{11} + \sigma_{33})] \tag{2.2.2}$$

$$\varepsilon_{33} = \frac{1}{E}\,[\sigma_{33} - \mu(\sigma_{11} + \sigma_{22})] \tag{2.2.3}$$

$$\varepsilon_{12} = \frac{\sigma_{12}}{G} \tag{2.2.4}$$

$$\varepsilon_{13} = \frac{\sigma_{13}}{G} \tag{2.2.5}$$

$$\varepsilon_{23} = \frac{\sigma_{23}}{G} \tag{2.2.6}$$

where σ_{11}, σ_{22}, and σ_{33} are normal stresses and σ_{12}, σ_{13}, and σ_{23} are shear stresses as shown in Fig. 2.2.1. Note that

$$\sigma_{12} = \sigma_{21} \quad \sigma_{13} = \sigma_{31} \quad \sigma_{23} = \sigma_{32} \tag{2.2.7}$$

We will later assume that transverse shear deflections can be neglected. This implies that

$$\varepsilon_{13} = 0 \quad \varepsilon_{23} = 0 \tag{2.2.8}$$

However, we will not neglect the integrated effect of the transverse shear stresses σ_{13} and σ_{23}. This will be discussed later.

The normal stress σ_{33}, which acts in normal direction to the neutral surface, will be neglected:

$$\sigma_{33} = 0 \tag{2.2.9}$$

This is based on the argument that on an unloaded outer shell surface it is zero, or if a load acts on the shell, it is equivalent in magnitude to the external load on the shell, which is a relatively small value in most cases. Only in the close vicinity of a concentrated load do we reach magnitudes that would make the consideration of σ_{33} worthwhile. Our equation system reduces therefore to

$$\varepsilon_{11} = \frac{1}{E}\,(\sigma_{11} - \mu\sigma_{22}) \tag{2.2.10}$$

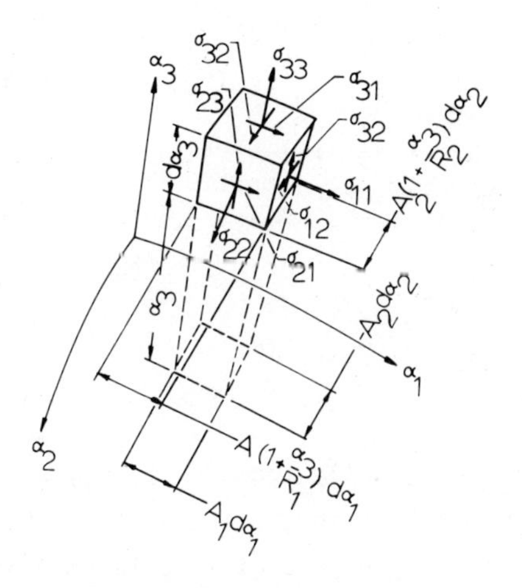

Figure 2.2.1

$$\varepsilon_{22} = \frac{1}{E}(\sigma_{22} - \mu\sigma_{11}) \tag{2.2.11}$$

$$\varepsilon_{12} = \frac{\sigma_{12}}{G} \tag{2.2.12}$$

and

$$\varepsilon_{33} = -\frac{\mu}{E}(\sigma_{11} + \sigma_{22}) \tag{2.2.13}$$

Only the first three relationships will be of importance in the following. Equation (2.2.13) can later be used to calculate the constriction of the shell thickness during vibration, which is of some interest to acousticians since it is an additional noise generating mechanism, along with transverse deflection.

2.3 STRAIN-DISPLACEMENT RELATIONSHIPS

We have seen that the infinitesimal distance between two points P_1 and P_1' of an undeflected shell is given by Eq. (2.1.34). Defining, for the purpose of a short notation,

$$A_1^2\left(1 + \frac{\alpha_3}{R_1}\right)^2 = g_{11}(\alpha_1, \alpha_2, \alpha_3) \tag{2.3.1}$$

$$A_2^2\left(1+\frac{\alpha_3}{R_2}\right)^2 = g_{22}(\alpha_1,\alpha_2,\alpha_3) \tag{2.3.2}$$

$$1 = g_{33}(\alpha_1,\alpha_2,\alpha_3) \tag{2.3.3}$$

we may write Eq. (2.1.34) as

$$(ds)^2 = \sum_{i=1}^{3} g_{ii}(\alpha_1,\alpha_2,\alpha_3)(d\alpha_i)^2 \tag{2.3.4}$$

If point P_1, originally located at $(\alpha_1,\alpha_2,\alpha_3)$, is deflected in the α_1 direction by U_1, in the α_2 direction by U_2, and in the α_3 (normal) direction by U_3, it will be located at $(\alpha_1 + \xi_1,\ \alpha_2 + \xi_2,\ \alpha_3 + \xi_3)$. Deflections U_i and coordinate changes ξ_i are related by

$$U_i = \sqrt{g_{ii}(\alpha_1,\alpha_2,\alpha_3)}\,\xi_i \tag{2.3.5}$$

Point P_1', originally at $(\alpha_1 + d\alpha_1,\ \alpha_2 + d\alpha_2,\ \alpha_3 + d\alpha_3)$, will be located at $(\alpha_1 + d\alpha_1 + \xi + d\xi_1,\ \alpha_2 + d\alpha_2 + \xi_2 + d\xi_2,\ \alpha_3 + d\alpha_3 + \xi_3 + d\xi_3)$ after deflection (Fig. 2.3.1). The distance ds' between P_1 and P_1' in the deflected state will therefore be

$$(ds')^2 = \sum_{i=1}^{3} g_{ii}(\alpha_1 + \xi_1,\ \alpha_2 + \xi_2,\ \alpha_3 + \xi_3)(d\alpha_i + d\xi_i)^2 \tag{2.3.6}$$

Since $g_{ii}(\alpha_1,\alpha_2,\alpha_3)$ varies in a continuous fashion as α_1, α_2, and α_3 change, we may utilize as an approximation the first few terms of a Taylor series expansion of $g_{ii}(\alpha_1 + \xi_1,\ \alpha_2 + \xi_2,\ \alpha_3 + \xi_3)$ about the point $(\alpha_1,\alpha_2,\alpha_3)$:

$$g_{ii}(\alpha_1 + \xi_1,\ \alpha_2 + \xi_2,\ \alpha_3 + \xi_3) = g_{ii}(\alpha_1,\alpha_2,\alpha_3) + \sum_{j=1}^{3} \frac{\partial g_{ii}(\alpha_1,\alpha_2,\alpha_3)}{\partial \alpha j}\,\xi j + \ldots \tag{2.3.7}$$

For the special case of an arch that deflects only in the

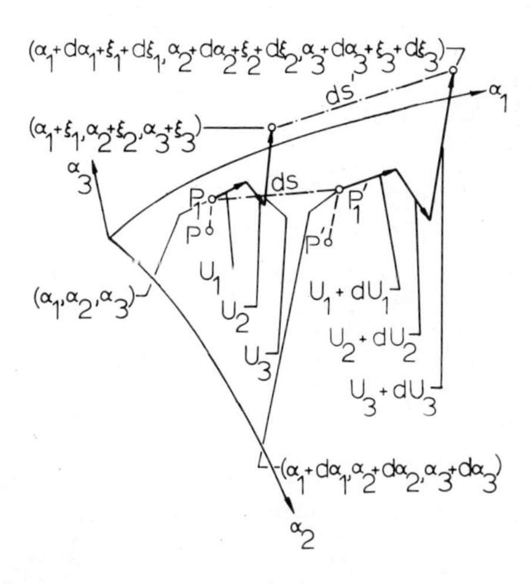

Figure 2.3.1

plane of its curvature, the Taylor series expansion is illustrated in Fig. 2.3.2. In this example, $g_{22}(\alpha_1,\alpha_2,\alpha_3) = 0$, $g_{33}(\alpha_1,\alpha_2,\alpha_3) = 0$, and $g_{11}(\alpha_1,\alpha_2,\alpha_3) = g_{11}(\alpha_1)$. Equation (2.3.7) becomes

$$g_{11}(\alpha_1 + \xi_1) = g_{11}(\alpha_1) + \frac{\partial g_{11}(\alpha_1)}{\partial \alpha_1} \xi_1 \tag{2.3.8}$$

Next, continuing with the general case, we may write

$$(d\alpha_i + d\xi_i)^2 = (d\alpha_i)^2 + 2d\alpha_i\, d\xi_i + (d\xi_i)^2 \tag{2.3.9}$$

In the order of approximation consistent with linear theory, $(d\xi_i)^2$ can be neglected. Thus

$$(d\alpha_i + d\xi_i)^2 = (d\alpha_i)^2 + 2d\alpha_i\, d\xi_i \tag{2.3.10}$$

The differential $d\xi_i$ is

$$d\xi_i = \sum_{j=1}^{3} \frac{\partial \xi_i}{\partial \alpha_j} d\alpha_j \tag{2.3.11}$$

Therefore

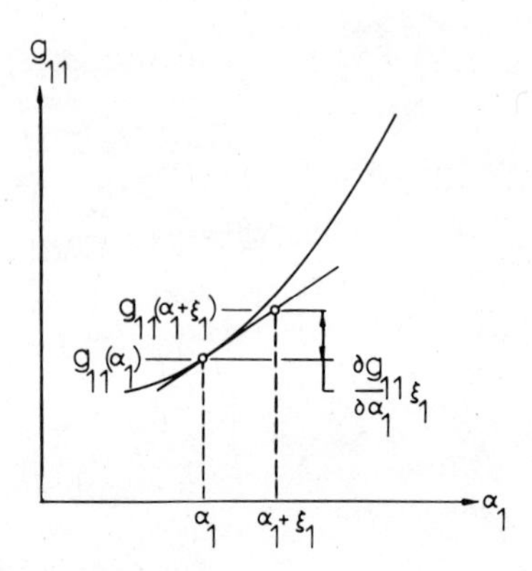

Figure 2.3.2

$$(d\alpha_i + d\xi_i)^2 = (d\alpha_i)^2 + 2d\alpha_i \sum_{j=1}^{3} \frac{\partial \xi_i}{\partial \alpha_j} d\alpha_j \tag{2.3.12}$$

Substituting Eqs. (2.3.12) and (2.3.7) in Eq. (2.3.6) gives

$$(ds')^2 = \sum_{i=1}^{3} \left[g_{ii}(\alpha_1,\alpha_2,\alpha_3) + \sum_{j=1}^{3} \frac{\partial g_{ii}(\alpha_1,\alpha_2,\alpha_3)}{\partial \alpha_j} \xi_j \right]$$

$$\left[(d\alpha_i)^2 + 2d\alpha_i \sum_{j=1}^{3} \frac{\partial \xi_i}{\partial \alpha_j} d\alpha_j \right] \tag{2.3.13}$$

Expanding this equation and writing

$$g_{ii}(\alpha_1,\alpha_2,\alpha_3) = g_{ii} \tag{2.3.14}$$

gives

$$(ds')^2 = \sum_{i=1}^{3} [(g_{ii} + \sum_{j=1}^{3} \frac{\partial g_{ii}}{\partial \alpha_j} \xi_j)(d\alpha_i)^2 + 2d\alpha_i\, g_{ii} \sum_{j=1}^{3} \frac{\partial \xi_i}{\partial \alpha_j} d\alpha_j$$

$$+ 2d\alpha_i \sum_{j=1}^{3} \frac{\partial g_{ii}}{\partial \alpha_j} \xi_j \sum_{j=1}^{3} \frac{\partial \xi_i}{\partial \alpha_j} d\alpha_j] \tag{2.3.15}$$

The last term is negligible except for cases where high initial stresses exist in the shell. We have therefore, replacing j by k in the first term,

$$(ds')^2 = \sum_{i=1}^{3} \left(g_{ii} + \sum_{k=1}^{3} \frac{\partial g_{ii}}{\partial \alpha_k} \xi_k\right)(d\alpha_i)^2$$
$$+ \sum_{i=1}^{3} \sum_{j=1}^{3} g_{ii} \frac{\partial \xi_i}{\partial \alpha_j} d\alpha_j d\alpha_i + \sum_{i=1}^{3} \sum_{j=1}^{3} g_{ii} \frac{\partial \xi_i}{\partial \alpha_j} d\alpha_j d\alpha_i \tag{2.3.16}$$

Utilizing the Kronecker delta notation

$$\delta_{ij} = \begin{cases} 1 & i = j \\ 0 & i \neq j \end{cases} \tag{2.3.17}$$

We may write the first term of Eq. (2.3.16) as

$$\sum_{i=1}^{3} \sum_{j=1}^{3} \left(g_{ii} + \sum_{k=1}^{3} \frac{\partial g_{ii}}{\partial \alpha_k} \xi_k\right) \delta_{ij} d\alpha_i d\alpha_j \tag{2.3.18}$$

The last two terms of Eq. (2.3.16) we may write in symmetric fashion by noting that

$$\sum_{i=1}^{3} \sum_{j=1}^{3} g_{ii} \frac{\partial \xi_i}{\partial \alpha_j} d\alpha_j d\alpha_i = \sum_{i=1}^{3} \sum_{j=1}^{3} g_{jj} \frac{\partial \xi_j}{\partial \alpha_i} d\alpha_i d\alpha_j \tag{2.3.19}$$

Thus

$$(ds')^2 = \sum_{i=1}^{3} \sum_{j=1}^{3} \left[\left(g_{ii} + \sum_{k=1}^{3} \frac{\partial g_{ii}}{\partial \alpha_k} \xi_k\right) \delta_{ij} + g_{ii} \frac{\partial \xi_i}{\partial \alpha_j} + g_{jj} \frac{\partial \xi_j}{\partial \alpha_i}\right] d\alpha_i d\alpha_j \tag{2.3.20}$$

Denoting

$$G_{ij} = \left(g_{ii} + \sum_{k=1}^{3} \frac{\partial g_{ii}}{\partial \alpha_k} \xi_k\right) \delta_{ij} + g_{ii} \frac{\partial \xi_i}{\partial \alpha_j} + g_{jj} \frac{\partial \xi_j}{\partial \alpha_i} \tag{2.3.21}$$

gives

$$(ds')^2 = \sum_{i=1}^{3} \sum_{j=1}^{3} G_{ij}\, d\alpha_i d\alpha_j \tag{2.3.22}$$

Note that

$$G_{ij} = G_{ji} \tag{2.3.23}$$

Equation (2.3.22) defines the distance between two points P and P' after deflection, where point P was originally located at $(\alpha_1, \alpha_2, \alpha_3)$ and point P' at $(\alpha_1 + d\alpha_1,\ \alpha_2 + d\alpha_2,\ \alpha_3 + d\alpha_3)$. For instance, if P' was originally located at $(\alpha_1 + d\alpha_1,\ \alpha_2,\ \alpha_3)$, that is, $d\alpha_2 = 0$ and $d\alpha_3 = 0$,

$$(ds)^2 = g_{11}(d\alpha_1)^2 = (ds)_{11}^2 \tag{2.3.24}$$

$$(ds')^2 = G_{11}(d\alpha_1)^2 = (ds')_{11}^2 \tag{2.3.25}$$

If point P' was originally located at $(\alpha_1,\ \alpha_2 + d\alpha_2,\ \alpha_3)$, that is, $d\alpha_1 = 0$ and $d\alpha_3 = 0$,

$$(ds)^2 = g_{22}(d\alpha_2)^2 = (ds)_{22}^2 \tag{2.3.26}$$

$$(ds')^2 = G_{22}(d\alpha_2)^2 = (ds')_{22}^2 \tag{2.3.27}$$

Now let us investigate the case shown in Fig. 2.3.3, where P

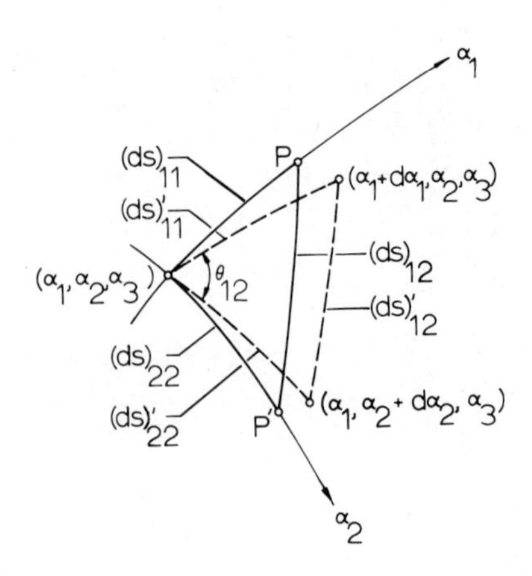

Figure 2.3.3

was originally located at $(\alpha_1 + d\alpha_1, \alpha_2, \alpha_3)$ and P' was originally located at $(\alpha_1, \alpha_2 + d\alpha_2, \alpha_3)$. This is equivalent to saying that P was originally located at $(\alpha_1, \alpha_2, \alpha_3)$ and P' at $(\alpha_1 - d\alpha_1, \alpha_2 + d\alpha_2, \alpha_3)$. We get then

$$(ds)^2 = g_{11}(d\alpha_1)^2 + g_{22}(d\alpha_2)^2 = (ds)^2_{12} \tag{2.3.28}$$

$$(ds')^2 = G_{11}(d\alpha_1)^2 + G_{22}(d\alpha_2)^2 - 2G_{12}d\alpha_1 d\alpha_2 = (ds')^2_{12} \tag{2.3.29}$$

In general,

$$(ds)^2_{ii} = g_{ii}(d\alpha_i)^2 \tag{2.3.30}$$

$$(ds')^2_{ii} = G_{ii}(d\alpha_i)^2 \tag{2.3.31}$$

and

$$(ds)^2_{ij} = g_{ii}(d\alpha_i)^2 + g_{jj}(d\alpha_j)^2 \tag{2.3.32}$$

$$(ds')^2_{ij} = G_{ii}(d\alpha_i)^2 + G_{jj}(d\alpha_j)^2 - 2G_{ij}d\alpha_i d\alpha_j \tag{2.3.33}$$

We are ready now to formulate strains. The normal strains are

$$\varepsilon_{ii} = \frac{(ds')_{ii} - (ds)_{ii}}{(ds)_{ii}} = \sqrt{\frac{G_{ii}}{g_{ii}}} - 1 = \sqrt{1 + \frac{G_{ii} - g_{ii}}{g_{ii}}} - 1 \tag{2.3.34}$$

Noting that since

$$\frac{G_{ii} - g_{ii}}{g_{ii}} << 1 \tag{2.3.35}$$

we have the expansion

$$\sqrt{1 + \frac{G_{ii} - g_{ii}}{g_{ii}}} = 1 + \frac{1}{2} \frac{G_{ii} - g_{ii}}{g_{ii}} - \cdots \tag{2.3.36}$$

Thus

$$\varepsilon_{ii} = \frac{1}{2} \frac{G_{ii} - g_{ii}}{g_{ii}} \tag{2.3.37}$$

Shear strains ε_{ij} $(i \neq j)$ are defined as the angular change of an infinitesimal element

$$\varepsilon_{ij} = \frac{\pi}{2} - \theta_{ij} \tag{2.3.38}$$

θ_{ij} for $i = 1$ and $j = 2$ is shown in Fig. 2.3.3. Utilizing the cosine law, we may compute this angle

$$(ds')^2_{ij} = (ds')^2_{ii} + (ds')^2_{jj} - 2(ds')_{ii}(ds')_{jj} \cos \theta_{ij} \tag{2.3.39}$$

Substituting Eqs. (2.3.31) and (2.3.33) and solving for $\cos \theta_{ij}$ gives

$$\cos \theta_{ij} = \frac{G_{ij}}{\sqrt{G_{ii} G_{jj}}} \tag{2.3.40}$$

Substituting Eq. (2.3.38) results in

$$\cos\left(\frac{\pi}{2} - \varepsilon_{ij}\right) = \sin \varepsilon_{ij} = \frac{G_{ij}}{\sqrt{G_{ii} G_{jj}}} \tag{2.3.41}$$

and since for reasonable shear strain magnitudes

$$\sin \varepsilon_{ij} \cong \varepsilon_{ij} \tag{2.3.42}$$

and
$$\frac{G_{ij}}{\sqrt{G_{ii} G_{jj}}} \cong \frac{G_{ij}}{\sqrt{g_{ii} g_{jj}}} \tag{2.3.43}$$

we may express the shear strain as

$$\varepsilon_{ij} = \frac{G_{ij}}{\sqrt{g_{ii} g_{jj}}} \tag{2.3.44}$$

Substituting Eqs. (2.3.21),(2.3.5), and (2.3.1) to (2.3.3) in Eq. (2.3.37) gives, for instance for $i = 1$,

$$\varepsilon_{11} = \frac{1}{2A_1^{\,2}}\left(1 + \frac{\alpha_3}{R_1}\right)^2 \left(\frac{\partial[A_1^{\,2}(1 + \frac{\alpha_3}{R_1})^2]}{\partial\alpha_1}\;\frac{U_1}{A_1(1 + \frac{\alpha_3}{R_1})}\right.$$

$$\left. + \frac{\partial[A_1^{\,2}(1 + \frac{\alpha_3}{R_1})^2]}{\partial\alpha_2}\;\frac{U_2}{A_2(1 + \frac{\alpha_3}{R_2})} + \frac{\partial[A_1^{\,2}(1 + \frac{\alpha_3}{R_1})^2]}{\partial\alpha_3}\;U_3\right)$$

$$+ \frac{\partial}{\partial\alpha_1}\left(\frac{U_1}{A_1(1 + \frac{\alpha_3}{R_1})}\right)$$

$$= \frac{1}{A_1(1 + \frac{\alpha_3}{R_1})}\left\{\frac{\partial[A_1(1 + \frac{\alpha_3}{R_1})]}{\partial\alpha_1}\;\frac{U_1}{A_1(1 + \frac{\alpha_3}{R_1})}\right.$$

$$\left. + \frac{\partial[A_1(1 + \frac{\alpha_3}{R_1})]}{\partial\alpha_2}\;\frac{U_2}{A_2(1 + \frac{\alpha_3}{R_2})} + \frac{A_1}{R_1}\,U_3\right\}$$

$$+ \frac{1}{A_1(1 + \frac{\alpha_3}{R_1})}\;\frac{\partial U_1}{\partial\alpha_1} - \frac{\partial[A_1(1 + \frac{\alpha_3}{R_1})]}{\partial\alpha_1}\;\frac{U_1}{A_1^{\,2}(1 + \frac{\alpha_3}{R_1})^2} \tag{2.3.45}$$

Next, we will utilize the equalities

$$\frac{\partial[A_1(1 + \frac{\alpha_3}{R_1})]}{\partial\alpha_2} = \left(1 + \frac{\alpha_3}{R_2}\right)\frac{\partial A_1}{\partial\alpha_2} \tag{2.3.46}$$

and

$$\frac{\partial[A_2(1 + \frac{\alpha_3}{R_2})]}{\partial\alpha_1} = \left(1 + \frac{\alpha_3}{R_1}\right)\frac{\partial A_2}{\partial\alpha_1} \tag{2.3.47}$$

These relations are named after Codazzi.

Substituting them in Eq. (2.3.45), we get

$$\varepsilon_{11} = \frac{1}{A_1(1 + \frac{\alpha_3}{R_1})}\left[\frac{\partial U_1}{\partial\alpha_1} + \frac{U_2}{A_2}\;\frac{\partial A_1}{\partial\alpha_2} + U_3\,\frac{A_1}{R_1}\right] \tag{2.3.48}$$

Similarly

$$\varepsilon_{22} = \frac{1}{A_2(1 + \frac{\alpha_3}{R_2})} \left[\frac{\partial U_2}{\partial \alpha_2} + \frac{U_1}{A_1} \frac{\partial A_2}{\partial \alpha_1} + U_3 \frac{A_2}{R_2} \right] \tag{2.3.49}$$

$$\varepsilon_{33} = \frac{\partial U_3}{\partial \alpha_3} \tag{2.3.50}$$

Substituting Eqs. (2.3.21), (2.3.5), and (2.3.1) to (2.3.3) in Eq. (2.3.44) gives, for instance for i=1, j=2,

$$\varepsilon_{12} = \frac{A_1(1 + \frac{\alpha_3}{R_1})}{A_2(1 + \frac{\alpha_3}{R_2})} \frac{\partial}{\partial \alpha_2} \left(\frac{U_1}{A_1(1 + \frac{\alpha_3}{R_1})} \right)$$

$$+ \frac{A_2(1 + \frac{\alpha_3}{R_2})}{A_1(1 + \frac{\alpha_3}{R_1})} \frac{\partial}{\partial \alpha_1} \left(\frac{U_2}{A_2(1 + \frac{\alpha_3}{R_2})} \right) \tag{2.3.51}$$

Similarly

$$\varepsilon_{13} = A_1\left(1 + \frac{\alpha_3}{R_1}\right) \frac{\partial}{\partial \alpha_3} \left(\frac{U_1}{A_1(1 + \frac{\alpha_3}{R_1})} \right) + \frac{1}{A_1(1 + \frac{\alpha_3}{R_1})} \frac{\partial U_3}{\partial \alpha_1} \tag{2.3.52}$$

$$\varepsilon_{23} = A_2\left(1 + \frac{\alpha_3}{R_2}\right) \frac{\partial}{\partial \alpha_3} \left(\frac{U_2}{A_2(1 + \frac{\alpha_3}{R_2})} \right) + \frac{1}{A_2(1 + \frac{\alpha_3}{R_2})} \frac{\partial U_3}{\partial \alpha_2} \tag{2.3.53}$$

2.4 THE LOVE SIMPLIFICATIONS

If the shell is thin, we may assume that the displacements in the α_1 and α_2 direction vary linearly through the shell thickness while displacements in the α_3 direction are independent of α_3:

$$U_1(\alpha_1,\alpha_2,\alpha_3) = u_1(\alpha_1,\alpha_2) + \alpha_3\beta_1(\alpha_1,\alpha_2) \tag{2.4.1}$$

$$U_2(\alpha_1,\alpha_2,\alpha_3) = u_2(\alpha_1,\alpha_2) + \alpha_3\beta_2(\alpha_1,\alpha_2) \tag{2.4.2}$$

$$U_3(\alpha_1,\alpha_2,\alpha_3) = u_3(\alpha_1,\alpha_2) \tag{2.4.3}$$

where β_1 and β_2 represent angles. If we assume that we may neglect shear deflection, which implies that the normal shear strains are zero

$$\varepsilon_{13} = 0 \tag{2.4.4}$$

$$\varepsilon_{23} = 0 \tag{2.4.5}$$

we obtain, for example, a definition of β_1 from Eq. (2.3.52),

$$0 = A_1\left(1 + \frac{\alpha_3}{R_1}\right)\frac{\partial}{\partial\alpha_3}\left(\frac{u_1 + \alpha_3\beta_1}{A_1(1 + \frac{\alpha_3}{R_1})}\right) + \frac{1}{A_1(1 + \frac{\alpha_3}{R_1})}\frac{\partial u_3}{\partial\alpha_1}$$

$$= \beta_1 - \frac{u_1}{R_1} + \frac{1}{A_1}\frac{\partial u_3}{\partial\alpha_1} \tag{2.4.6}$$

or $$\beta_1 = \frac{u_1}{R_1} - \frac{1}{A_1}\frac{\partial u_3}{\partial\alpha_1} \tag{2.4.7}$$

Similarly, we get

$$\beta_2 = \frac{u_2}{R_2} - \frac{1}{A_2}\frac{\partial u_3}{\partial\alpha_2} \tag{2.4.8}$$

Substituting Eqs. (2.4.1) to (2.4.3) into Eqs. (2.3.48) to (2.3.51), and recognizing that

$$\frac{\alpha_3}{R_1} \ll 1 \qquad \frac{\alpha_3}{R_2} \ll 1 \tag{2.4.9}$$

we get

$$\varepsilon_{11} = \frac{1}{A_1}\frac{\partial}{\partial\alpha_1}(u_1 + \alpha_3\beta_1) + \frac{1}{A_1A_2}\frac{\partial A_1}{\partial\alpha_2}(u_2 + \alpha_3\beta_2) + \frac{u_3}{R_1} \tag{2.4.10}$$

$$\varepsilon_{22} = \frac{1}{A_2}\frac{\partial}{\partial\alpha_2}(u_2 + \alpha_3\beta_2) + \frac{1}{A_1A_2}\frac{\partial A_2}{\partial\alpha_1}(u_1 + \alpha_3\beta_1) + \frac{u_3}{R_2} \tag{2.4.11}$$

$$\varepsilon_{33} = 0 \tag{2.4.12}$$

$$\varepsilon_{13} = 0 \tag{2.4.13}$$

$$\varepsilon_{23} = 0 \tag{2.4.14}$$

$$\varepsilon_{12} = \frac{A_2}{A_1}\frac{\partial}{\partial\alpha_1}\left(\frac{u_2 + \alpha_3\beta_2}{A_2}\right) + \frac{A_1}{A_2}\frac{\partial}{\partial\alpha_2}\left(\frac{u_1 + \alpha_3\beta_1}{A_1}\right) \tag{2.4.15}$$

It is convenient to express Eqs. (2.4.10) to (2.4.15) in a form where so-called membrane strains (independent of α_3) and so-called bending strains (proportional to α_3) are separated

$$\varepsilon_{11} = \varepsilon_{11}^{\circ} + \alpha_3 k_{11} \tag{2.4.16}$$

$$\varepsilon_{22} = \varepsilon_{22}^{\circ} + \alpha_3 k_{22} \tag{2.4.17}$$

$$\varepsilon_{12} = \varepsilon_{12}^{\circ} + \alpha_3 k_{12} \tag{2.4.18}$$

where the membrane strains are

$$\varepsilon_{11}^{\circ} = \frac{1}{A_1}\frac{\partial u_1}{\partial \alpha_1} + \frac{u_2}{A_1 A_2}\frac{\partial A_1}{\partial \alpha_2} + \frac{u_3}{R_1} \tag{2.4.19}$$

$$\varepsilon_{22}^{\circ} = \frac{1}{A_2}\frac{\partial u_2}{\partial \alpha_2} + \frac{u_1}{A_1 A_2}\frac{\partial A_2}{\partial \alpha_1} + \frac{u_3}{R_2} \tag{2.4.20}$$

$$\varepsilon_{12}^{\circ} = \frac{A_2}{A_1}\frac{\partial}{\partial \alpha_1}\left(\frac{u_2}{A_2}\right) + \frac{A_1}{A_2}\frac{\partial}{\partial \alpha_2}\left(\frac{u_1}{A_1}\right) \tag{2.4.21}$$

and where the change in curvature terms (bending strains) are

$$k_{11} = \frac{1}{A_1}\frac{\partial \beta_1}{\partial \alpha_1} + \frac{\beta_2}{A_1 A_2}\frac{\partial A_1}{\partial \alpha_2} \tag{2.4.22}$$

$$k_{22} = \frac{1}{A_2}\frac{\partial \beta_2}{\partial \alpha_2} + \frac{\beta_1}{A_1 A_2}\frac{\partial A_2}{\partial \alpha_1} \tag{2.4.23}$$

$$k_{12} = \frac{A_2}{A_1}\frac{\partial}{\partial \alpha_1}\left(\frac{\beta_2}{A_2}\right) + \frac{A_1}{A_2}\frac{\partial}{\partial \alpha_2}\left(\frac{\beta_1}{A_1}\right) \tag{2.4.24}$$

The displacement relations of Eqs. (2.4.1) and (2.4.7) are illustrated in Fig. 2.4.1.

2.5 MEMBRANE FORCES AND BENDING MOMENTS

In the following we will integrate all stresses acting on a shell element whose dimensions are infinitesimal in the α_1 and α_2 directions and equal to the shell thickness in the normal direction. Solving Eqs. (2.2.10) to (2.2.12) for stresses yields

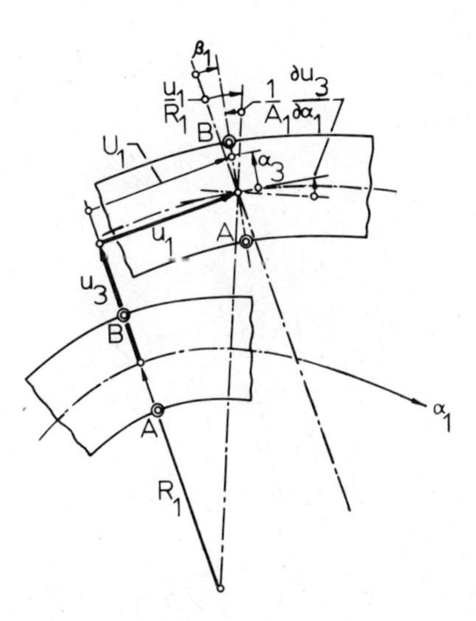

Figure 2.4.1

$$\sigma_{11} = \frac{E}{1 - \mu^2} (\varepsilon_{11} + \mu\varepsilon_{22}) \tag{2.5.1}$$

$$\sigma_{22} = \frac{E}{1 - \mu^2} (\varepsilon_{22} + \mu\varepsilon_{11}) \tag{2.5.2}$$

$$\sigma_{12} = \varepsilon_{12} G \tag{2.5.3}$$

Substituting Eqs. (2.4.16) to (2.4.18) gives

$$\sigma_{11} = \frac{E}{1 - \mu^2} [\varepsilon_{11}^{\circ} + \mu\varepsilon_{22}^{\circ} + \alpha_3(k_{11} + \mu k_{22})] \tag{2.5.4}$$

$$\sigma_{22} = \frac{E}{1 - \mu^2} [\varepsilon_{22}^{\circ} + \mu\varepsilon_{11}^{\circ} + \alpha_3(k_{22} + \mu k_{11})] \tag{2.5.5}$$

$$\sigma_{12} = G(\varepsilon_{12}^{\circ} + \alpha_3 k_{12}) \tag{2.5.6}$$

For instance, referring to Fig. 2.5.1, the force in the α_1 direction acting on a strip of the element face of height $d\alpha_3$ and width $A_2(1 + \frac{\alpha 3}{R_2})d\alpha_2$ is

$$\sigma_{11} A_2 \left(1 + \frac{\alpha_3}{R_2}\right) d\alpha_2 \; d\alpha_3$$

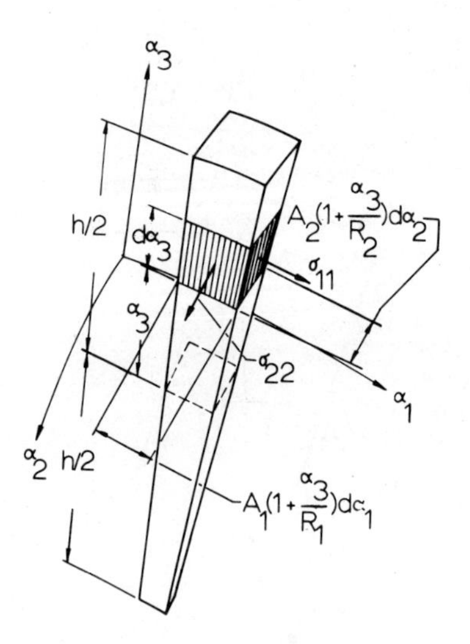

Figure 2.5.1

Thus, the total force acting on the element in the α_1 direction is

$$\int_{\alpha_3=\frac{h}{2}}^{\alpha_3=\frac{h}{2}} \sigma_{11} A_2 \left(1 + \frac{\alpha_3}{R_2}\right) d\alpha_2 \, d\alpha_3$$

and the force per unit length of neutral surface $A_2 \, d\alpha_2$ is

$$N_{11} = \int_{-\frac{h}{2}}^{\frac{h}{2}} \sigma_{11} \left(1 + \frac{\alpha_3}{R_2}\right) d\alpha_3 \tag{2.5.7}$$

Neglecting the second term in the parentheses, we obtain

$$N_{11} = \int_{-\frac{h}{2}}^{\frac{h}{2}} \sigma_{11} \, d\alpha_3 \tag{2.5.8}$$

Substituting Eq. (2.5.4) results in

$$N_{11} = K(\varepsilon_{11}^{\circ} + \mu\varepsilon_{22}^{\circ}) \tag{2.5.9}$$

where

$$K = \frac{Eh}{1 - \mu^2} \tag{2.5.10}$$

K is the so-called membrane stiffness. Similarly, integrating σ_{22} on the α_2 face of the element with the shear stresses $\sigma_{12} = \sigma_{21}$ gives

$$N_{22} = K(\varepsilon_{22}^{\circ} + \mu\varepsilon_{11}^{\circ}) \tag{2.5.11}$$

$$N_{12} = N_{21} = \frac{K(1 - \mu)}{2} \varepsilon_{12}^{\circ} \tag{2.5.12}$$

To obtain bending moments, we express first the bending moment about the neutral surface due to the element strip $A_2(1 + \frac{\alpha_3}{R_2})\, d\alpha_2\, d\alpha_3$:

$$\sigma_{11}\alpha_3 A_2\left(1 + \frac{\alpha_3}{R_2}\right) d\alpha_2\, d\alpha_3$$

Thus, the total bending moment acting on the element in the α_1 direction is

$$\int_{\alpha_3 = \frac{h}{2}}^{\alpha_3 = \frac{h}{2}} \sigma_{11}\alpha_3 A_2\left(1 + \frac{\alpha_3}{R_2}\right) d\alpha_2\, d\alpha_3$$

and the bending moment per unit length neutral surface is

$$M_{11} = \int_{-\frac{h}{2}}^{\frac{h}{2}} \sigma_{11}\alpha_3\left(1 + \frac{\alpha_3}{R_2}\right) d\alpha_3 \tag{2.5.13}$$

Neglecting the second term in the parentheses, we have

$$M_{11} = \int_{-\frac{h}{2}}^{\frac{h}{2}} \sigma_{11}\alpha_3\, d\alpha_3 \tag{2.5.14}$$

Substituting Eq. (2.5.4) results in

$$M_{11} = D(k_{11} + \mu k_{22}) \tag{2.5.15}$$

where

$$D = \frac{Eh^3}{12(1 - \mu^2)} \tag{2.5.16}$$

D is the so-called bending stiffness. Similarly, integrating σ_{22} and $\sigma_{12} = \sigma_{21}$ gives

$$M_{22} = D(k_{22} + \mu k_{11}) \tag{2.5.17}$$

$$M_{12} = M_{21} = \frac{D(1 - \mu)}{2} k_{12} \tag{2.5.18}$$

While we have assumed that strains ε_{13} and ε_{23} due to transverse shear stresses σ_{13} and σ_{23} are negligible, we will by no means neglect the transverse shear forces in the following:

$$Q_{13} = \int_{-\frac{h}{2}}^{\frac{h}{2}} \sigma_{13}\, d\alpha_3 \tag{2.5.19}$$

and
$$Q_{23} = \int_{-\frac{h}{2}}^{\frac{h}{2}} \sigma_{23}\, d\alpha_3 \tag{2.5.20}$$

These forces will be defined by the resulting equations themselves.

Lastly, if we solve Eqs. (2.5.9), (2.5.11), (2.5.12), (2.5.15), (2.5.17), and (2.5.18) for the strains, we may write Eqs. (2.5.4) to (2.5.6) as

$$\sigma_{11} = \frac{N_{11}}{h} + \frac{12M_{11}}{h^3}\alpha_3 \tag{2.5.21}$$

$$\sigma_{22} = \frac{N_{22}}{h} + \frac{12M_{22}}{h^3}\alpha_3 \tag{2.5.22}$$

$$\sigma_{12} = \frac{N_{12}}{h} + \frac{12M_{12}}{h^3}\alpha_3 \tag{2.5.23}$$

2.6 ENERGY EXPRESSIONS

The strain energy stored in one infinitesimal element that is acted on by stresses σ_{ij} is

$$dU = \frac{1}{2}(\sigma_{11}\varepsilon_{11} + \sigma_{22}\varepsilon_{22} + \sigma_{12}\varepsilon_{12} + \sigma_{13}\varepsilon_{13} + \sigma_{23}\varepsilon_{23} + \sigma_{33}\varepsilon_{33})\, dV \tag{2.6.1}$$

The last term is neglected in line with assumption (2.4.3). We do however have to keep the transverse shear terms, even though we have previously assumed ε_{13} and ε_{23} to be negligible, in order to obtain expressions for β_1 and β_2. The infinitesimal volume is given by

$$dV = A_1A_2\left(1 + \frac{\alpha_3}{R_1}\right)\left(1 + \frac{\alpha_3}{R_2}\right) d\alpha_1 d\alpha_2 d\alpha_3 \tag{2.6.2}$$

Integrating Eq. (2.6.1) over the volume of the shell gives

$$U = \iiint_{\alpha_1\alpha_2\alpha_3} F\, dV \tag{2.6.3}$$

where

$$F = \frac{1}{2}(\sigma_{11}\varepsilon_{11} + \sigma_{22}\varepsilon_{22} + \sigma_{12}\varepsilon_{12} + \sigma_{13}\varepsilon_{13} + \sigma_{23}\varepsilon_{23}) \tag{2.6.4}$$

The kinetic energy of one infinitesimal element is given by

$$dK = \frac{1}{2}\rho(\dot{U}_1^2 + \dot{U}_2^2 + \dot{U}_3^2)\, dV \tag{2.6.5}$$

The dot indicates a time derivative.

Substituting Eqs. (2.4.1) to (2.4.3) and considering all the elements of the shell gives

$$K = \frac{\rho}{2}\iiint_{\alpha_1\alpha_2\alpha_3} [\dot{u}_1^2 + \dot{u}_2^2 + \dot{u}_3^2 + \alpha_3^2(\dot{\beta}_1^2 + \dot{\beta}_2^2) + 2\alpha_3(\dot{u}_1\dot{\beta}_1 + \dot{u}_2\dot{\beta}_2)]A_1A_2\left(1 + \frac{\alpha_3}{R_1}\right)\left(1 + \frac{\alpha_3}{R_2}\right) d\alpha_1 d\alpha_2 d\alpha_3 \tag{2.6.6}$$

Neglecting the α_3/R_1 and α_3/R_2 terms, we integrate over the thickness of the shell ($\alpha_3 = -\frac{h}{2}$ to $\alpha_3 = \frac{h}{2}$). This gives

$$K = \frac{\rho h}{2} \int_{\alpha_1}\int_{\alpha_2} \left[\dot{u}_1^2 + \dot{u}_2^2 + \dot{u}_3^2 + \frac{h^2}{12}(\dot{\beta}_1^2 + \dot{\beta}_2^2)\right] A_1 A_2 \, d\alpha_1 \, d\alpha_2 \tag{2.6.7}$$

The energy put into the shell by possible applied boundary force resultants and moment resultants is, along typical α_2 = constant and α_1 = constant lines,

$$E_B = \int_{\alpha_1} (u_2 N_{22}^* + u_1 N_{21}^* + u_3 Q_{23}^* + \beta_2 M_{22}^* + \beta_1 M_{21}^*) A_1 d\alpha_1$$
$$+ \int_{\alpha_2} (u_1 N_{11}^* + u_2 N_{12}^* + u_3 Q_{13}^* + \beta_1 M_{11}^* + \beta_2 M_{12}^*) A_2 \, d\alpha_2 \tag{2.6.8}$$

Taking, for example, the α_2 = constant edge, N_{22}^* is the boundary force normal to the boundary in the tangent plane to the neutral surface. The units are [N/m]. Q_{23}^* is a shear force acting on the boundary normal to the shell surface, while N_{21}^* is a shear force acting along the boundary in the tangent plane. M_{22}^* is a moment in the α_2 direction and M_{21}^* is a twisting moment in the α_1 direction. Figure 2.6.1 illustrates this.

The energy introduced into the shell by distributed load components in the α_1, α_2, and α_3 directions, namely q_1, q_2, and q_3 [N/m^2], is

$$E_L = \int_{\alpha_1}\int_{\alpha_2} (q_1 u_1 + q_2 u_2 + q_3 u_3) A_1 A_2 \, d\alpha_1 \, d\alpha_2 \tag{2.6.9}$$

All loads are assumed to act on the neutral surface of the shell. The components are shown in Fig. 2.6.2.

2.7 LOVE'S EQUATIONS BY WAY OF HAMILTON'S PRINCIPLE

Hamilton's principle is given as

$$\delta \int_{t_o}^{t_1} (\Pi - K)\, dt = 0 \tag{2.7.1}$$

where Π is the total potential energy. In our case

$$\Pi = U - E_B - E_L \tag{2.7.2}$$

The times t_1 and t_o are arbitrary, except that at $t = t_1$ and $t = t_o$, all variations are zero. The symbol δ is the variational symbol and is treated mathematically like a differential symbol. Variational displacements are arbitrary.

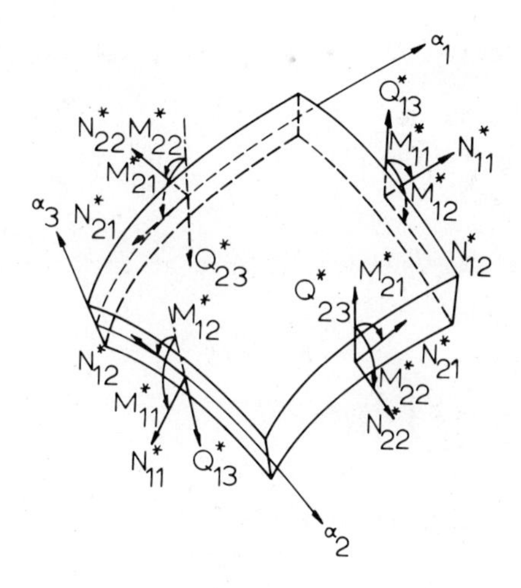

Figure 2.6.1

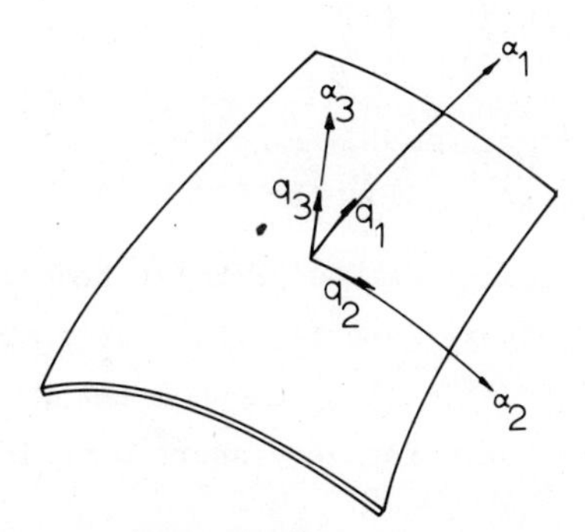

Figure 2.6.2

Substituting Eq. (2.7.2) for Eq. (2.7.1) and taking the variational operator inside the integral, we obtain

$$\int_{t_o}^{t_o} (\delta U - \delta E_B - \delta E_L - \delta K)\, dt = 0 \qquad (2.7.3)$$

let us examine these variations one by one. First, from Eq. (2.6.6),

$$\int_{t_o}^{t_1} \delta K dt = \rho h \int_{t_o}^{t_1} \int_{\alpha_1}\int_{\alpha_2} \left[\dot{u}_1 \delta\dot{u}_1 + \dot{u}_2 \delta\dot{u}_2 + \dot{u}_3 \delta u_3 + \frac{h^2}{12}(\dot{\beta}_1 \delta\dot{\beta}_1 + \dot{\beta}_2 \delta\dot{\beta}_2)\right] A_1 A_2\, d\alpha_1\, d\alpha_2\, dt \qquad (2.7.4)$$

Integrating by parts, for instance, the first term becomes

$$\int_{t_o}^{t_1} \dot{u}_1 \delta\dot{u}_1\, dt = [\dot{u}_1 \delta u_1]_{t_o}^{t_1} - \int_{t_o}^{t_1} \ddot{u}_1 \delta u_1\, dt \qquad (2.7.5)$$

Since the virtual displacement is zero at $t=t_o$ and $t=t_1$, we are left with

$$\int_{t_o}^{t_1} \dot{u}_1 \delta\dot{u}_1\, dt = -\int_{t_o}^{t_1} \ddot{u}_1 \delta u_1\, dt \qquad (2.7.6)$$

Proceeding similarly with the other term in the integral, we get

$$\int_{t_o}^{t_1} \delta K\, dt = -\rho h \int_{t_o}^{t_1} \int_{\alpha_1}\int_{\alpha_2} [\ddot{u}_1 \delta u_1 + \ddot{u}_2 \delta u_2 + \ddot{u}_3 \delta u_3 + \frac{h^2}{12}(\ddot{\beta}_1 \delta\beta_1 + \ddot{\beta}_2 \delta\beta_2)]\, A_1 A_2\, d\alpha_1\, d\alpha_2\, dt \qquad (2.7.7)$$

As in the classical Bernoulli-Euler beam theory, we neglect the influence of rotatory inertia, which we recognize as the terms involving $\ddot{\beta}_1$ and $\ddot{\beta}_2$. It will be shown later that rotatory inertia has to be considered only for very short wave lengths of vibration and even then shear deformation is a more important effect.

$$\int_{t_o}^{t_1} \delta K \, dt = -\rho h \int_{t_o}^{t_1} \int_{\alpha_1}\int_{\alpha_2} (\ddot{u}_1 \, \delta u_1 + \ddot{u}_2 \, \delta u_2 + \ddot{u}_3 \, \delta u_3) \, A_1 A_2 \, d\alpha_1 \, d\alpha_2 \, dt \tag{2.7.8}$$

Next, let us evaluate the variation of the energy due to the load. From Eq. (2.6.9),

$$\int_{t_o}^{t_1} \delta E_L \, dt = \int_{t_o}^{t_1} \int_{\alpha_1}\int_{\alpha_2} (q_1 \, \delta u_1 + q_2 \, \delta u_2 + q_3 \, \delta u_3) \, A_1 A_2 \, d\alpha_1 \, d\alpha_2 \, dt \tag{2.7.9}$$

The integral of the variation of boundary energy is, using Eq. (2.6.8),

$$\int_{t_o}^{t_1} \delta E_B \, dt = \int_{t_o}^{t_1} \int_{\alpha_1} (N_{22}^* \, \delta u_2 + N_{21}^* \, \delta u_1 + Q_{23}^* \, \delta u_3 + M_{22}^* \delta\beta_2 + M_{21}^* \delta\beta_1) \, A_1 d\alpha_1 dt + \int_{t_o}^{t_1} \int_{\alpha_2} (N_{11}^* \, \delta u_1 + N_{12}^* \, \delta u_2 + Q_{13}^* \, \delta u_3 + M_{11}^* \, \delta\beta_1 + M_{12}^* \, \delta\beta_2) \, A_2 \, d\alpha_2 \, dt \tag{2.7.10}$$

It is more complicated to evaluate the integral of the variation in strain energy. Starting with Eq. (2.6.3), we have

$$\int_{t_o}^{t_1} \delta U \, dt = \int_{t_o}^{t_1} \int_{\alpha_1}\int_{\alpha_2}\int_{\alpha_3} \delta F \, dV \, dt \tag{2.7.11}$$

where

$$\delta F = \frac{\partial F}{\partial \varepsilon_{11}} \delta\varepsilon_{11} + \frac{\partial F}{\partial \varepsilon_{22}} \delta\varepsilon_{22} + \frac{\partial F}{\partial \varepsilon_{12}} \delta\varepsilon_{12} + \frac{\partial F}{\partial \varepsilon_{13}} \delta\varepsilon_{13} + \frac{\partial F}{\partial \varepsilon_{23}} \delta\varepsilon_{23} \tag{2.7.12}$$

Examining the first term of this equation, we see that

$$\frac{\partial F}{\partial \varepsilon_{11}} \delta\varepsilon_{11} = \frac{1}{2}\left(\frac{\partial \sigma_{11}}{\partial \varepsilon_{11}} \varepsilon_{11} + \sigma_{11} + \varepsilon_{22} \frac{\partial \sigma_{22}}{\partial \varepsilon_{11}}\right) \delta\varepsilon_{11} \tag{2.7.13}$$

Substituting Eqs. (2.5.1) and (2.5.2) gives

$$\frac{\partial F}{\partial \varepsilon_{11}} \delta\varepsilon_{11} = \sigma_{11}\delta\varepsilon_{11} \tag{2.7.14}$$

Thus, we can show that

$$\int_{t_o}^{t_1} \delta U \, dt = \int_{t_o}^{t_1} \iiint_{\alpha_1\alpha_2\alpha_3} (\sigma_{11}\delta\varepsilon_{11} + \sigma_{22}\delta\varepsilon_{22} + \sigma_{12}\delta\varepsilon_{12}$$
$$+ \sigma_{13}\delta\varepsilon_{13} + \sigma_{23}\delta\varepsilon_{23}) \, A_1A_2 \left(1 + \frac{\alpha_3}{R_1}\right)\left(1 + \frac{\alpha_3}{R_2}\right) d\alpha_1 \, d\alpha_2 \, d\alpha_3 \, dt \tag{2.7.15}$$

We neglect the α_3/R_1 and α_3/R_2 terms as small. Substituting Eqs. (2.4.10) to (2.4.15) allows us to express the strain variations in terms of displacement variations. Integration with respect to α_3 allows us to introduce force resultants and moment resultants. Integration by parts will put the integral into a manageable form. Let us illustrate all this on the first term of Eq. (2.7.15):

$$\int_{t_o}^{t_1} \iiint_{\alpha_1\alpha_2\alpha_3} \sigma_{11} \, \delta\varepsilon_{11} \, A_1A_2 \, d\alpha_1 \, d\alpha_2 \, d\alpha_3 \, dt$$
$$= \int_{t_o}^{t_1} \iiint_{\alpha_1\alpha_2\alpha_3} \left[\sigma_{11}\left(A_2\frac{\partial(\delta u_1)}{\partial \alpha_1} + \delta u_2 \frac{\partial A_1}{\partial \alpha_2} + \frac{A_1A_2}{R_1} \delta u_3\right)\right.$$
$$\left. + \alpha_3\sigma_{11}\left(A_2\frac{\partial(\delta\beta_1)}{\partial \alpha_1} + \delta\beta_2\frac{\partial A_1}{\partial \alpha_2}\right)\right] d\alpha_1 \, d\alpha_2 \, d\alpha_3 \, dt =$$
$$\int_{t_o}^{t_1} \iint_{\alpha_1\alpha_2} \left[N_{11}\left(A_2\frac{\partial(\delta u_1)}{\partial \alpha_1} + \delta u_2 \frac{\partial A_1}{\partial \alpha_2} + \frac{A_1A_2}{R_1} \delta u_3\right)\right.$$
$$\left. + M_{11}\left(A_2 \frac{\partial(\delta\beta_1)}{\partial \alpha_1} + \delta\beta_2 \frac{\partial A_1}{\partial \alpha_2}\right)\right] d\alpha_1 \, d\alpha_2 \, dt \tag{2.7.16}$$

Now we illustrate the integration by parts on the first term of Eq. (2.7.16):

$$\iint_{\alpha_2 \alpha_1} N_{11} A_2 \frac{\partial(\delta u_1)}{\partial \alpha_1} \, d\alpha_1 \, d\alpha_2 = \int_{\alpha_2} N_{11} A_2 \, \delta u_1 \, d\alpha_2 - \iint_{\alpha_1 \alpha_2} \frac{\partial(N_{11} A_2)}{\partial \alpha_1} \delta u_1 \, d\alpha_1 \, d\alpha_2 \tag{2.7.17}$$

Proceeding with all terms of Eq. (2.7.15) in this fashion, we get

$$\begin{aligned}
\int_{t_o}^{t_1} \delta U \, dt = \int_{t_o}^{t_1} \iint_{\alpha_1 \alpha_2} \Bigg[& \left(- \frac{\partial(N_{11} A_2)}{\partial \alpha_1} - \frac{\partial(N_{21} A_1)}{\partial \alpha_2} - N_{12} \frac{\partial A_1}{\partial \alpha_2} \right. \\
& \left. + N_{22} \frac{\partial A_2}{\partial \alpha_1} - Q_{13} \frac{A_1 A_2}{R_1} \right) \delta u_1 + \left(- \frac{\partial(N_{12} A_2)}{\partial \alpha_1} \right. \\
& \left. - \frac{\partial(N_{22} A_1)}{\partial \alpha_2} + N_{11} \frac{\partial A_1}{\partial \alpha_2} - N_{21} \frac{\partial A_2}{\partial \alpha_1} - Q_{23} \frac{A_1 A_2}{R_2} \right) \delta u_2 \\
& + \left(N_{11} \frac{A_1 A_2}{R_1} + N_{22} \frac{A_1 A_2}{R_2} - \frac{\partial(Q_{13} A_2)}{\partial \alpha_1} - \frac{\partial(Q_{23} A_1)}{\partial \alpha_2} \right) \delta u_3 \\
& + \left(- \frac{\partial(M_{21} A_1)}{\partial \alpha_2} - M_{12} \frac{\partial A_1}{\partial \alpha_2} + M_{22} \frac{\partial A_2}{\partial \alpha_1} - \frac{\partial(M_{11} A_2)}{\partial \alpha_1} \right. \\
& \left. + Q_{13} A_1 A_2 \right) \delta \beta_1 + \left(- \frac{\partial(M_{12} A_2)}{\partial \alpha_1} - \frac{\partial(M_{22} A_1)}{\partial \alpha_2} - M_{21} \frac{\partial A_2}{\partial \alpha_1} \right. \\
& \left. + M_{11} \frac{\partial A_1}{\partial \alpha_2} + Q_{23} A_1 A_2 \right) \delta \beta_2 \Bigg] \, d\alpha_1 d\alpha_2 dt \\
& + \int_{t_o}^{t_1} \int_{\alpha_2} \left(N_{11} \delta u_1 + M_{11} \delta \beta_1 + N_{12} \delta u_2 + M_{12} \delta \beta_2 \right. \\
& \left. + Q_{13} \delta u_3 \right) A_2 \, d\alpha_2 dt + \int_{t_o}^{t_1} \int_{\alpha_1} \left(N_{22} \delta u_2 \right. \\
& \left. + M_{22} \delta \beta_2 + N_{21} \delta u_1 + M_{21} \delta \beta_1 + Q_{23} \delta u_3 \right) A_1 \, d\alpha_1 dt
\end{aligned} \tag{2.7.18}$$

We are now ready to substitute Eqs. (2.7.18), (2.7.10), (2.7.9), and (2.7.8) in Eq. (2.7.3). This gives

$$\int_{t_o}^{t_1}\iint_{\alpha_1\alpha_2}\left\{\left[\frac{\partial(N_{11}A_2)}{\partial\alpha_1}+\frac{\partial(N_{21}A_1)}{\partial\alpha_2}+N_{12}\frac{\partial A_1}{\partial\alpha_2}-N_{22}\frac{\partial A_2}{\partial\alpha_1}\right.\right.$$

$$\left.+Q_{13}\frac{A_1A_2}{R_1}+(q_1-\rho h\ddot{u}_1)A_1A_2\right]\delta u_1+\left[\frac{\partial(N_{12}A_2)}{\partial\alpha_1}\right.$$

$$+\frac{\partial(N_{22}A_1)}{\partial\alpha_2}+N_{21}\frac{\partial A_2}{\partial\alpha_1}-N_{11}\frac{\partial A_1}{\partial\alpha_2}+Q_{23}\frac{A_1A_2}{R_2}$$

$$\left.+(q_2-\rho h\ddot{u}_2)A_1A_2\right]\delta u_2+\left[\frac{\partial(Q_{13}A_2)}{\partial\alpha_1}+\frac{\partial(Q_{23}A_1)}{\partial\alpha_2}\right.$$

$$\left.-\left(\frac{N_{11}}{R_1}+\frac{N_{22}}{R_2}\right)A_1A_2+(q_3-\rho h\ddot{u}_3)A_1A_2\right]\delta u_3+\left(\frac{\partial(M_{11}A_2)}{\partial\alpha_1}\right.$$

$$\left.+\frac{\partial(M_{21}A_1)}{\partial\alpha_2}+M_{12}\frac{\partial A_1}{\partial\alpha_2}-M_{22}\frac{\partial A_2}{\partial\alpha_1}-Q_{13}A_1A_2\right)\delta\beta_1+\left(\frac{\partial(M_{12}A_2)}{\partial\alpha_1}\right.$$

$$\left.\left.+\frac{\partial(M_{22}A_1)}{\partial\alpha_2}+M_{21}\frac{\partial A_2}{\partial\alpha_1}-M_{11}\frac{\partial A_1}{\partial\alpha_2}-Q_{23}A_1A_2\right)\delta\beta_2\right\}d\alpha_1 d\alpha_2 dt$$

$$+\int_{t_o}^{t_1}\int_{\alpha_1}\left[(N_{22}^*-N_{22})\,\delta u_2+(N_{21}^*-N_{21})\,\delta u_1+(Q_{23}^*-Q_{23})\,\delta u_3\right.$$

$$\left.+(M_{22}^*-M_{22})\,\delta\beta_2+(M_{21}^*-M_{21})\,\delta\beta_1\right]A_1\,d\alpha_1\,dt$$

$$+\int_{t_o}^{t_1}\int_{\alpha_2}\left[(N_{11}^*-N_{11})\,\delta u_1+(N_{12}^*-N_{12})\,\delta u_2+(Q_{13}^*-Q_{13})\,\delta u_3\right.$$

$$\left.+(M_{11}^*-M_{11})\,\delta\beta_1+(M_{12}^*-M_{12})\,\delta\beta_2\right]A_2\,d\alpha_2\,dt=0 \qquad (2.7.19)$$

The equation can only be satisfied if each of the triple and double integral parts are zero individually. Moreover, since the variational displacements are arbitrary, each integral equation can only be satisfied if the coefficients of the variational displacements are zero. Thus, the coefficients of the triple integral set to zero give the following five equations:

$$-\frac{\partial(N_{11}A_2)}{\partial\alpha_1} - \frac{\partial(N_{21}A_1)}{\partial\alpha_2} - N_{12}\frac{\partial A_1}{\partial\alpha_2} + N_{22}\frac{\partial A_2}{\partial\alpha_1} - A_1A_2\frac{Q_{13}}{R_1} + A_1A_2\,\rho h\ddot{u}_1 = A_1A_2q_1 \tag{2.7.20}$$

$$-\frac{\partial(N_{12}A_2)}{\partial\alpha_1} - \frac{\partial(N_{22}A_1)}{\partial\alpha_2} - N_{21}\frac{\partial A_2}{\partial\alpha_1} + N_{11}\frac{\partial A_1}{\partial\alpha_2} - A_1A_2\frac{Q_{23}}{R_2} + A_1A_2\,\rho h\ddot{u}_2 = A_1A_2q_2 \tag{2.7.21}$$

$$-\frac{\partial(Q_{13}A_2)}{\partial\alpha_1} - \frac{\partial(Q_{23}A_1)}{\partial\alpha_2} + A_1A_2\left(\frac{N_{11}}{R_1} + \frac{N_{22}}{R_2}\right) + A_1A_2\,\rho h\ddot{u}_3 = A_1A_2q_3 \tag{2.7.22}$$

where Q_{13} and Q_{23} are defined by

$$\frac{\partial(M_{11}A_2)}{\partial\alpha_1} + \frac{\partial(M_{21}A_1)}{\partial\alpha_2} + M_{12}\frac{\partial A_1}{\partial\alpha_2} - M_{22}\frac{\partial A_2}{\partial\alpha_1} - Q_{13}A_1A_2 = 0 \tag{2.7.23}$$

$$\frac{\partial(M_{12}A_2)}{\partial\alpha_1} + \frac{\partial(M_{22}A_1)}{\partial\alpha_2} + M_{21}\frac{\partial A_2}{\partial\alpha_1} - M_{11}\frac{\partial A_1}{\partial\alpha_2} - Q_{23}A_1A_2 = 0 \tag{2.7.24}$$

These five equations are known as *Love's equations*. They define the motion (or static deflection, for all it matters) due to any type of pressure load. Shear deflection and rotatory inertia is not included.

2.8 BOUNDARY CONDITIONS

Examining Love's equations, the stress strain and the strain displacement relations, we see that the equations are eighth order partial differential equations in space. This means, that we can accommodate at most four boundary conditions on each edge.

However, the two line integrals of Eq. (2.7.19), when set to zero, are satisfied only if the five coefficients in each are zero or if the virtual displacements are zero. This would define as necessary boundary conditions five. A similar problem was encountered by Kirchhoff [2.3] in the nineteenth century, when he investigated the boundary conditions of a plate. It appeared as if three conditions at each edge were needed, but the forth order equation would allow only two. Kirchhoff combined the three conditions into two by noting that the twisting moment and shear boundary conditions were related.

Following Kirchhoff's lead, the two line integrals are rewritten utilizing the definitions of Eqs. (2.4.7) and (2.4.8). For instance, for the first line integral equation we get

$$\int_{t_o}^{t_1}\int_{\alpha_1} [(N^*_{22} - N_{22})\,\delta u_2 + (N^*_{21} - N_{21})\,\delta u_1 + (Q^*_{23} - Q_{23})\,\delta u_3$$

$$+ (M^*_{22} - M_{22})\,\delta\beta_2 + (M^*_{21} - M_{21})(\frac{\delta u_1}{R_1} - \frac{1}{A_1}\frac{\partial}{\partial\alpha_1}(\delta u_3))]\,A_1 d\alpha_1 dt = 0 \qquad (2.8.1)$$

Before collecting coefficients of δu_3, we have to perform an integration by parts

$$\int_{t_o}^{t_1}\int_{\alpha_1} (M^*_{21} - M_{21})\,\frac{\partial}{\partial\alpha_1}\,(\delta u_3)\,d\alpha_1\,dt$$

$$= \left|\int_{t_o}^{t_1} (M^*_{21} - M_{21})\,\delta u_3\,dt\,\right|_{\alpha_1} - \int_{t_o}^{t_1}\int_{\alpha_1} \frac{\partial}{\partial\alpha_1}\,(M^*_{21} - M_{21})\,d\alpha_1\,\delta u_3\,dt \qquad (2.8.2)$$

Since $M_{21} = M^*_{21}$ along the entire edge, the term in parentheses is zero. Thus, substituting Eq. (2.8.2) in Eq. (2.8.1) and collecting coefficients of virtual displacements,

$$\int_{t_o}^{t_1}\int_{\alpha_1}\left\{(N^*_{22} - N_{22})\ \delta u_2 + \left[\left(N^*_{21} + \frac{M^*_{21}}{R_1}\right) - \left(N_{21} + \frac{M_{21}}{R_1}\right)\right]\delta u_1\right.$$

$$+ (M^*_{22} - M_{22})\ \delta\beta_2 + \left[\left(Q^*_{23} + \frac{1}{A_1}\frac{\partial M^*_{21}}{\partial\alpha_1}\right)\right.$$

$$\left.\left. - \left(Q_{23} + \frac{1}{A_1}\frac{\partial M_{21}}{\partial\alpha_1}\right)\right]\delta u_3\right\} A_1 d\alpha_1 dt = 0 \tag{2.8.3}$$

Similarly, for the second line integral, we get

$$\int_{t_o}^{t_1}\int_{\alpha_2}\left\{(N^*_{11} - N_{11})\ \delta u_1 + \left[\left(N^*_{12} + \frac{M^*_{12}}{R_2}\right) - \left(N_{12} + \frac{M_{12}}{R_2}\right)\right]\delta u_2\right.$$

$$+ (M^*_{11} - M_{11})\ \delta\beta_1 + \left[\left(Q^*_{13} + \frac{1}{A_2}\frac{\partial M^*_{12}}{\partial\alpha_2}\right)\right.$$

$$\left.\left. - \left(Q_{13} + \frac{1}{A_2}\frac{\partial M_{12}}{\partial\alpha_2}\right)\right]\delta u_3\right\} A_2\ d\alpha_2\ dt = 0 \tag{2.8.4}$$

These equation are satisfied if either the virtual displacements vanish or the coefficients of the virtual displacements vanish. Defining, in memory of Kirchhoff, the so-called Kirchhoff effective shear stress resultants of the first kind

$$V_{13} = Q_{13} + \frac{1}{A_2}\frac{\partial M_{12}}{\partial\alpha_2} \tag{2.8.5}$$

and
$$V_{23} = Q_{23} + \frac{1}{A_1}\frac{\partial M_{21}}{\partial\alpha_1} \tag{2.8.6}$$

and as the Kirchhoff effective shear stress resultants of the second kind

$$T_{12} = N_{12} + \frac{M_{12}}{R_2} \tag{2.8.7}$$

and
$$T_{21} = N_{21} + \frac{M_{21}}{R_1} \tag{2.8.8}$$

we may write the integrals as

$$\int_{t_o}^{t_1}\int_{\alpha_1} [(N^*_{22} - N_{22})\ \delta u_2 + (T^*_{21} - T_{21})\ \delta u_1 + (M^*_{22} - M_{22})\ \delta\beta_2$$
$$+ (V^*_{23} - V_{23})\ \delta u_3]\ A_1\ d\alpha_1\ dt = 0 \tag{2.8.9}$$

and

$$\int_{t_o}^{t_1}\int_{\alpha_2} [(N^*_{11} - N_{11})\ \delta u_1 + (T^*_{12} - T_{12})\ \delta u_2 + (M^*_{11} - M_{11})\ \delta\beta_1$$
$$+ (V^*_{13} - V_{13})\ \delta u_3]\ A_2\ d\alpha_2\ dt = 0 \tag{2.8.10}$$

Now we may argue that each of these integrals can be satisfied only if the coefficients of the virtual displacements, the virtual displacements, or one of the two for each term are zero. Since virtual displacements are only zero at all times when the boundary displacements are prescribed, this translates into the following possible boundary conditions for an α_1 = constant edge [Eq. (2.8.10)].

$$N_{11} = N^*_{11} \quad \text{or} \quad u_1 = u^*_1 \tag{2.8.11}$$
$$M_{11} = M^*_{11} \quad \text{or} \quad \beta_1 = \beta^*_1 \tag{2.8.12}$$
$$V_{13} = V^*_{13} \quad \text{or} \quad u_3 = u^*_3 \tag{2.8.13}$$
$$T_{12} = T_{12} \quad \text{or} \quad u_2 = u^*_2 \tag{2.8.14}$$

This states the intuitively obvious fact that we have to prescribe at a boundary either forces (moments) or displacements (angular displacements). However, four conditions have to be identified per edge. In a later chapter we will see that under certain simplifying assumptions we may reduce this number even further. Similarly, examining Eq. (2.8.3) that describes an α_2 = constant edge, the four boundary conditions have to be

$$N_{22} = N^*_{22} \quad \text{or} \quad u_2 = u^*_2 \tag{2.8.15}$$
$$M_{22} = M^*_{22} \quad \text{or} \quad \beta_2 = \beta^*_2 \tag{2.8.16}$$
$$V_{23} = V^*_{23} \quad \text{or} \quad u_3 = u^*_3 \tag{2.8.17}$$
$$T_{21} = T^*_{21} \quad \text{or} \quad u_1 = u^*_1 \tag{2.8.18}$$

We can therefore state in general that if n denotes the subscript that defines the normal direction to the edge and if t denotes the subscript that defines the tangential direction to the edge, the necessary boundary conditions are

$$N_{nn} = N^*_{nn} \quad \text{or} \quad u_n = u^*_n \tag{2.8.19}$$

$$M_{nn} = M^*_{nn} \quad \text{or} \quad \beta_n = \beta^*_n \tag{2.8.20}$$

$$V_{n3} = V^*_{n3} \quad \text{or} \quad u_3 = u^*_3 \tag{2.8.21}$$

$$T_{nt} = T^*_{nt} \quad \text{or} \quad u_t = u^*_t \tag{2.8.22}$$

Let us consider a few examples. First there is the case where the edge is completely free. This means that no forces or moments act on this edge:

$$N_{nn} = 0 \tag{2.8.23}$$

$$M_{nn} = 0 \tag{2.8.24}$$

$$V_{n3} = 0 \tag{2.8.25}$$

$$T_{nt} = 0 \tag{2.8.26}$$

The other extreme is the case where the edge is completely prevented from deflecting by being clamped:

$$u_n = 0 \tag{2.8.27}$$

$$u_t = 0 \tag{2.8.28}$$

$$u_3 = 0 \tag{2.8.29}$$

$$\beta_n = 0 \tag{2.8.30}$$

If the edge is supported on knife edges such that it is free to rotate in normal direction, but is prevented from having any transverse deflection, clearly the two conditions

$$u_3 = 0 \tag{2.8.31}$$

$$M_{nn} = 0 \tag{2.8.32}$$

apply. If the knife edges are such that the shell is free to slide between them, then the other two conditions are

$$N_{nn} = 0 \tag{2.8.33}$$

$$T_{nt} = 0 \tag{2.8.34}$$

If the shell is somehow prevented from sliding, the conditions

$$u_n = 0 \tag{2.8.35}$$

$$u_t = 0 \tag{2.8.36}$$

should be used.

REFERENCES

2.1 A.E.N. Love, "On the small free vibrations and deformations of thin elastic shells," *Phil. Trans. Royal Society (London), Vol. 179A*, 1888, pp. 491-546.

2.2 J.W.S. Rayleigh, *The Theory of Sound*, Dover, New York, 1945 (1896).

2.3 G.R. Kirchhoff, "Über das Gleichgewicht und die Bewegung einer elastischen Scheibe," *J. Math. (Crelle), vol. 40*, 1850.

2.4 E. Reissner, "A new derivation of the equations for the deformation of elastic shells," *Amer. J. Math., vol. 63*, 1941, pp. 177-184.

2.5 H. Kraus, *Thin Elastic Shells*, John Wiley, New York, 1967.

3

EQUATIONS OF MOTION FOR COMMONLY OCCURRING GEOMETRIES

In the following we will derive the general shell of revolution equations by reduction from the general Love equations. The shell of revolution equations are then further reduced to specific cases like the conical shell, the circular cylindrical shell, etc. Note that one can obtain the specific cases directly, without going through the general shell of revolution case, by directly substituting the proper values for α_1, α_2, A_1, A_2, R_1, and R_2 into Love's equations. For literature that uses reduction see references 3.1 to 3.5. For literature where equations for specific geometries are derived directly see references 3.6 to 3.8.

3.1 SHELLS OF REVOLUTION

Consider a shell whose neutral surface is a surface of revolution. For such a shell, the lines of principal curvature are its meridians and its parallel circles as shown in Fig. 3.1.1. Thus

$$\alpha_1 = \phi \tag{3.1.1}$$

$$\alpha_2 = \theta \tag{3.1.2}$$

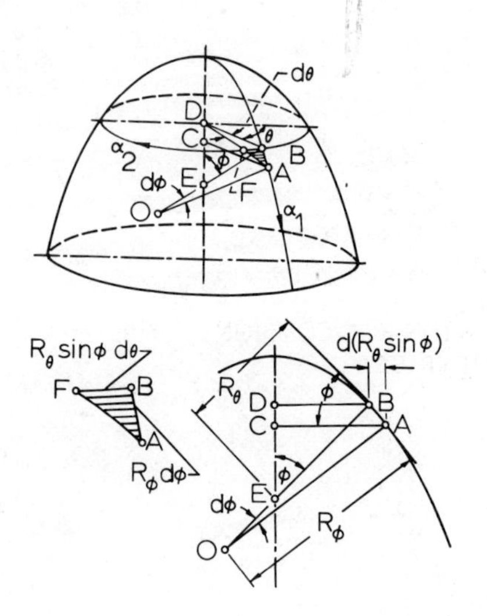

Figure 3.1.1

$$R_1 = R_\phi \tag{3.1.3}$$

$$R_2 = R_\theta \tag{3.1.4}$$

The infinitesimal distances $\overline{BA}$ and $\overline{BF}$ are

$$\overline{BA} = R_\phi \, d\phi \tag{3.1.5}$$

$$\overline{BF} = R_\theta \sin \phi \, d\theta \tag{3.1.6}$$

The fundamental form is therefore

$$(ds)^2 = R_\phi{}^2(d\phi)^2 + R_\theta{}^2\sin^2\phi \; (d\theta)^2 \tag{3.1.7}$$

and therefore

$$A_1 = R_\phi \tag{3.1.8}$$

$$A_2 = R_\theta \sin \phi \tag{3.1.9}$$

Inserting Eqs. (3.1.1) to (3.1.4) and Eqs. (3.1.8) and (3.1.9) in Eqs. (2.7.20) to (2.7.24) gives, with subscripts 1 and 2 changed to ϕ and θ, and with $R_\phi \cos\phi \, d\phi = d(R_\theta \sin \phi)$ from Fig. 3.1.1,

$$\frac{\partial}{\partial\phi}(N_{\phi\phi}R_\theta \sin\phi) + R_\phi \frac{\partial N_{\theta\phi}}{\partial\theta} - N_{\theta\theta}R_\phi \cos\phi$$

$$+ R_\phi R_\theta \sin\phi\left(\frac{Q_{\phi 3}}{R_\phi} + q_\phi\right) = R_\phi R_\theta \sin\phi\rho h \frac{\partial^2 u_\phi}{\partial t^2} \tag{3.1.10}$$

$$\frac{\partial}{\partial\phi}(N_{\phi\theta}R_\theta \sin\phi) + R_\phi \frac{\partial N_{\theta\theta}}{\partial\theta} + N_{\theta\phi}R_\phi \cos\phi$$

$$+ R_\phi R_\theta \sin\phi\left(\frac{Q_{\theta 3}}{R_\theta} + q_\theta\right) = R_\phi R_\theta \sin\phi\rho h \frac{\partial^2 u_\theta}{\partial t^2} \tag{3.1.11}$$

$$\frac{\partial}{\partial\phi}(Q_{\phi 3}R_\theta \sin\phi) + R_\phi \frac{\partial Q_{\theta 3}}{\partial\theta} - \left(\frac{N_{\phi\phi}}{R_\phi} + \frac{N_{\theta\theta}}{R_\theta}\right)R_\phi R_\theta \sin\phi$$

$$+ q_3 R_\phi R_\theta \sin\phi = R_\phi R_\theta \sin\phi\rho h \frac{\partial^2 u_3}{\partial t^2} \tag{3.1.12}$$

where

$$Q_{\phi 3} = \frac{1}{R_\phi R_\theta \sin\phi}\left[\frac{\partial}{\partial\phi}(M_{\phi\phi}R_\theta \sin\phi) + R_\phi \frac{\partial M_{\theta\phi}}{\partial\theta} - M_{\theta\theta}R_\phi \cos\phi\right] \tag{3.1.13}$$

$$Q_{\theta 3} = \frac{1}{R_\phi R_\theta \sin\phi}\left[\frac{\partial}{\partial\phi}(M_{\phi\theta}R_\theta \sin\phi) + R_\phi \frac{\partial M_{\theta\theta}}{\partial\theta} + M_{\theta\phi}R_\phi \cos\phi\right] \tag{3.1.14}$$

The strain-displacement relations (2.4.19) to (2.4.24) become

$$\varepsilon^\circ_{\phi\phi} = \frac{1}{R_\phi}\left(\frac{\partial u_\phi}{\partial\phi} + u_3\right) \tag{3.1.15}$$

$$\varepsilon^\circ_{\theta\theta} = \frac{1}{R_\theta \sin\phi}\left(\frac{\partial u_\theta}{\partial\theta} + u_\phi \cos\phi + u_3 \sin\phi\right) \tag{3.1.16}$$

$$\varepsilon^\circ_{\phi\theta} = \frac{R_\theta}{R_\phi} \sin\phi \frac{\partial}{\partial\phi}\left(\frac{u_\theta}{R_\theta \sin\phi}\right) + \frac{1}{R_\theta \sin\phi} \frac{\partial u_\phi}{\partial\theta} \tag{3.1.17}$$

$$k_{\phi\phi} = \frac{1}{R_\phi} \frac{\partial\beta_\phi}{\partial\phi} \tag{3.1.18}$$

$$k_{\theta\theta} = \frac{1}{R_\theta \sin\phi}\left(\frac{\partial\beta_\theta}{\partial\theta} + \beta_\phi \cos\phi\right) \tag{3.1.19}$$

$$k_{\phi\theta} = \frac{R_\theta}{R_\phi} \sin \phi \frac{\partial}{\partial \phi}\left(\frac{\beta_\theta}{R_\theta \sin \phi}\right) + \frac{1}{R_\theta \sin \phi} \frac{\partial \beta_\phi}{\partial \theta} \tag{3.1.20}$$

and β_ϕ and β_θ are, from Eqs. (2.4.7) and (2.4.8),

$$\beta_\phi = \frac{1}{R_\phi}\left(u_\phi - \frac{\partial u_3}{\partial \phi}\right) \tag{3.1.21}$$

$$\beta_\theta = \frac{1}{R_\theta \sin \phi}\left(u_\theta \sin \phi - \frac{\partial u_3}{\partial \theta}\right) \tag{3.1.22}$$

The relations between the force and moment resultants and the strains are given by Eqs. (2.5.9), (2.5.11), (2.5.12), (2.5.15), (2.5.17), and (2.5.18):

$$N_{\phi\phi} = K(\varepsilon^\circ_{\phi\phi} + \mu\varepsilon^\circ_{\theta\theta}) \tag{3.1.23}$$

$$N_{\theta\theta} = K(\varepsilon^\circ_{\theta\theta} + \mu\varepsilon^\circ_{\phi\phi}) \tag{3.1.24}$$

$$N_{\phi\theta} = N_{\theta\phi} = \frac{K(1 - \mu)}{2} \varepsilon^\circ_{\phi\theta} \tag{3.1.25}$$

$$M_{\phi\phi} = D(k_{\phi\phi} + \mu k_{\theta\theta}) \tag{3.1.26}$$

$$M_{\theta\theta} = D(k_{\theta\theta} + \mu k_{\phi\phi}) \tag{3.1.27}$$

$$M_{\phi\theta} = M_{\theta\phi} = \frac{D(1 - \mu)}{2} k_{\phi\theta} \tag{3.1.28}$$

3.2 CIRCULAR CONICAL SHELL

For the case of the circular conical shell, as shown in Fig. 3.2.1, we see that

$$\frac{1}{R_\phi} = 0 \tag{3.2.1}$$

$$R_\theta = x \tan \alpha \tag{3.2.2}$$

and since

$$\phi = \frac{\pi}{2} - \alpha \tag{3.2.3}$$

we get

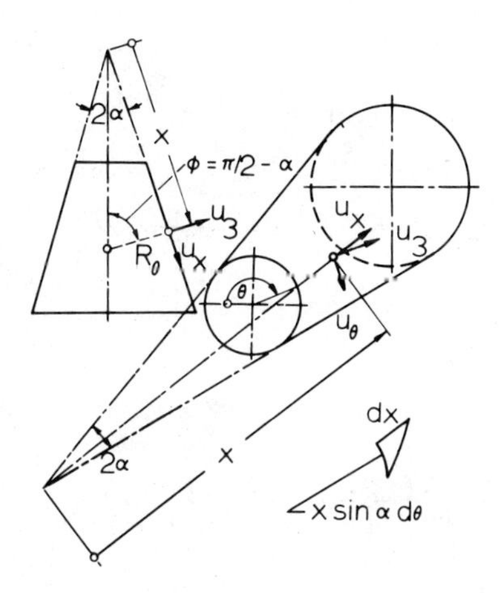

Figure 3.2.1

$$\sin \phi = \cos \alpha \tag{3.2.4}$$

$$\cos \phi = \sin \alpha \tag{3.2.5}$$

Furthermore, the fundamental form is now

$$(ds)^2 = (dx)^2 + x^2 \sin^2 \alpha \, (d\theta)^2 \tag{3.2.6}$$

Comparing this to Eq. (3.1.7), we obtain

$$R_\phi d\phi = dx \tag{3.2.7}$$

and $$R_\theta \sin \phi = x \sin \alpha \tag{3.2.8}$$

The subscript ϕ is replaced by x and Eqs. (3.1.10) to (3.1.14) read

$$\frac{\partial N_{xx}}{\partial x} + \frac{1}{x \sin \alpha} \frac{\partial N_{\theta x}}{\partial \theta} + \frac{1}{x}(N_{xx} - N_{\theta\theta}) + q_x = \rho h \frac{\partial^2 u_x}{\partial t^2} \tag{3.2.9}$$

$$\frac{\partial N_{x\theta}}{\partial x} + \frac{2}{x} N_{\theta x} + \frac{1}{x \sin \alpha} \frac{\partial N_{\theta\theta}}{\partial \theta} + \frac{1}{x \tan \alpha} Q_{\theta 3} + q_\theta = \rho h \frac{\partial^2 u_\theta}{\partial t^2} \tag{3.2.10}$$

$$\frac{\partial Q_{x3}}{\partial x} + \frac{1}{x} Q_{x3} + \frac{1}{x \sin \alpha} \frac{\partial Q_{\theta 3}}{\partial \theta} - \frac{1}{x \tan \alpha} N_{\theta\theta} + q_3 = \rho h \frac{\partial^2 u_3}{\partial t^2} \tag{3.2.11}$$

where

$$Q_{x3} = \frac{\partial M_{xx}}{\partial x} + \frac{M_{xx}}{x} + \frac{1}{x \sin \alpha} \frac{\partial M_{\theta x}}{\partial \theta} - \frac{M_{\theta\theta}}{x} \tag{3.2.12}$$

$$Q_{\theta 3} = \frac{\partial M_{x\theta}}{\partial x} + \frac{2}{x} M_{\theta x} + \frac{1}{x \sin \alpha} \frac{\partial M_{\theta\theta}}{\partial \theta} \tag{3.2.13}$$

The strain-displacement relations become

$$\varepsilon_{xx}^{\circ} = \frac{\partial u_x}{\partial x} \tag{3.2.14}$$

$$\varepsilon_{\theta\theta}^{\circ} = \frac{1}{x \sin \alpha} \frac{\partial u_\theta}{\partial \theta} + \frac{1}{x} u_x + \frac{1}{x \tan \alpha} u_3 \tag{3.2.15}$$

$$\varepsilon_{x\theta}^{\circ} = \frac{\partial u_\theta}{\partial x} - \frac{1}{x} u_\theta + \frac{1}{x \sin \alpha} \frac{\partial u_x}{\partial \theta} \tag{3.2.16}$$

$$k_{xx} = \frac{\partial \beta_x}{\partial x} \tag{3.2.17}$$

$$k_{\theta\theta} = \frac{1}{x \sin \alpha} \frac{\partial \beta_\theta}{\partial \theta} + \frac{1}{x} \beta_x \tag{3.2.18}$$

$$k_{x\theta} = x \frac{\partial}{\partial x} \left(\frac{\beta_\theta}{x} \right) + \frac{1}{x \sin \alpha} \frac{\partial \beta_x}{\partial \theta} \tag{3.2.19}$$

and β_1 and β_2 become

$$\beta_x = - \frac{\partial u_3}{\partial x} \tag{3.2.20}$$

$$\beta_\theta = \frac{1}{x \tan \alpha} u_\theta - \frac{1}{x \sin \alpha} \frac{\partial u_3}{\partial \theta} \tag{3.2.21}$$

Note that as $\alpha \to 0$, a circular cylindrical shell results, with

$$x \tan \alpha \to a \qquad \sin \alpha \to 0 \qquad \cos \alpha \to 1$$

As $\alpha \to \frac{\pi}{2}$, we approach the circular plate equations.

3.3 CIRCULAR CYLINDRICAL SHELL

An important subcase of the circular conical shell is the circular cylindrical shell (Fig. 3.3.1), which has the fundamental form

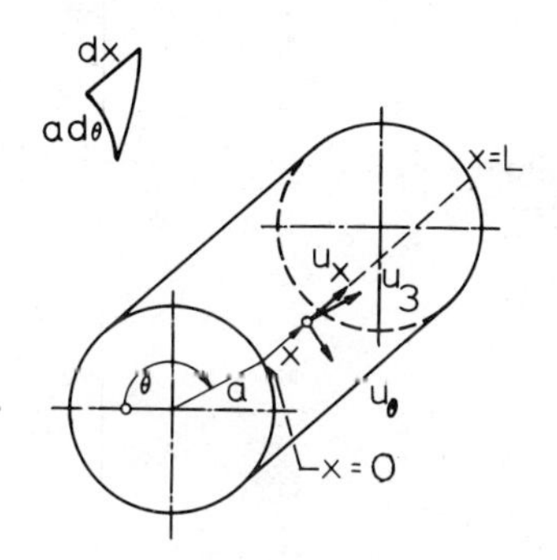

Figure 3.3.1

$$(ds)^2 = (dx)^2 + a^2(d\theta)^2 \tag{3.3.1}$$

Letting α approach zero with x sin α and x tan α approaching a and 1/x approaching zero, from Eqs. (3.2.9) to (3.2.13) we have

$$\frac{\partial N_{xx}}{\partial x} + \frac{1}{a}\frac{\partial N_{\theta x}}{\partial \theta} + q_x = \rho h \frac{\partial^2 u_x}{\partial t^2} \tag{3.3.2}$$

$$\frac{\partial N_{x\theta}}{\partial x} + \frac{1}{a}\frac{\partial N_{\theta\theta}}{\partial \theta} + \frac{Q_{\theta 3}}{a} + q_\theta = \rho h \frac{\partial^2 u_\theta}{\partial t^2} \tag{3.3.3}$$

$$\frac{\partial Q_{x3}}{\partial x} + \frac{1}{a}\frac{\partial Q_{\theta 3}}{\partial \theta} - \frac{N_{\theta\theta}}{a} + q_3 = \rho h \frac{\partial^2 u_3}{\partial t^2} \tag{3.3.4}$$

where

$$Q_{x3} = \frac{\partial M_{xx}}{\partial x} + \frac{1}{a}\frac{\partial M_{\theta x}}{\partial \theta} \tag{3.3.5}$$

$$Q_{\theta 3} = \frac{\partial M_{x\theta}}{\partial x} + \frac{1}{a}\frac{\partial M_{\theta\theta}}{\partial \theta} \tag{3.3.6}$$

The strain-displacement relations become

$$\varepsilon_{xx}^{\circ} = \frac{\partial u_x}{\partial x} \tag{3.3.7}$$

$$\varepsilon_{\theta\theta}^{\circ} = \frac{1}{a}\frac{\partial u_\theta}{\partial \theta} + \frac{u_3}{a} \tag{3.3.8}$$

$$\varepsilon_{x\theta}^{\circ} = \frac{\partial u_\theta}{\partial x} + \frac{1}{a}\frac{\partial u_x}{\partial \theta} \tag{3.3.9}$$

$$k_{xx} = \frac{\partial \beta_x}{\partial x} \tag{3.3.10}$$

$$k_{\theta\theta} = \frac{1}{a}\frac{\partial \beta_\theta}{\partial \theta} \tag{3.3.11}$$

$$k_{x\theta} = \frac{\partial \beta_\theta}{\partial x} + \frac{1}{a}\frac{\partial \beta_x}{\partial \theta} \tag{3.3.12}$$

and β_1 and β_2 become

$$\beta_x = -\frac{\partial u_3}{\partial x} \tag{3.3.13}$$

$$\beta_\theta = \frac{u_\theta}{a} - \frac{1}{a}\frac{\partial u_3}{\partial \theta} \tag{3.3.14}$$

3.4 SPHERICAL SHELL

In this particular case, from Fig. 3.4.1,

$$R_\phi = a \tag{3.4.1}$$

$$R_\theta = a \tag{3.4.2}$$

and the fundamental form becomes

$$(ds)^2 = a^2(d\phi)^2 + a^2\sin^2\phi\,(d\theta)^2 \tag{3.4.3}$$

From Eqs. (3.1.10) to (3.1.14) we get

$$\frac{\partial}{\partial \phi}(N_{\phi\phi}\sin\phi) + \frac{\partial N_{\theta\phi}}{\partial \theta} - N_{\theta\theta}\cos\phi + Q_{\phi 3}\sin\phi + aq_\phi\sin\phi = a\sin\phi\,\rho h\frac{\partial^2 u_\phi}{\partial t^2} \tag{3.4.4}$$

$$\frac{\partial}{\partial \phi}(N_{\phi\theta}\sin\phi) + \frac{\partial N_{\theta\theta}}{\partial \theta} + N_{\theta\phi}\cos\phi + Q_{\theta 3}\sin\phi + aq_\theta\sin\phi = a\sin\phi\,\rho h\frac{\partial^2 u_\theta}{\partial t^2} \tag{3.4.5}$$

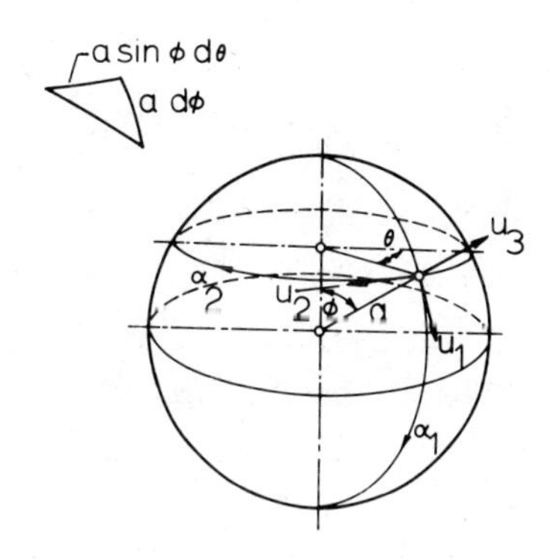

Figure 3.4.1

$$\frac{\partial}{\partial\phi}(Q_{\phi 3}\sin\phi) + \frac{\partial Q_{\theta 3}}{\partial\theta} - (N_{\phi\phi} + N_{\theta\theta})\sin\phi$$

$$+ aq_3\sin\phi = a\sin\phi\,\rho h\,\frac{\partial^2 u_3}{\partial t^2} \tag{3.4.6}$$

and
$$Q_{\phi 3} = \frac{1}{a\sin\phi}\left[\frac{\partial}{\partial\phi}(M_{\phi\phi}\sin\phi) + \frac{\partial M_{\theta\phi}}{\partial\theta} - M_{\theta}\cos\phi\right] \tag{3.4.7}$$

$$Q_{\theta 3} = \frac{1}{a\sin\phi}\left[\frac{\partial}{\partial\phi}(M_{\phi\theta}\sin\phi) + \frac{\partial M_{\theta\theta}}{\partial\theta} + M_{\theta\phi}\cos\phi\right] \tag{3.4.8}$$

The strain-displacement relations become

$$\varepsilon^{\circ}_{\phi\phi} = \frac{1}{a}\left(\frac{\partial u_\phi}{\partial\phi} + u_3\right) \tag{3.4.9}$$

$$\varepsilon^{\circ}_{\theta\theta} = \frac{1}{a\sin\phi}\left(\frac{\partial u_\theta}{\partial\theta} + u_\phi\cos\phi + u_3\sin\phi\right) \tag{3.4.10}$$

$$\varepsilon^{\circ}_{\phi\theta} = \frac{1}{a}\left(\frac{\partial u_\theta}{\partial\phi} - u_\theta\cot\phi + \frac{1}{\sin\phi}\frac{\partial u_\phi}{\partial\theta}\right) \tag{3.4.11}$$

$$k_{\phi\phi} = \frac{1}{a}\frac{\partial\beta_\phi}{\partial\phi} \tag{3.4.12}$$

$$k_{\theta\theta} = \frac{1}{a}\left(\frac{1}{\sin\phi}\frac{\partial\beta_\theta}{\partial\theta} + \beta_\phi\cot\phi\right) \tag{3.4.13}$$

$$k_{\phi\theta} = \frac{1}{a}\left(\frac{\partial\beta_\theta}{\partial\phi} - \beta_\theta\cot\phi + \frac{1}{\sin\phi}\frac{\partial\beta_\phi}{\partial\theta}\right) \tag{3.4.14}$$

and where

$$\beta_\phi = \frac{1}{a}\left(u_\phi - \frac{\partial u_3}{\partial \phi}\right) \tag{3.4.15}$$

$$\beta_\theta = \frac{1}{a}\left(u_\theta - \frac{1}{\sin\phi}\frac{\partial u_3}{\partial \theta}\right) \tag{3.4.16}$$

REFERENCES

3.1 H. Kraus, *Thin Elastic Shells*, John Wiley, New York, 1967.

3.2 W. Nowacki, *Dynamics of Elastic Systems*, John Wiley, New York, 1963.

3.3 V. Z. Vlasov, *General Theory of Shells and Its Applications in Engineering*, NASA-TT-F99, Washington, 1964.

3.4 V. V. Novozhilov, *Thin Shell Theory*, Noordhoff, Groningen, 1965.

3.5 N.A. Kilchevskiy, *Fundamentals of the Analytical Mechanics of Shells*, NASA-TT-F292, Washington, 1965.

3.6 W. Flügge, *Theory of Shells*, Springer-Verlag, Berlin, 1932.

3.7 S. Timoshenko and S. Woinowsky-Krieger, *Theory of Shells*, McGraw-Hill, New York, 1959.

3.8 L. H. Donnell, *Beams, Plates, and Shells*, McGraw-Hill, New York, 1976.

4

NON-SHELL-TYPE STRUCTURES

In the following discussion we will treat rings and beams as special cases of arches. The arch equation is derived by reduction from Love's equations for shells. Also by direct reduction from Love's equation we obtain the plate equation. In literature, reduction is usually not used for these relatively simple structures and each special case is derived from basic principles [4.1-4.4].

4.1 THE ARCH

The arch is a curved beam where all curvature is in one plane only, as shown in Fig. 4.1.1. Vibratory motion is assumed to occur only in that plane. Designating s as the coordinate along the neutral axis of the arch and y as the coordinate perpendicular to the neutral axis, the fundamental form becomes

$$(ds)^2 = (ds)^2 + (dy)^2 \tag{4.1.1}$$

Thus, $A_1 = 1$, $A_2 = 1$, $d\alpha_1 = ds$, and $d\alpha_2 = dy$. Furthermore,

$$R_1 = R_s \tag{4.1.2}$$

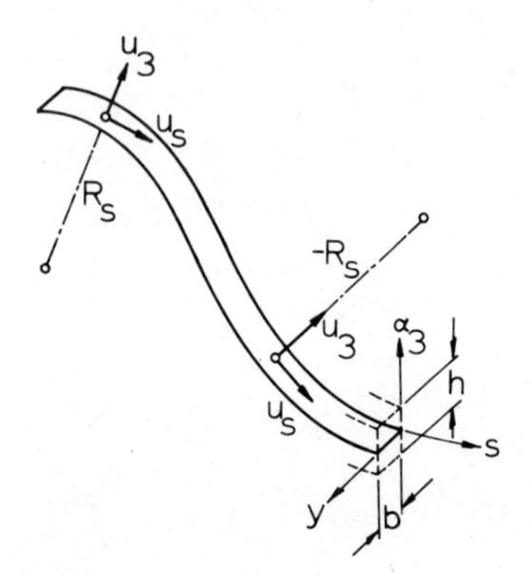

Figure 4.1.1

$$\frac{1}{R_2} = \frac{1}{R_y} = 0 \tag{4.1.3}$$

and

$$\frac{\partial(\ )}{\partial\alpha_2} = \frac{\partial(\ )}{\partial y} = 0 \tag{4.1.4}$$

Also, stresses on the arch in the y direction can be assumed to be zero. From the fact that there are no deflections in the y direction, shear stresses are zero.

$$\begin{aligned} &N_{yy} = 0 \quad M_{yy} = 0 \\ &N_{ys} = N_{sy} = 0 \quad M_{ys} = M_{sy} = 0 \end{aligned} \tag{4.1.5}$$

Love's equations become

$$\frac{\partial N_{ss}}{\partial s} + \frac{Q_{s3}}{R_s} + q_s = \rho h \frac{\partial^2 u_s}{\partial t^2} \tag{4.1.6}$$

$$\frac{\partial Q_{s3}}{\partial s} - \frac{N_{ss}}{R_s} + q_3 = \rho h \frac{\partial^2 u_3}{\partial t^2} \tag{4.1.7}$$

where

$$Q_{s3} = \frac{\partial M_{ss}}{\partial s} \tag{4.1.8}$$

The strain-displacement equations become

$$\varepsilon_{ss}^{\circ} = \frac{\partial u_s}{\partial s} + \frac{u_3}{R_s} \tag{4.1.9}$$

$$k_{ss} = \frac{\partial \beta_s}{\partial s} \tag{4.1.10}$$

where

$$\beta_s = \frac{u_s}{R_s} - \frac{\partial u_3}{\partial s} \tag{4.1.11}$$

Inserting Eqs. (4.1.9) to (4.1.11) in the stress resultant and stress couple relations, we get

$$N_{ss} = K\left(\frac{\partial u_s}{\partial s} + \frac{u_3}{R_s}\right) \tag{4.1.12}$$

$$M_{ss} = D\,\frac{\partial}{\partial s}\left(\frac{u_s}{R_s} - \frac{\partial u_3}{\partial s}\right) \tag{4.1.13}$$

and inserting this in Eqs. (4.1.6) to (4.1.8) results in

$$\frac{D}{R_s}\left[\frac{\partial^2}{\partial s^2}\left(\frac{u_s}{R_s}\right) - \frac{\partial^3 u_3}{\partial s^3}\right] + K\left[\frac{\partial^2 u_s}{\partial s^2} + \frac{\partial}{\partial s}\left(\frac{u_3}{R_s}\right)\right] + q_s = \rho h\,\frac{\partial^2 u_s}{\partial t^2} \tag{4.1.14}$$

$$D\left[\frac{\partial^3}{\partial s^3}\left(\frac{u_s}{R_s}\right) - \frac{\partial^4 u_3}{\partial s^4}\right] - \frac{K}{R_s}\left(\frac{\partial u_s}{\partial s} + \frac{u_3}{R_s}\right) + q_3 = \rho h\,\frac{\partial^2 u_3}{\partial t^2} \tag{4.1.15}$$

Instead of four boundary conditions, we need now only three at each edge. Examining Eqs. (2.8.11) to (2.8.14), we obtain the following necessary boundary conditions that are to be specified at each end of the arch:

$$N_{ss} = N_{ss}^* \quad \text{or} \quad u_s = u_s^* \tag{4.1.16}$$

$$M_{ss} = M_{ss}^* \quad \text{or} \quad \beta_s = \beta_s^* \tag{4.1.17}$$

$$Q_{s3} = Q_{s3}^* \quad \text{or} \quad u_3 = u_3^* \tag{4.1.18}$$

4.2 BEAM AND ROD

For a thin straight beam

$$\frac{1}{R_s} = 0 \tag{4.2.1}$$

and Eqs. (4.1.14) and (4.1.15) reduce to

$$K \frac{\partial^2 u_s}{\partial s^2} + q_s = \rho h \frac{\partial^2 u_s}{\partial t^2} \tag{4.2.2}$$

$$- D \frac{\partial^4 u_3}{\partial s^4} + q_3 = \rho h \frac{\partial^2 u_3}{\partial t^2} \tag{4.2.3}$$

Note that Eqs. (4.2.2) and (4.2.3) are independent from each other. Equation (4.2.2) describes longitudinal vibrations along the axis (commonly called *rod vibrations*), and Eq. (4.2.3) describes vibrations transverse to the beam neutral axis. For the rod vibration of Eq. (4.2.2) we have to specify one boundary condition at each end of the rod

$$N_{ss} = N_{ss}^* \quad \text{or} \quad u_s = u_s^* \tag{4.2.4}$$

For the transverse vibration equation of the beam we must specify two boundary conditions at each end

$$M_{ss} = M_{ss}^* \quad \text{or} \quad \frac{\partial u_3}{\partial s} = \left(\frac{\partial u_3}{\partial s}\right)^* \tag{4.2.5}$$

$$Q_{s3} = Q_{s3}^* \quad \text{or} \quad u_3 = u_3^* \tag{4.2.6}$$

To bring Eqs. (4.2.2) and (4.2.3) to the conventional form, we multiply through by the width b of the beam. Neglecting the Poisson effect, since the width b is of the order of h, we obtain $(s = x)$

$$-EA \frac{\partial^2 u_x}{\partial x^2} + \rho A \ddot{u}_x = q_x' \tag{4.2.7}$$

$$EI \frac{\partial^4 u_3}{\partial x^4} + \rho A \ddot{u}_3 = q_3' \tag{4.2.8}$$

where $A = bh$, $I = bh^3/12$, $q_x' = bq_x$, and $q_3' = bq_3$. Equations (4.2.7) and (4.2.8) apply of course to any cross section, but this fact cannot in general be obtained by the process of reduction. The boundary conditions become for the longitudinal case

$$EA \frac{\partial u_x}{\partial x} = N_x^* \quad \text{or} \quad u_x = u_x^* \tag{4.2.9}$$

where $N_x^* = bN_{xx}^*$ [N]. For the transverse vibration case we have

$$-EI \frac{\partial^2 u_3}{\partial x^2} = M_x^* \quad \text{or} \quad \frac{\partial u_3}{\partial x} = \left(\frac{\partial u_3}{\partial x}\right)^* \tag{4.2.10}$$

$$-EI \frac{\partial^3 u_3}{\partial x^3} = Q_x^* \quad \text{or} \quad u_3 = u_3^* \tag{4.2.11}$$

where $M_x^* = bM_{xx}^*$ [Nm] and $Q_x^* = bQ_{x3}^*$ [N].

4.3 THE CIRCULAR RING

In the case of a circular ring, the radius of arch curvature is constant (Figure 4.3.1),

$$R_s = a \tag{4.3.1}$$

and the coordinate s is commonly expressed as

$$s = a\theta \tag{4.3.2}$$

Therefore, Eqs. (4.1.14) and (4.1.15) reduce to

$$\frac{D}{a^4}\left(\frac{\partial^2 u_\theta}{\partial \theta^2} - \frac{\partial^3 u_3}{\partial \theta^3}\right) + \frac{K}{a^2}\left(\frac{\partial^2 u_\theta}{\partial \theta^2} + \frac{\partial u_3}{\partial \theta}\right) + q_\theta = \rho h \frac{\partial^2 u_\theta}{\partial t^2} \tag{4.3.3}$$

$$\frac{D}{a^4}\left(\frac{\partial^3 u_\theta}{\partial \theta^3} - \frac{\partial^4 u_3}{\partial \theta^4}\right) - \frac{K}{a^2}\left(\frac{\partial u_\theta}{\partial \theta} + u_3\right) + q_3 = \rho h \frac{\partial^2 u_3}{\partial t^2} \tag{4.3.4}$$

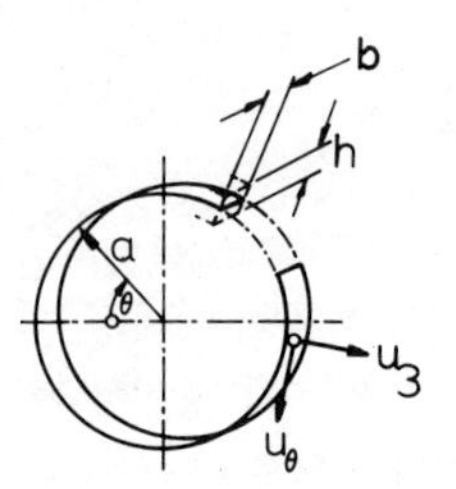

Figure 4.3.1

If there is no circumferential forcing and if the circumferential interia term can be assumed to be negligible, it is possible to eliminate u_θ and obtain Prescott's equation [4.5]:

$$\frac{D}{a^4}\left(\frac{\partial^4 u_3}{\partial \theta^4} + 2\frac{\partial^2 u_3}{\partial \theta^2} + u_3\right) + \rho h \frac{\partial^2 u_3}{\partial t^2} = q_3 \tag{4.3.5}$$

4.4 THE PLATE

A plate is a shell of zero curvatures. Thus

$$\frac{1}{R_1} = 0 \tag{4.4.1}$$

$$\frac{1}{R_2} = 0 \tag{4.4.2}$$

Love's equations become, after substituting Eqs. (4.4.1) and (4.4.2),

$$-\frac{\partial(N_{11}A_2)}{\partial\alpha_1} - \frac{\partial(N_{21}A_1)}{\partial\alpha_2} - N_{12}\frac{\partial A_1}{\partial\alpha_2} + N_{22}\frac{\partial A_2}{\partial\alpha_1} + A_1A_2\rho h\ddot{u}_1 = A_1A_2q_1 \tag{4.4.3}$$

$$-\frac{\partial(N_{12}A_2)}{\partial\alpha_1} - \frac{\partial(N_{22}A_1)}{\partial\alpha_2} - N_{21}\frac{\partial A_2}{\partial\alpha_1} + N_{11}\frac{\partial A_1}{\partial\alpha_2} + A_1A_2\rho h\ddot{u}_2 = A_1A_2q_2 \tag{4.4.4}$$

and, uncoupled from these two equations,

$$-\frac{\partial(Q_{13}A_2)}{\partial\alpha_1} - \frac{\partial(Q_{23}A_1)}{\partial\alpha_2} + A_1A_2\rho h\ddot{u}_3 = A_1A_2q_3 \tag{4.4.5}$$

where

$$Q_{13} = \frac{1}{A_1A_2}\left(\frac{\partial(M_{11}A_2)}{\partial\alpha_1} + \frac{\partial(M_{21}A_1)}{\partial\alpha_2} + M_{12}\frac{\partial A_1}{\partial\alpha_2} - M_{22}\frac{\partial A_2}{\partial\alpha_1}\right) \tag{4.4.6}$$

$$Q_{23} = \frac{1}{A_1A_2}\left(\frac{\partial(M_{12}A_2)}{\partial\alpha_1} + \frac{\partial(M_{22}A_1)}{\partial\alpha_2} + M_{21}\frac{\partial A_2}{\partial\alpha_1} - M_{11}\frac{\partial A_1}{\partial\alpha_2}\right) \tag{4.4.7}$$

Equation (4.4.5) describes the transverse vibrations of plates. Equations (4.4.3) and (4.4.4) describe the in-plane oscillations. For small amplitudes of vibration, these oscillations are independent from each other.

The strain-displacement relationships of Eqs. (2.4.19) to (2.4.21) become

$$\varepsilon_{11}^{\circ} = \frac{1}{A_1}\frac{\partial u_1}{\partial\alpha_1} + \frac{u_2}{A_1A_2}\frac{\partial A_1}{\partial\alpha_2} \tag{4.4.8}$$

$$\varepsilon_{22}^{\circ} = \frac{1}{A_2}\frac{\partial u_2}{\partial\alpha_2} + \frac{u_1}{A_1A_2}\frac{\partial A_2}{\partial\alpha_1} \tag{4.4.9}$$

$$\varepsilon_{12}^{\circ} = \frac{A_2}{A_1}\frac{\partial}{\partial\alpha_1}\left(\frac{u_2}{A_2}\right) + \frac{A_1}{A_2}\frac{\partial}{\partial\alpha_2}\left(\frac{u_1}{A_1}\right) \tag{4.4.10}$$

Since

$$\beta_1 = -\frac{1}{A_1}\frac{\partial u_3}{\partial\alpha_1} \tag{4.4.11}$$

$$\beta_2 = -\frac{1}{A_2}\frac{\partial u_3}{\partial\alpha_2} \tag{4.4.12}$$

we get for Eqs. (2.4.22) to (2.4.24)

$$k_{11} = -\frac{1}{A_1}\frac{\partial}{\partial\alpha_1}\left(\frac{1}{A_1}\frac{\partial u_3}{\partial\alpha_1}\right) - \frac{1}{A_1A_2^{\,2}}\frac{\partial u_3}{\partial\alpha_2}\frac{\partial A_1}{\partial\alpha_2} \tag{4.4.13}$$

$$k_{22} = -\frac{1}{A_2}\frac{\partial}{\partial\alpha_2}\left(\frac{1}{A_2}\frac{\partial u_3}{\partial\alpha_2}\right) - \frac{1}{A_1^{\,2}A_2}\frac{\partial u_3}{\partial\alpha_1}\frac{\partial A_2}{\partial\alpha_1} \tag{4.4.14}$$

$$k_{12} = -\frac{A_2}{A_1}\frac{\partial}{\partial\alpha_1}\left(\frac{1}{A_1^2}\frac{\partial u_3}{\partial\alpha_2}\right) - \frac{A_1}{A_2}\frac{\partial}{\partial\alpha_2}\left(\frac{1}{A_2^2}\frac{\partial u_3}{\partial\alpha_1}\right) \tag{4.4.15}$$

Inserting Eqs. (4.4.13) to (4.4.15) in Eqs. (2.5.15), (2.5.17), and (2.5.18) yields

$$M_{11} = -D\left\{\frac{1}{A_1}\frac{\partial}{\partial\alpha_1}\left(\frac{1}{A_1}\frac{\partial u_3}{\partial\alpha_1}\right) + \frac{1}{A_1A_2^2}\frac{\partial u_3}{\partial\alpha_2}\frac{\partial A_1}{\partial\alpha_2} + \mu\left[\frac{1}{A_2}\frac{\partial}{\partial\alpha_2}\left(\frac{1}{A_2}\frac{\partial u_3}{\partial\alpha_2}\right) + \frac{1}{A_1^2A_2}\frac{\partial u_3}{\partial\alpha_1}\frac{\partial A_2}{\partial\alpha_1}\right]\right\} \tag{4.4.16}$$

$$M_{22} = -D\left\{\frac{1}{A_2}\frac{\partial}{\partial\alpha_2}\left(\frac{1}{A_2}\frac{\partial u_3}{\partial\alpha_2}\right) + \frac{1}{A_1^2A_2}\frac{\partial u_3}{\partial\alpha_1}\frac{\partial A_2}{\partial\alpha_1} + \mu\left[\frac{1}{A_1}\frac{\partial}{\partial\alpha_1}\left(\frac{1}{A_1}\frac{\partial u_3}{\partial\alpha_1}\right) + \frac{1}{A_1A_2^2}\frac{\partial u_3}{\partial\alpha_2}\frac{\partial A_1}{\partial\alpha_2}\right]\right\} \tag{4.4.17}$$

$$M_{12} = M_{21} = -\frac{D(1-\mu)}{2}\left[\frac{A_2}{A_1}\frac{\partial}{\partial\alpha_1}\left(\frac{1}{A_2^2}\frac{\partial u_3}{\partial\alpha_2}\right) + \frac{A_1}{A_2}\frac{\partial}{\partial\alpha_2}\left(\frac{1}{A_1^2}\frac{\partial u_3}{\partial\alpha_1}\right)\right] \tag{4.4.18}$$

Inserting Eqs. (4.4.16) to (4.4.18) in Eqs. (4.4.6) and (4.4.7) and then the resulting expressions in Eq. (4.4.5) gives

$$D\nabla^4 u_3 + \rho h\ddot{u}_3 = q_3 \tag{4.4.19}$$

where

$$\nabla^4(\cdot) = \nabla^2\nabla^2(\cdot) \tag{4.4.20}$$

$$\nabla^2(\cdot) = \frac{1}{A_1A_2}\left[\frac{\partial}{\partial\alpha_1}\left(\frac{A_2}{A_1}\frac{\partial(\cdot)}{\partial\alpha_1}\right) + \frac{\partial}{\partial\alpha_2}\left(\frac{A_1}{A_2}\frac{\partial(\cdot)}{\partial\alpha_2}\right)\right] \tag{4.4.21}$$

The operator ∇^2 is the Laplacian operator. Since it is expressed in curvilinear coordinates, it is now very easy to express it in the coordinate system of our choice. For instance, for cartesian coordinates, $A_1 = 1$, $d\alpha_1 = dx$, $A_2 = 1$, $d\alpha_2 = dy$. Thus

$$\nabla^2(\cdot) = \frac{\partial^2(\cdot)}{\partial x^2} + \frac{\partial^2(\cdot)}{\partial y^2} \tag{4.4.22}$$

and therefore

$$\nabla^4(\cdot) = \frac{\partial^4(\cdot)}{\partial x^4} + 2\frac{\partial^4(\cdot)}{\partial x^2 \partial y^2} + \frac{\partial^4(\cdot)}{\partial y^4} \tag{4.4.23}$$

Thus, Eq. (4.4.19) becomes

$$D\left(\frac{\partial^4 u_3}{\partial x^4} + 2\frac{\partial^4 u_3}{\partial x^2 \partial y^2} + \frac{\partial^4 u_3}{\partial y^4}\right) + \rho h \ddot{u}_3 = q_3 \tag{4.4.24}$$

For circular plates, it is of advantage to employ polar coordinates. In this case, $A_1 = 1$, $d\alpha_1 = dr$, $A_2 = r$, $d\alpha_2 = d\theta$, and

$$\nabla^2(\cdot) = \frac{\partial^2(\cdot)}{\partial r^2} + \frac{1}{r}\frac{\partial(\cdot)}{\partial r} + \frac{1}{r^2}\frac{\partial^2(\cdot)}{\partial \theta^2} \tag{4.4.25}$$

Elliptical plates are defined by elliptical coordinates: $A_1 = A_2 = (a^2 - b^2)(\sin^2 v + \sinh^2 u)$, $d\alpha_1 = du$, $d\alpha_2 = dv$, where a and b are the major halfaxes of the ellipse. The Laplacian operator becomes

$$\nabla^2(\cdot) = \frac{1}{(a^2 - b^2)(\sin^2 v + \sinh^2 u)}\left(\frac{\partial^2(\cdot)}{\partial u^2} + \frac{\partial^2(\cdot)}{\partial v^2}\right) \tag{4.4.26}$$

In general, there are two boundary conditions that are required on each edge. They are, by reduction from Eqs. (2.8.20) and (2.8.21),

$$M_{nn} = M_{nn}^* \quad \text{or} \quad \beta_n = \beta_n^* \tag{4.4.27}$$

$$V_{n3} = V_{n3}^* \quad \text{or} \quad u_3 = u_3^* \tag{4.4.28}$$

For example, for cartesian coordinates these boundary conditions are, at an x = constant edge

$$-D\left(\frac{\partial^2 u_3}{\partial x^2} + \mu\frac{\partial^2 u_3}{\partial y^2}\right) = M_{xx}^* \quad \text{or} \quad \frac{\partial u_3}{\partial x} = \left(\frac{\partial u_3}{\partial x}\right)^* \tag{4.4.29}$$

$$-D\left[\frac{\partial^3 u_3}{\partial x^3} + (2 - \mu)\frac{\partial^3 u_3}{\partial y^2 \partial x}\right] = V^*_{x3} \quad \text{or} \quad u_3 = u^*_3 \tag{4.4.30}$$

REFERENCES

4.1 S. Timoshenko, *Vibration Problems in Engineering*, Van Nostrand, Princeton, N.J., 1955.

4.2 C.B. Biezeno, and R. Grammel, *Engineering Dynamics*, Van Nostrand, Princeton, N.J., 1954.

4.3 W.T. Thomson, *Theory of Vibration with Applications*, Prentice-Hall, Englewood Cliffs, N.J., 1972.

4.4 L. Meirovitch, *Analytical Methods in Vibrations*, Macmillan, London, 1967.

4.5 J. Prescott, *Applied Elasticity*, Dover, New York, 1961 (1924).

5

NATURAL FREQUENCIES AND MODES

Not only is knowledge of natural frequencies and modes important from a design viewpoint (to avoid resonance conditions, for instance), but it is also the foundation for forced response calculations.

In the following, first some generalities are outlined and then specific examples are given.

5.1 THE GENERAL APPROACH

Love's equations can be written, after the substitution of the strain-displacement relations, as

$$L_1\{u_1,u_2,u_3\} + q_1 = \rho h \frac{\partial^2 u_1}{\partial t^2} \quad (5.1.1)$$

$$L_2\{u_1,u_2,u_3\} + q_2 = \rho h \frac{\partial^2 u_2}{\partial t^2} \quad (5.1.2)$$

$$L_3\{u_1,u_2,u_3\} + q_3 = \rho h \frac{\partial^2 u_3}{\partial t^2} \quad (5.1.3)$$

or in short,

$$L_i\{u_1,u_2,u_3\} + q_i = \rho h \frac{\partial^2 u_i}{\partial t^2} \tag{5.1.4}$$

Setting $q_i = 0$ ($i = 1,2,3$) and recognizing that at a natural frequency every point in the elastic system moves harmonically, we may assume that

$$u_1(\alpha_1,\alpha_2,t) = U_1(\alpha_1,\alpha_2)e^{j\omega t} \tag{5.1.5}$$

$$u_2(\alpha_1,\alpha_2,t) = U_2(\alpha_1,\alpha_2)e^{j\omega t} \tag{5.1.6}$$

$$u_3(\alpha_1,\alpha_2,t) = U_3(\alpha_1,\alpha_2)e^{j\omega t} \tag{5.1.7}$$

or, in short,

$$u_i(\alpha_1,\alpha_2,t) = U_i(\alpha_1,\alpha_2)e^{j\omega t} \tag{5.1.8}$$

All three of the $U_i(\alpha_1,\alpha_2)$ functions together constitute a natural mode. Substituting (5.1.8) in (5.1.4) gives, with $q_i = 0$,

$$L_i\{U_1,U_2,U_3\} + \rho h\omega^2 U_i = 0 \tag{5.1.9}$$

Boundary conditions can in general be written

$$B_K\{u_1,u_2,u_3\} = 0 \tag{5.1.10}$$

where $K = 1, 2, \ldots, N$ and where N is the total number of boundary conditions. In the most general case of a four-sided shell segment, we have $N = 16$. For a beam, $N = 4$. For a plate, $N = 8$.

After the substitution of (5.1.8), Eq. (5.1.10) becomes

$$B_K\{U_1,U_2,U_3\} = 0 \tag{5.1.11}$$

The next step is to try separation of variables on Eqs. (5.1.9) and (5.1.11).

$$U_i(\alpha_1,\alpha_2) = R_i(\alpha_1)S_i(\alpha_2) \tag{5.1.12}$$

If this is possible, a set of ordinary differential equations results. Solutions of these equations have N unknown coefficients. Substitution of these solutions into the separated boundary conditions will give a homogeneous set of N equations. The determinant of these equations will furnish the so-called characteristic equation. The roots of this equation will give the natural frequencies.

Often it is not possible to obtain a general solution that is valid for all boundary condition combinations, but solutions for certain boundary conditions can be guessed. Obviously, if the guess satisfies the equations of motion and the particular set of boundary conditions, it is a valid solution, even while we cannot be sure that it is the complete solution. However, experimental evidence usually takes care of this objection.

5.2 TRANSVERSELY VIBRATING BEAMS

The equation of motion is

$$EI \frac{\partial^4 u_3}{\partial x^4} + \rho' \frac{\partial^2 u_3}{\partial t^2} = 0 \tag{5.2.1}$$

where I is the area moment and ρ' is the mass per unit length, having multiplied Eq. (4.2.3) by the width of the beam. Substituting

$$u_3(x,t) = U_3(x)e^{j\omega t} \tag{5.2.2}$$

gives

$$\frac{d^4U_3}{dx^4} - \lambda^4 U_3 = 0 \tag{5.2.3}$$

where

$$\lambda^4 = \frac{\omega^2 \rho'}{EI} \tag{5.2.4}$$

We approach the solution utilizing the Laplace transform. We get

$$(s^4 - \lambda^4)U_3(s) - s^3 U_3(o) - s^2 \frac{dU_3}{dx}(o) - s\frac{d^2U_3}{dx^2}(o) - \frac{d^3U_3}{dx^3}(o) = 0 \tag{5.2.5}$$

Thus

$$U_3(s) = \frac{1}{s^4 - \lambda^4}\left[s^3 U_3(o) + s^2 \frac{dU_3}{dx}(o) + s\frac{d^2U_3}{dx^2}(o) + \frac{d^3U_3}{dx^3}(o)\right] \tag{5.2.6}$$

Taking the inverse transformation yields

$$U_3(x) = U_3(o)A(\lambda x) + \frac{1}{\lambda}\frac{dU_3}{dx}(o)\ B(\lambda x) + \frac{1}{\lambda^2}\frac{d^2U_3}{dx^2}(o)\ C(\lambda x) + \frac{1}{\lambda^3}\frac{d^3U_3}{dx^3}(o)\ D(\lambda x) \tag{5.2.7}$$

where

$$A(\lambda x) = \frac{1}{2}(\cosh \lambda x + \cos \lambda x) \tag{5.2.8}$$

$$B(\lambda x) = \frac{1}{2}(\sinh \lambda x + \sin \lambda x) \tag{5.2.9}$$

$$C(\lambda x) = \frac{1}{2}(\cosh \lambda x - \cos \lambda x) \tag{5.2.10}$$

$$D(\lambda x) = \frac{1}{2}(\sinh \lambda x - \sin \lambda x) \tag{5.2.11}$$

Note that $A(o) = 1$, $B(o) = 0$, $C(o) = 0$, and $D(o) = 0$. Since we will need, for application to specific boundary conditions, the derivatives of Eq. (5.2.7), they are given in the following:

$$\frac{dU_3}{dx}(x) = \lambda U_3(o)D(\lambda x) + \frac{dU_3}{dx}(o)A(\lambda x) + \frac{1}{\lambda}\frac{d^2U_3}{dx^2}(o)\ B(\lambda x) + \frac{1}{\lambda^2}\frac{d^3U_3}{dx^3}(o)\ C(\lambda x) \tag{5.2.12}$$

$$\frac{d^2U_3}{dx^2}(x) = \lambda^2 U_3(o)C(\lambda x) + \lambda\frac{dU_3}{dx}(o)\ D(\lambda x) + \frac{d^2U_3}{dx^2}(o)\ A(\lambda x) + \frac{1}{\lambda}\frac{d^3U_3}{dx^3}(o)\ B(\lambda x) \tag{5.2.13}$$

$$\frac{d^3U_3}{dx^3}(x) = \lambda^3 U_3(o)B(\lambda x) + \lambda^2 \frac{dU_3}{dx}(o)\ C(\lambda x)$$

$$+ \lambda \frac{d^2U_3}{dx^2}(o)\ D(\lambda x) + \frac{d^3U_3}{dx^3}(o)\ A(\lambda x) \tag{5.2.14}$$

Let us treat the clamped-free beam as an example. From Eqs. (4.2.5) and (4.2.6) we formulate the boundary conditions for the clamped end at $x = 0$ as

$$u_3(x = 0,\ t) = 0 \tag{5.2.15}$$

$$\frac{\partial u_3}{\partial x}(x = 0,\ t) = 0 \tag{5.2.16}$$

and at the free end $(x = L)$

$$M_{xx}(x = L,\ t) = 0 \tag{5.2.17}$$

$$Q_{x3}(x = L,\ t) = 0 \tag{5.2.18}$$

Substituting the strain-displacement relations and substituting Eq. (5.2.2) gives

$$U_3(x = 0) = 0 \tag{5.2.19}$$

$$\frac{dU_3}{dx}(x = 0) = 0 \tag{5.2.20}$$

$$\frac{d^2U_3}{dx^2}(x = L) = 0 \tag{5.2.21}$$

$$\frac{d^3U_3}{dx^3}(x = L) = 0 \tag{5.2.22}$$

Substituting Eqs. (5.2.7) and (5.2.12) to (5.2.14) in these conditions gives

$$0 = \frac{d^2U_3}{dx^2}(o)\ A(\lambda L) + \frac{1}{\lambda}\ \frac{d^3U_3}{dx^3}(o)\ B(\lambda L) \tag{5.2.23}$$

$$0 = \lambda \frac{d^2U_3}{dx^2}(o)\ D(\lambda L) + \frac{d^3U_3}{dx^3}(o)\ A(\lambda L) \tag{5.2.24}$$

or

$$\begin{bmatrix} A(\lambda L) & \frac{1}{\lambda} B(\lambda L) \\ \lambda D(\lambda L) & A(\lambda L) \end{bmatrix} \begin{Bmatrix} \frac{d^2U_3}{dx^2}(o) \\ \frac{d^3U_3}{dx^3}(o) \end{Bmatrix} = 0 \tag{5.2.25}$$

Since

$$\begin{Bmatrix} \frac{d^2U_3}{dx^2}(o) \\ \frac{d^3U_3}{dx^3}(o) \end{Bmatrix} \neq 0 \tag{5.2.26}$$

it must be that

$$\begin{vmatrix} A(\lambda L) & \frac{1}{\lambda} B(\lambda L) \\ \lambda D(\lambda L) & A(\lambda L) \end{vmatrix} = 0 \tag{5.2.27}$$

or

$$A^2(\lambda L) - D(\lambda L)B(\lambda L) = 0 \tag{5.2.28}$$

Substituting Eqs. (5.2.8) to (5.2.11) gives

$$\cosh \lambda L \cos \lambda L + 1 = 0 \tag{5.2.29}$$

The roots of this equation are

$$\begin{aligned} \lambda_1 L &= 1.875 \\ \lambda_2 L &= 4.694 \\ \lambda_3 L &= 7.855 \\ \lambda_4 L &= 10.996 \\ \lambda_5 L &= 14.137 \\ &\ldots\ldots\ldots\ldots \end{aligned} \tag{5.2.30}$$

From Eq. (5.2.4)

$$\omega_m = \frac{(\lambda_m L)^2}{L^2} \sqrt{\frac{EI}{\rho'}} \tag{5.2.31}$$

The natural mode is given by Eq. (5.2.7)

$$U_{3m}(x) = \frac{1}{\lambda_m^2} \frac{d^2U_{3m}}{dx^2}(o) \left[C(\lambda_m x) + \frac{1}{\lambda_m} \frac{\dfrac{d^3U_{3m}}{dx^3}(o)}{\dfrac{d^2U_{3m}}{dx^2}(o)} D(\lambda_m x) \right] \tag{5.2.32}$$

From Eq. 5.2.25 we get

$$\frac{\dfrac{d^3U_{3m}}{dx^3}(o)}{\dfrac{d^2U_{3m}}{dx^2}(o)} = -\frac{\lambda_m D(\lambda_m L)}{A(\lambda_m L)} = -\frac{\lambda_m A(\lambda_m L)}{B(\lambda_m L)} \tag{5.2.33}$$

Thus

$$U_{3m}(x) = \frac{1}{\lambda_m^2} \frac{d^2U_{3m}}{dx^2}(o) \left[C(\lambda_m x) - \frac{A(\lambda_m L)}{B(\lambda_m L)} D(\lambda_m x) \right] \tag{5.2.34}$$

The so-called mode shape is determined by the bracketed quantity. The magnitude of the coefficient

$$\frac{1}{\lambda_m^2} \frac{d^2U_{3m}}{dx^2}(o) \tag{5.2.35}$$

is arbitrary as far as the mode shape is concerned and is a function of the excitation.

As a final example, let us look at the simply supported beam. The boundary conditions are

$$u_3(x = 0) = 0 \tag{5.2.36}$$

$$u_3(x = L) = 0 \tag{5.2.37}$$

$$M_{xx}(x = 0) = 0 \tag{5.2.38}$$

$$M_{xx}(x = L) = 0 \tag{5.2.39}$$

Substituting the strain-displacement relations and substituting Eq. (5.2.2) in Eqs. (5.2.36) to (5.2.39) gives

$$U_3(x = 0) = 0 \tag{5.2.40}$$

$$U_3(x = L) = 0 \tag{5.2.41}$$

$$\frac{d^2U_3}{dx^2}(x = 0) = 0 \tag{5.2.42}$$

$$\frac{d^2U_3}{dx^2}(x = L) = 0 \tag{5.2.43}$$

Substituting Eqs. (5.2.7) and (5.2.13) in these relations gives

$$0 = \frac{1}{\lambda}\frac{dU_3}{dx}(o)\ B(\lambda L) + \frac{1}{\lambda^3}\frac{d^3U_3}{dx^3}(o)\ D(\lambda L) \tag{5.2.44}$$

$$0 = \lambda\frac{dU_3}{dx}(o)\ D(\lambda L) + \frac{1}{\lambda}\frac{d^3U_3}{dx^3}(o)\ B(\lambda L) \tag{5.2.45}$$

or

$$\begin{bmatrix} \frac{1}{\lambda} B(\lambda L) & \frac{1}{\lambda^3} D(\lambda L) \\ \lambda D(\lambda L) & \frac{1}{\lambda} B(\lambda L) \end{bmatrix} \begin{Bmatrix} \frac{dU_3}{dx}(o) \\ \frac{d^3U_3}{dx^3}(o) \end{Bmatrix} = 0 \tag{5.2.46}$$

and therefore, following the same argument as before,

$$\frac{1}{\lambda^2}\left(B^2(\lambda L) - D^2(\lambda L)\right) = 0 \tag{5.2.47}$$

or $\sinh \lambda L \sin \lambda L = 0$ (5.2.48)

Since

$$\lambda \neq 0 \tag{5.2.49}$$

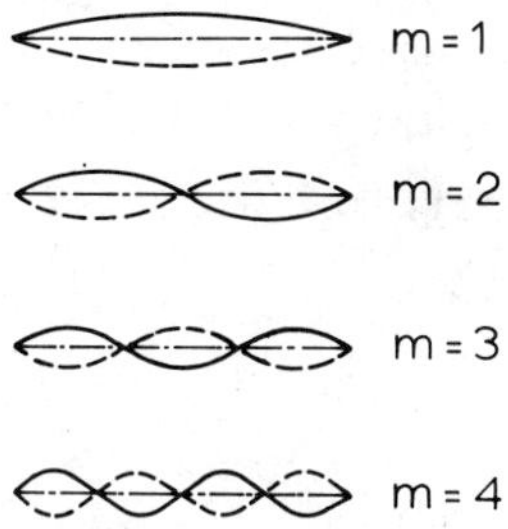

Figure 5.2.1

this equation reduces to

$$\sin \lambda L = 0 \tag{5.2.50}$$

or

$$\lambda_m L = m\pi \quad (m = 1, 2, \ldots) \tag{5.2.51}$$

and

$$\omega_m = \frac{m^2\pi^2}{L^2}\sqrt{\frac{EI}{\rho'}} \tag{5.2.52}$$

The natural mode is, from Eq. (5.2.7),

$$U_{3m}(x) = \frac{1}{\lambda_m}\frac{dU_{3m}}{dx}(o)\left[B(\lambda_m x) + \frac{1}{\lambda_m^2}\frac{\frac{d^3U_{3m}}{dx^3}(o)}{\frac{dU_{3m}}{dx}(o)}D(\lambda_m x)\right] \tag{5.2.53}$$

From Eq. (5.2.46)

$$\frac{\frac{d^3U_{3m}}{dx^3}(o)}{\frac{dU_{3m}}{dx}(o)} = -\lambda_m^2\frac{B(\lambda_m L)}{D(\lambda_m L)} = -\lambda_m^2\frac{D(\lambda_m L)}{B(\lambda_m L)} = -\lambda_m^2\frac{\sinh m\pi}{\sinh m\pi} = -\lambda_m^2 \tag{5.2.54}$$

and Eq. (4.2.53) becomes

$$U_{3m}(x) = \frac{1}{\lambda_m}\frac{dU_{3m}}{dx}(o)\sin \lambda_m x \tag{5.2.55}$$

The modes consist of sine waves as shown in Fig. 5.2.1. Note for later reference that we could have guessed the modes for this particular case and by substituting the guess in Eq. (5.2.3) could have found the natural frequencies.

5.3 THE CIRCULAR RING

Equations governing the vibrations of a circular ring in its plane of curvature are given in Sec. 4.3. For no load, substituting

$$u_\theta(\theta,t) = U_{\theta n}(\theta)e^{j\omega_n t} \tag{5.3.1}$$

$$u_3(\theta,t) = U_{3n}(\theta)e^{j\omega_n t} \tag{5.3.2}$$

gives

$$\frac{D}{a^4}\left(\frac{d^2U_{\theta n}}{d\theta^2} - \frac{d^3U_{3n}}{d\theta^3}\right) + \frac{K}{a^2}\left(\frac{d^2U_{\theta n}}{d\theta^2} + \frac{dU_{3n}}{d\theta}\right) + \rho h\omega_n^{\ 2}U_{\theta n} = 0 \tag{5.3.3}$$

$$\frac{D}{a^4}\left(\frac{d^3U_{\theta n}}{d\theta^3} - \frac{d^4U_{3n}}{d\theta^4}\right) - \frac{K}{a^2}\left(\frac{dU_{\theta n}}{d\theta} + U_{3n}\right) + \rho h\omega_n^{\ 2}U_{3n} = 0 \tag{5.3.4}$$

It is possible to approach the solution using the Laplace transformation in a manner similar to that in the previous chapter. However, it is possible, in certain cases, to take a shortcut. The approach in these cases is an inspired guess. Let us take, for example, the free floating closed ring. (See Figure 5.3.1 for a

$$U_{3n}(\theta) = A_n \cos n(\theta - \phi) \tag{5.3.5}$$

$$U_{\theta n}(\theta) = B_n \sin n(\theta - \phi) \tag{5.3.6}$$

physical interpretation of the following assumption.) We assume that ϕ is an arbitrary phase angle which has to be included since the ring does not show a preference for the orientation of its modes; rather, the orientation is determined later by the distribution of the external forces. In Figure 5.3.1. we have sketched the n = 2 mode, setting $\phi = 0$. Note that physical intuition confirms Eqs.

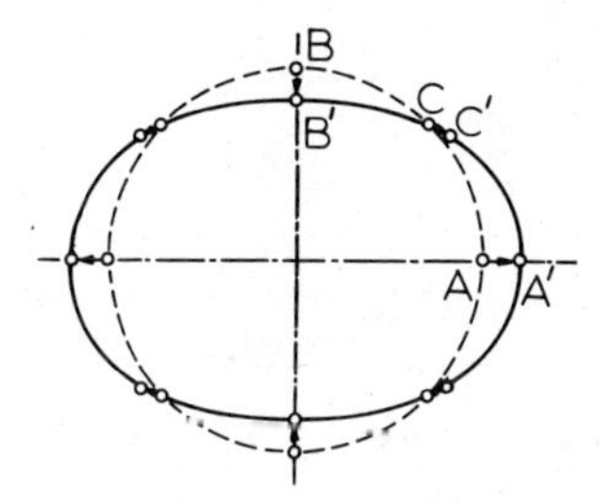

Figure 5.3.1

(5.3.5) and (5.3.6) since obviously point A will move along the x axis to the point A' $[U_{32}(o) = A_2;\ U_{\theta 2}(o) = 0]$, point B will move along the y axis to the point B'$[U_{32}(\frac{\pi}{2}) = -A_2;\ U_{\theta 2}(\frac{\pi}{2}) = 0]$, while point C will not move in the normal direction, but rather will move in the circumferential direction $[U_{32}(\frac{\pi}{4}) = 0;\ U_{\theta 2}(\frac{\pi}{4}) = B_2]$.

Substituting Eqs. (5.3.5) and (5.3.6) in Eqs. (5.3.3) and (5.3.4) gives

$$\begin{bmatrix} \rho h \omega_n^2 - \frac{n^4 D}{a^4} - \frac{K}{a^2} & -\frac{n^3 D}{a^4} - \frac{nK}{a^2} \\ -\frac{n^3 D}{a^4} - \frac{nK}{a^2} & \rho h \omega_n^2 - \frac{n^2 D}{a^4} - \frac{n^2 K}{a^2} \end{bmatrix} \begin{Bmatrix} A_n \\ B_n \end{Bmatrix} = 0 \tag{5.3.7}$$

Since, in general

$$\begin{Bmatrix} A_n \\ B_n \end{Bmatrix} \neq 0 \tag{5.3.8}$$

It must be that the determinant is zero. Thus

$$\omega_n^4 - K_1 \omega_n^2 + K_2 = 0 \tag{5.3.9}$$

where

$$K_1 = \frac{n^2 + 1}{a^2 \rho h} \left(\frac{n^2 D}{a^2} + K \right) \tag{5.3.10}$$

$$K_2 = \frac{n^2(n^2 - 1)^2}{a^6(\rho h)^2} DK \tag{5.3.11}$$

Therefore

$$\omega_n^2 = \frac{K_1}{2}\left(1 \pm \sqrt{1 - 4\frac{K_2}{K_1^2}}\right) \tag{5.3.12}$$

We encounter, therefore, for each value of n, a frequency

$$\omega_{n1}^2 = \frac{K_1}{2}\left(1 - \sqrt{1 - 4\frac{K_2}{K_1^2}}\right) \tag{5.3.13}$$

and a frequency

$$\omega_{n2}^2 = \frac{K_1}{2}\left(1 + \sqrt{1 - 4\frac{K_2}{K_1^2}}\right) \tag{5.3.14}$$

As it turns out, for typical rings,

$$\omega_{n2} >> \omega_{n1} \tag{5.3.15}$$

From Eqs. (5.3.7) we get

$$\frac{A_{ni}}{B_{ni}} = \frac{\frac{n}{a^2}\left(\frac{n^2}{a^2} D + K\right)}{\rho h \omega_{ni}^2 - \frac{1}{a^2}\left(\frac{n^4 D}{a^2} + K\right)} = \frac{\rho h \omega_{ni}^2 - \frac{n^2}{a^2}\left(\frac{D}{a^2} + K\right)}{\frac{n}{a^2}\left(\frac{n^2 D}{a^2} + K\right)} \tag{5.3.16}$$

where i = 1, 2. To gain an intuitive feeling of this ratio, let us look at lower n numbers where

$$\frac{n^2 D}{a^2} << K \tag{5.3.17}$$

We get

$$\rho h \omega_{n1}^2 << \frac{K}{a^2} \tag{5.3.18}$$

$$\omega_{n2}^2 \cong K_1 \cong \frac{n^2 + 1}{a^2 \rho h} K \tag{5.3.19}$$

Thus

$$\frac{A_{n1}}{B_{n1}} \cong -n \tag{5.3.20}$$

and
$$\frac{A_{n2}}{B_{n2}} \cong \frac{1}{n} \tag{5.3.21}$$

The conclusion is that at ω_{n1} frequencies, transverse deflections dominate: The ring is essentially vibrating in bending, analogous to the transverse bending vibration of a beam. At the ω_{n2} frequencies, circumferential deflections dominate, analogous to the longitudinal vibrations of a beam.

Now let us look at some specific values of n. At n = 0,

$$K_1 = \frac{K}{a^2 \rho h} \tag{5.3.22}$$

$$K_2 = 0 \tag{5.3.23}$$

and therefore

$$\omega_{o1}^2 = 0 \tag{5.3.24}$$

$$\omega_{o2}^2 = K_1 = \frac{K}{a^2 \rho h} \tag{5.3.25}$$

and
$$\frac{A_{01}}{B_{01}} = 0 \qquad \frac{B_{02}}{A_{02}} = 0 \tag{5.3.26}$$

The mode shape is shown in Fig. 5.3.2 and is sometimes called the *breathing mode* of the ring.

At n = 1, we get

$$K_1 = \frac{2}{a^2 \rho h}\left(\frac{D}{a^2} + K\right) \tag{5.3.27}$$

$$K_2 = 0 \tag{5.3.28}$$

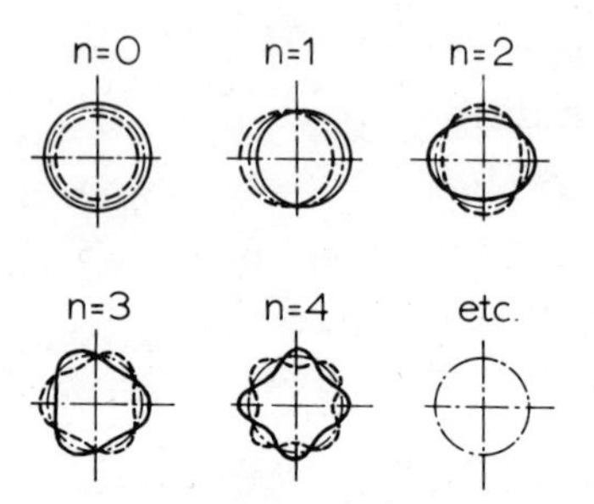

Figure 5.3.2

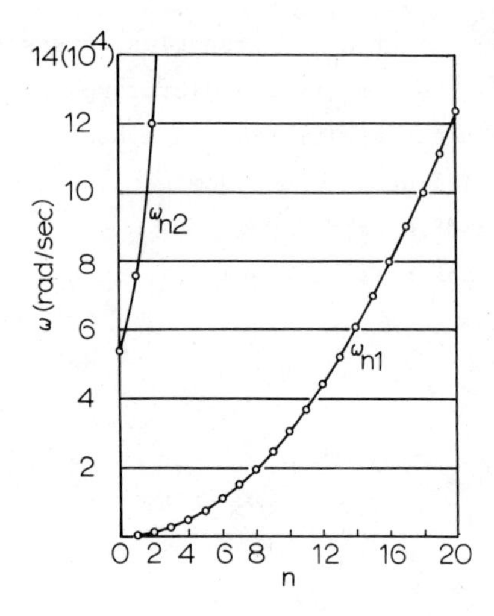

Figure 5.3.3

and therefore

$$\omega_{11}^{\ 2} = 0 \tag{5.3.29}$$

Thus, a bending vibration still does not exist; we have to think of the ring as being simply displaced in a rigid body motion as shown in Fig. 5.3.2. However, a ω_{n2} frequency does occur and the mode is one compression and one tension region around the ring.

Starting with n = 2, two nonzero sets of natural frequencies and modes exist. Frequencies as function of n value are plotted in Fig. 5.3.3 ($E = 20.6 \times 10^4$ N/mm^2, $\rho = 7.85 \times 10^{-9}$ Nsec^2/mm^4, $\mu = 0.3$, h = 2 mm, a = 100 mm).

5.4 RECTANGULAR PLATES THAT ARE SIMPLY SUPPORTED ALONG TWO OPPOSING EDGES

Let us suppose that the simply supported edges occur always along the x = 0 and x = a edge as shown in Fig. 5.4.1. The boundary conditions on these edges are therefore

$$u_3(o,y,t) = 0 \tag{5.4.1}$$

$$u_3(a,y,t) = 0 \tag{5.4.2}$$

$$M_{xx}(o,y,t) = 0 \tag{5.4.3}$$

$$M_{xx}(a,y,t) = 0 \tag{5.4.4}$$

The equation of motion is, from Eq. (4.4.24),

$$D\left(\frac{\partial^4 u_3}{\partial x^4} + 2\frac{\partial^4 u_3}{\partial x^2 \partial y^2} + \frac{\partial^4 u_3}{\partial y^4}\right) + \rho h \frac{\partial^2 u_3}{\partial t^2} = 0 \tag{5.4.5}$$

Substituting

$$u_3(x,y,t) = U_3(x,y)e^{j\omega t} \tag{5.4.6}$$

gives

$$D\left(\frac{\partial^4 U_3}{\partial x^4} + 2\frac{\partial^4 U_3}{\partial x^2 \partial y^2} + \frac{\partial^4 U_3}{\partial y^4}\right) - \rho h\omega^2 U_3 = 0 \tag{5.4.7}$$

Substituting the strain-displacement relations and Eq. (5.4.6) in Eqs. (5.4.1) to (5.4.4) gives

$$U_3(o,y) = 0 \tag{5.4.8}$$

$$U_3(a,y) = 0 \tag{5.4.9}$$

$$\frac{\partial^2 U_3}{\partial x^2}(o,y) = 0 \tag{5.4.10}$$

$$\frac{\partial^2 U_3}{\partial x^2}(a,y) = 0 \tag{5.4.11}$$

Variables become separated and the boundary conditions (5.4.8)

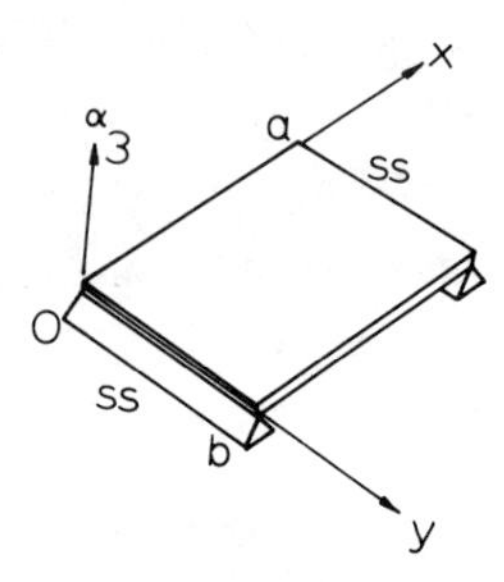

Figure 5.4.1

to (5.4.11) are satisfied if we assume a solution of the form

$$U_3(x,y) = Y(y)\sin\frac{m\pi x}{a} \tag{5.4.12}$$

Substituting Eq. (5.4.12) in Eq. (5.4.7) gives

$$\frac{d^4Y}{dy^4} - 2\left(\frac{m\pi}{a}\right)^2\frac{d^2Y}{dy^2} + \left[\left(\frac{m\pi}{a}\right)^4 - \frac{\rho h}{D}\omega^2\right]Y = 0 \tag{5.4.13}$$

The solution to this ordinary fourth-order differential equation must satisfy four boundary condtions. Substituting

$$Y(y) = \sum_{i=1}^{4} C_i e^{\lambda_i \frac{y}{b}} \tag{5.4.14}$$

gives

$$\left(\frac{\lambda_i}{b}\right)^4 - 2\left(\frac{m\pi}{a}\right)^2\left(\frac{\lambda_i}{b}\right)^2 + \left[\left(\frac{m\pi}{a}\right)^4 - \frac{\rho h}{D}\omega^2\right] = 0 \tag{5.4.15}$$

or $$\lambda_i = \pm\frac{b}{a}m\pi\sqrt{1 \pm K} \tag{5.4.16}$$

where

$$K = \frac{\omega}{\frac{m^2\pi^2}{a^2}\sqrt{\frac{D}{\rho h}}} \tag{5.4.17}$$

Since the frequency of a simple supported beam of length a is

$$\omega_b = \frac{m^2\pi^2}{a^2}\sqrt{\frac{D}{\rho h}} \tag{5.4.18}$$

we get

$$K = \frac{\omega}{\omega_b} \tag{5.4.19}$$

Thus

$$K > 1 \tag{5.4.20}$$

and thus

$$\lambda_1 = +\rho_1 \quad \lambda_2 = -\rho_1 \quad \lambda_3 = +j\rho_2 \quad \lambda_4 = -j\rho_2 \tag{5.4.21}$$

where

$$\rho_1 = \frac{b}{a} m\pi\sqrt{K + 1} \quad \rho_2 = \frac{b}{a} m\pi\sqrt{K - 1} \tag{5.4.22}$$

and solution (5.4.14) reads

$$Y(y) = C_1 e^{\rho_1 \frac{y}{b}} + C_2 e^{-\rho_1 \frac{y}{b}} + C_3 e^{j\rho_2 \frac{y}{b}} + C_4 e^{-j\rho_2 \frac{y}{b}} \tag{5.4.23}$$

Now let

$$C_1 = \frac{A + B}{2} \quad C_2 = \frac{A - B}{2} \quad C_3 = \frac{C + D}{2} \quad C_4 = \frac{C - D}{2} \tag{5.4.24}$$

and we get

$$Y(y) = A\,\frac{e^{\rho_1 \frac{y}{b}} + e^{-\rho_1 \frac{y}{b}}}{2} + B\,\frac{e^{\rho_1 \frac{y}{b}} - e^{-\rho_1 \frac{y}{b}}}{2}$$

$$+ C\,\frac{e^{j\rho_2 \frac{y}{b}} + e^{-j\rho_2 \frac{y}{b}}}{2} + D\,\frac{e^{j\rho_2 \frac{y}{b}} - e^{-j\rho_2 \frac{y}{b}}}{2} \tag{5.4.25}$$

or

$$Y(y) = A \cosh \rho_1 \frac{y}{b} + B \sinh \rho_1 \frac{y}{b} + C \cos \rho_2 \frac{y}{b} + D \sin \rho_2 \frac{y}{b} \tag{5.4.26}$$

Let us now consider a few examples

5.4.1 Two Other Edges Are Clamped

In this case the additional four boundary conditions are

$$u_3(x,o,t) = 0 \tag{5.4.27}$$

$$u_3(x,b,t) = 0 \tag{5.4.28}$$

$$\frac{\partial u_3}{\partial y}(x,o,t) = 0 \tag{5.4.29}$$

$$\frac{\partial u_3}{\partial y}(x,b,t) = 0 \tag{5.4.30}$$

and substituting Eqs. (5.4.6) and (5.4.12), this gives

$$Y(o) = 0 \tag{5.4.31}$$

$$Y(b) = 0 \tag{5.4.32}$$

$$\frac{dY}{dy}(o) = 0 \tag{5.4.33}$$

$$\frac{dY}{dy}(b) = 0 \tag{5.4.34}$$

Substituting Eq. (5.4.26) in Eqs. (5.4.31) to (5.4.34) gives

$$0 = A + C \tag{5.4.35}$$

$$0 = A\cosh\rho_1 + B\sin\rho_1 + C\cos\rho_2 + D\sin\rho_2 \tag{5.4.36}$$

$$0 = B\frac{\rho_1}{b} + D\frac{\rho_2}{b} \tag{5.4.37}$$

$$0 = A\frac{\rho_1}{b}\sinh\rho_1 + B\frac{\rho_1}{b}\cosh\rho_1 - C\frac{\rho_2}{b}\sin\rho_2 + D\frac{\rho_2}{b}\cos\rho_2 \tag{5.4.38}$$

or

$$\begin{bmatrix} 1 & 0 & 1 & 0 \\ \cosh\rho_1 & \sinh\rho_1 & \cos\rho_2 & \sin\rho_2 \\ 0 & \rho_1 & 0 & \rho_2 \\ \rho_1\sinh\rho_1 & \rho_1\cosh\rho_1 & -\rho_2\sin\rho_2 & \rho_2\cos\rho_2 \end{bmatrix} \begin{Bmatrix} A \\ B \\ C \\ D \end{Bmatrix} = 0 \tag{5.4.39}$$

This equation is satisfied if the determinant is zero. Expanding the determinant, we obtain

$$2\rho_1\rho_2(\cosh\rho_1\cos\rho_2 - 1) + (\rho_2^2 - \rho_1^2)\sinh\rho_1\sin\rho_2 = 0 \tag{5.4.40}$$

Solving this equation for its roots K_n ($n = 1, 2, \ldots$) and substituting them into Eq. (5.4.17) gives

$$\omega_{mn}a^2\sqrt{\frac{\rho h}{D}} = m^2\pi^2K_n \tag{5.4.41}$$

For example, for a square plate where $a/b = 1.0$, for $m^2\pi^2K_n$ we get the values

m \ n	1	2	3
1	28.9	69.2	129.1
2	54.8	94.6	154.8
3	102.2	140.2	199.9

The mode shape is obtained by letting

$$\rho_{1n} = \frac{b}{a}m\pi\sqrt{K_n + 1} \tag{5.4.42}$$

$$\rho_{2n} = \frac{b}{a}m\pi\sqrt{K_n - 1} \tag{5.4.43}$$

and by solving for three of the four coefficients of Eq. (5.4.39) in terms of the fourth. Thus

$$\begin{bmatrix} 0 & 1 & 0 \\ \sinh\rho_{1n} & \cos\rho_{2n} & \sin\rho_{2n} \\ \rho_{1n} & 0 & \rho_{2n} \end{bmatrix} \begin{Bmatrix} B \\ C \\ D \end{Bmatrix} = -A \begin{Bmatrix} 1 \\ \cosh\rho_{1n} \\ 0 \end{Bmatrix} \tag{5.4.44}$$

Thus

$$\frac{B}{A} = -\frac{\begin{vmatrix} 1 & 1 & 0 \\ \cosh\rho_{1n} & \cos\rho_{2n} & \sin\rho_{2n} \\ 0 & 0 & \rho_{2n} \end{vmatrix}}{D'} \tag{5.4.45}$$

$$\frac{C}{A} = -\frac{\begin{vmatrix} 0 & 1 & 0 \\ \sinh \rho_{1n} & \cosh \rho_{1n} & \sin \rho_{2n} \\ \rho_{1n} & 0 & \rho_{2n} \end{vmatrix}}{D'} \tag{5.4.46}$$

$$\frac{D}{A} = -\frac{\begin{vmatrix} 0 & 1 & 1 \\ \sinh \rho_{1n} & \cos \rho_{2n} & \cosh \rho_{1n} \\ \rho_{1n} & 0 & 0 \end{vmatrix}}{D'} \tag{5.4.47}$$

$$D' = \begin{vmatrix} 0 & 1 & 0 \\ \sinh \rho_{1n} & \cos \rho_{2n} & \sin \rho_{2n} \\ \rho_{1n} & 0 & \rho_{2n} \end{vmatrix}$$

$$= \rho_{1n} \sin \rho_{2n} - \rho_{2n} \sinh \rho_{1n} \tag{5.4.48}$$

Thus

$$\frac{B}{A} = -\frac{\rho_{2n}(\cos \rho_{2n} - \cosh \rho_{1n})}{\rho_{1n} \sin \rho_{2n} - \rho_{2n} \sinh \rho_{1n}} \tag{5.4.49}$$

$$\frac{C}{A} = -\frac{\rho_{1n} \sin \rho_{2n} - \rho_{2n} \sinh \rho_{1n}}{\rho_{1n} \sin \rho_{2n} - \rho_{2n} \sinh \rho_{1n}} = -1 \tag{5.4.50}$$

$$\frac{D}{A} = -\frac{\rho_{1n}(\cosh \rho_{1n} - \cos \rho_{2n})}{\rho_{1n} \sin \rho_{2n} - \rho_{2n} \sinh \rho_{1n}} = \frac{\rho_{1n}}{\rho_{2n}} \frac{B}{A} \tag{5.4.51}$$

and from Eqs. (5.4.26) and (5.4.12) we have

$$U_{3mn}(x,y) = A\left[\left(\cosh \rho_{1n} \frac{y}{b} - \cos \rho_{2n} \frac{y}{b}\right) - \frac{\rho_{2n}(\cos \rho_{2n} - \cosh \rho_{1n})}{\rho_{1n} \sin \rho_{2n} - \rho_{2n} \sinh \rho_{1n}} \left(\sinh \rho_{1n} \frac{y}{b} + \frac{\rho_{1n}}{\rho_{2n}} \sin \rho_{2n} \frac{y}{b}\right)\right] \sin \frac{m\pi x}{a} \tag{5.4.52}$$

Lines where

$$U_{3mn}(x,y) = 0 \tag{5.4.53}$$

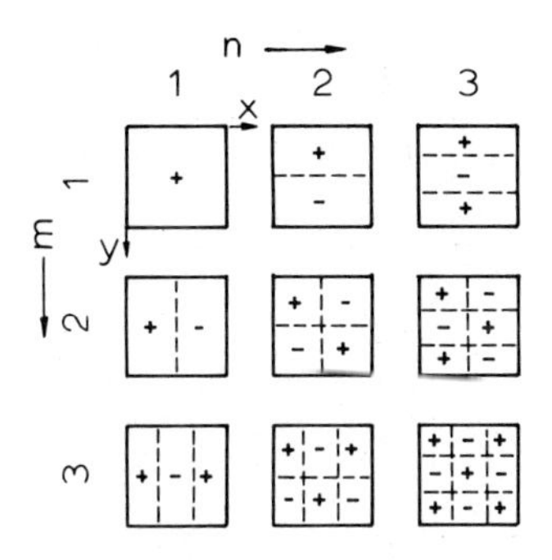

Figure 5.4.2

are node lines and can be obtained by searching for the x,y points that satisfy Eq. (5.4.53). Node lines for a square plate are shown in Fig. 5.4.2.

5.4.2 Two Other Edges Simply Supported

For this case the additional four boundary conditions are

$$u_3(x,o,t) = 0 \tag{5.4.54}$$

$$u_3(x,b,t) = 0 \tag{5.4.55}$$

$$M_{xx}(x,o,t) = 0 \tag{5.4.56}$$

$$M_{xx}(x,b,t) = 0 \tag{5.4.57}$$

Substituting the strain-displacement relations and Eqs. (5.4.6) and (5.4.12) results in

$$Y(o) = 0 \tag{5.4.58}$$

$$Y(b) = 0 \tag{5.4.59}$$

$$\frac{d^2Y}{dy^2}(o) = 0 \tag{5.4.60}$$

$$\frac{d^2Y}{dy^2}(b) = 0 \tag{5.4.61}$$

Substituting Eq. (5.4.26) gives

$$0 = A \qquad\qquad + C \tag{5.4.62}$$

$$0 = A \cosh \rho_1 + B \sinh \rho_1 + C \cos \rho_2 + D \sin \rho_2 \tag{5.4.63}$$

$$0 = A\rho_1^2 \qquad\qquad - C\rho_2^2 \tag{5.4.64}$$

$$0 = A\rho_1^2 \cosh \rho_1 + B\rho_1^2 \sinh \rho_1 - C\rho_2^2 \cos \rho_2 - D\rho_2^2 \sin \rho_2 \tag{5.4.65}$$

or

$$\begin{bmatrix} 1 & 0 & 1 & 0 \\ \cosh \rho_1 & \sinh \rho_1 & \cos \rho_2 & \sin \rho_2 \\ \rho_1^2 & 0 & -\rho_2^2 & 0 \\ \rho_1^2 \cosh \rho_1 & \rho_1^2 \sinh \rho_1 & -\rho_2^2 \cos \rho_2 & -\rho_2^2 \sin \rho_2 \end{bmatrix} \begin{Bmatrix} A \\ B \\ C \\ D \end{Bmatrix} = 0 \tag{5.4.66}$$

This equation is satisfied if the determinant is zero. Expanding the determinant gives

$$(\rho_1^2 + \rho_2^2)^2 \sinh \rho_1 \sin \rho_2 = 0 \tag{5.4.67}$$

Since neither $(\rho_1^2 + \rho_2^2)^2$ nor $\sinh \rho_1$ are zero for nontrivial solutions,

$$\sin \rho_2 = 0 \tag{5.4.68}$$

or

$$\rho_2 = n\pi \quad (n = 1, 2, \ldots) \tag{5.4.69}$$

or

$$K = \left(\frac{n}{m}\right)^2 \left(\frac{a}{b}\right)^2 + 1 \tag{5.4.70}$$

and therefore

$$\omega_{mn} = \pi^2 \left[\left(\frac{m}{a}\right)^2 + \left(\frac{n}{b}\right)^2\right] \sqrt{\frac{D}{\rho h}} \tag{5.4.71}$$

For example, for a square plate where $a/b = 1.0$, for $\omega_{mn} a^2 \sqrt{\frac{\rho h}{D}}$ we get the values

m \ n	1	2	3
1	19.72	49.30	98.60
2	49.30	78.88	128.17
3	98.60	128.17	177.47

To obtain the natural modes, we solve for three of the four coefficients of Eq. (5.4.66)

$$\begin{bmatrix} 1 & 0 & 1 \\ \cosh \rho_1 & \sinh \rho_1 & \cos \rho_2 \\ \rho_1^2 & 0 & -\rho_2^2 \end{bmatrix} \begin{Bmatrix} A \\ B \\ C \end{Bmatrix} = -D \begin{Bmatrix} 0 \\ \sin \rho_2 \\ 0 \end{Bmatrix} \tag{5.4.72}$$

This gives

$$\frac{A}{D} = 0 \tag{5.4.73}$$

$$\frac{B}{D} = 0 \tag{5.4.74}$$

$$\frac{C}{D} = 0 \tag{5.4.75}$$

and we get

$$Y(y) = D \sin \frac{n\pi y}{b} \tag{5.4.76}$$

or $$U_{3mn}(x,y) = D \sin \frac{n\pi y}{b} \sin \frac{m\pi x}{a} \tag{5.4.77}$$

This is an example of a result that could have been guessed.

5.5 CIRCULAR CYLINDRICAL SHELL SIMPLY SUPPORTED

The shell is shown in Fig. 5.5.1. It is assumed that the boundaries are such that

$$u_3(o,\theta,t) = 0 \tag{5.5.1}$$

$$u_\theta(o,\theta,t) = 0 \tag{5.5.2}$$

$$M_{xx}(o,\theta,t) = 0 \tag{5.5.3}$$

$$N_{xx}(o,\theta,t) = 0 \tag{5.5.4}$$

and $$u_3(L,\theta,t) = 0 \tag{5.5.5}$$

$$u_\theta(L,\theta,t) = 0 \tag{5.5.6}$$

$$M_{xx}(L,\theta,t) = 0 \tag{5.5.7}$$

$$N_{xx}(L,\theta,t) = 0 \tag{5.5.8}$$

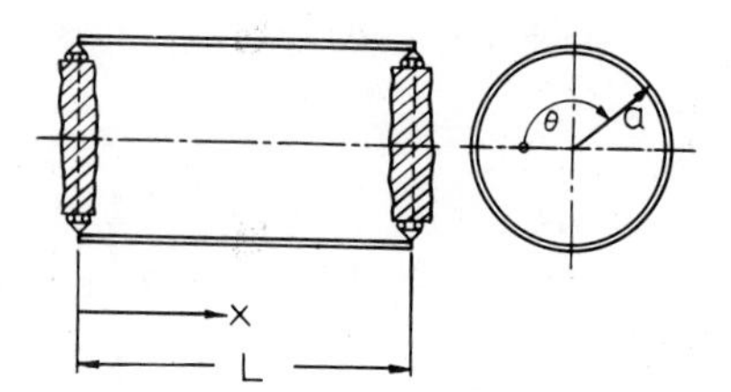

Figure 5.5.1

The equations of motion are, from Eqs. (3.3.2) to (3.3.4),

$$\frac{\partial N_{xx}}{\partial x} + \frac{1}{a}\frac{\partial N_{x\theta}}{\partial \theta} - \rho h \frac{\partial^2 u_x}{\partial t^2} = 0 \tag{5.5.9}$$

$$\frac{\partial N_{x\theta}}{\partial x} + \frac{1}{a}\frac{\partial N_{\theta\theta}}{\partial \theta} + \frac{Q_{\theta 3}}{a} - \rho h \frac{\partial^2 u_\theta}{\partial t^2} = 0 \tag{5.5.10}$$

$$\frac{\partial Q_{x3}}{\partial x} + \frac{1}{a}\frac{\partial Q_{\theta 3}}{\partial \theta} - \frac{N_{\theta\theta}}{a} - \rho h \frac{\partial^2 u_3}{\partial t^2} = 0 \tag{5.5.11}$$

with all terms defined by Eqs. (3.3.5) to (3.3.14). At a natural frequency

$$u_x(x,\theta,t) = U_x(x,\theta)e^{j\omega t} \tag{5.5.12}$$

$$u_\theta(x,\theta,t) = U_\theta(x,\theta)e^{j\omega t} \tag{5.5.13}$$

$$u_3(x,\theta,t) = U_3(x,\theta)e^{j\omega t} \tag{5.5.14}$$

This gives

$$\frac{\partial N'_{xx}}{\partial x} + \frac{1}{a}\frac{\partial N'_{x\theta}}{\partial \theta} + \rho h \omega^2 U_x = 0 \tag{5.5.15}$$

$$\frac{\partial N'_{x\theta}}{\partial x} + \frac{1}{a}\frac{\partial N'_{\theta\theta}}{\partial \theta} + \frac{1}{a}Q'_{\theta 3} + \rho h \omega^2 U_\theta = 0 \tag{5.5.16}$$

$$\frac{\partial Q'_{x3}}{\partial x} + \frac{1}{a}\frac{\partial Q'_{\theta 3}}{\partial \theta} - \frac{N'_{\theta\theta}}{a} + \rho h \omega^2 U_3 = 0 \tag{5.5.17}$$

where

$$Q'_{x3} = \frac{\partial M'_{xx}}{\partial x} + \frac{1}{a} \frac{\partial M'_{x\theta}}{\partial \theta} \tag{5.5.18}$$

$$Q'_{\theta 3} = \frac{\partial M'_{x\theta}}{\partial x} + \frac{1}{a} \frac{\partial M'_{\theta\theta}}{\partial \theta} \tag{5.5.19}$$

$$N'_{xx} = K(\varepsilon'^{o}_{xx} + \mu\varepsilon'^{o}_{\theta\theta}) \tag{5.5.20}$$

$$N'_{\theta\theta} = K(\varepsilon'^{o}_{\theta\theta} + \mu\varepsilon'^{o}_{xx}) \tag{5.5.21}$$

$$N'_{x\theta} = \frac{K(1 - \mu)}{2} \varepsilon'^{o}_{x\theta} \tag{5.5.22}$$

$$M'_{xx} = D(K'_{xx} + \mu k'_{\theta\theta}) \tag{5.5.23}$$

$$M'_{\theta\theta} = D(k'_{\theta\theta} + \mu k'_{xx}) \tag{5.5.24}$$

$$M'_{x\theta} = \frac{D(1 - \mu)}{2} k'_{x\theta} \tag{5.5.25}$$

$$\varepsilon'^{o}_{xx} = \frac{\partial U_x}{\partial x} \tag{5.5.26}$$

$$\varepsilon'^{o}_{\theta\theta} = \frac{1}{a} \frac{\partial U_\theta}{\partial \theta} + \frac{U_3}{a} \tag{5.5.27}$$

$$\varepsilon'^{o}_{x\theta} = \frac{\partial U_\theta}{\partial x} + \frac{1}{a} \frac{\partial U_x}{\partial \theta} \tag{5.5.28}$$

$$k'_{xx} = \frac{\partial \beta'_x}{\partial x} \tag{5.5.29}$$

$$k'_{\theta\theta} = \frac{1}{a} \frac{\partial \beta'_\theta}{\partial \theta} \tag{5.5.30}$$

$$k'_{x\theta} = \frac{\partial \beta'_\theta}{\partial x} + \frac{1}{a} \frac{\partial \beta'_x}{\partial \theta} \tag{5.5.31}$$

$$\beta'_x = -\frac{\partial U_3}{\partial x} \tag{5.5.32}$$

$$\beta'_\theta = \frac{U_\theta}{a} - \frac{1}{a} \frac{\partial U_3}{\partial \theta} \tag{5.5.33}$$

The boundary conditions become

$$U_3(o,\theta) = 0 \tag{5.5.34}$$

$$U_\theta(o,\theta) = 0 \tag{5.5.35}$$

$$M'_{xx}(o,\theta) = 0 \tag{5.5.36}$$

$$N'_{xx}(o,\theta) = 0 \tag{5.5.37}$$

$$U_3(L,\theta) = 0 \tag{5.5.38}$$

$$U_\theta(L,\theta) = 0 \tag{5.5.39}$$

$$M'_{xx}(L,\theta) = 0 \tag{5.5.40}$$

$$N'_{xx}(L,\theta) = 0 \tag{5.5.41}$$

Based on our experience with the ring and the simply supported beam, we assume the following solution:

$$U_x(x,\theta) = A \cos \frac{m\pi x}{L} \cos n(\theta-\phi) \tag{5.5.42}$$

$$U_\theta(x,\theta) = B \sin \frac{m\pi x}{L} \sin n(\theta-\phi) \tag{5.5.43}$$

$$U_3(x,\theta) = C \sin \frac{m\pi x}{L} \cos n(\theta-\phi) \tag{5.5.44}$$

While the assumptions for $U_\theta(x,\theta)$ and $U_3(x,\theta)$ are fairly obvious, the assumption for $U_x(x,\theta)$ needs some explanation. First of all, the term $\cos n(\theta-\phi)$ was chosen since it is to be expected that a longitudinal node line will not experience deflections in the x direction. Next, the term $\cos(\frac{m\pi x}{L})$ is based on the boundary condition requirement that

$$N'_{xx}(o,\theta) = 0 \tag{5.5.45}$$

$$N'_{xx}(L,\theta) = 0 \tag{5.5.46}$$

Substituting Eqs. (5.5.42) to (5.5.44) in all boundary conditions shows that all are satisfied.

Next, we subtitute these assumed solutions into Eqs. (5.5.15) to (5.5.33), starting with Eq. (5.5.33) and working backward. We get

$$\beta'_\theta = \frac{1}{a} (B + nC) \sin \frac{m\pi x}{L} \sin n(\theta-\phi) \tag{5.5.47}$$

$$\beta'_x = -\frac{m\pi}{L} C \cos \frac{m\pi x}{L} \cos n(\theta-\phi) \tag{5.5.48}$$

$$k'_{x\theta} = \frac{m\pi}{La} (B + 2nC) \cos \frac{m\pi x}{L} \sin n(\theta-\phi) \tag{5.5.49}$$

$$k'_{\theta\theta} = \frac{n}{a^2} (B + nC) \sin \frac{m\pi x}{L} \cos n(\theta-\phi) \tag{5.5.50}$$

$$k'_{xx} = \left(\frac{m\pi}{L}\right)^2 C \sin \frac{m\pi x}{L} \cos n(\theta-\phi) \tag{5.5.51}$$

$$\varepsilon'^{\circ}_{x\theta} = (\frac{m\pi}{L} B - \frac{n}{a} A) \cos \frac{m\pi x}{L} \sin n(\theta-\phi) \tag{5.5.52}$$

$$\varepsilon'^{\circ}_{\theta\theta} = \frac{1}{a} (Bn + C) \sin \frac{m\pi x}{L} \cos n(\theta-\phi) \tag{5.5.53}$$

$$\varepsilon'^{\circ}_{xx} = -A \frac{m\pi}{L} \sin \frac{m\pi x}{L} \cos n(\theta-\phi) \tag{5.5.54}$$

$$M'_{x\theta} = \frac{D(1 - \mu)}{2} \frac{m\pi}{La} (B + 2nC) \cos \frac{m\pi x}{L} \sin n(\theta-\phi) \tag{5.5.55}$$

$$M'_{\theta\theta} = D\left\{\frac{n}{a^2} B + \left[\frac{n^2}{a^2} + \mu\left(\frac{m\pi}{L}\right)^2\right] C\right\} \sin \frac{m\pi x}{L} \cos n(\theta-\phi) \tag{5.5.56}$$

$$M'_{xx} = D\left\{\frac{\mu n}{a^2} B + \left[\mu \frac{n^2}{a^2} + \left(\frac{m\pi}{L}\right)^2\right] C\right\} \sin \frac{m\pi x}{L} \cos n(\theta-\phi) \tag{5.5.57}$$

$$N'_{x\theta} = \frac{K(1 - \mu)}{2} \left(\frac{m\pi}{L} B - \frac{n}{a} A\right) \cos \frac{m\pi x}{L} \sin n(\theta-\phi) \tag{5.5.58}$$

$$N'_{\theta\theta} = K(\frac{n}{a} B + \frac{1}{a} C - \mu \frac{m\pi}{L} A) \sin \frac{m\pi x}{L} \cos n(\theta-\phi) \tag{5.5.59}$$

$$N'_{xx} = K(\frac{\mu n}{a} B + \frac{\mu}{a} C - \frac{m\pi}{L} A) \sin \frac{m\pi x}{L} \cos n(\theta-\phi) \tag{5.5.60}$$

$$Q'_{x3} = D \frac{m\pi}{L}\left\{\frac{n}{a^2} \frac{1 + \mu}{2} B + \left[\left(\frac{n}{a}\right)^2 + \left(\frac{m\pi}{L}\right)^2\right] C\right\} \cos \frac{m\pi x}{L} \cos n(\theta-\phi) \tag{5.5.61}$$

$$Q'_{\theta3} = - \frac{D}{a} \left\{\left[\frac{1 - \mu}{2} \left(\frac{m\pi}{L}\right)^2 + \left(\frac{n}{a}\right)^2\right] B + n\left[\left(\frac{m\pi}{L}\right)^2 + \left(\frac{n}{a}\right)^2\right] C\right\} \sin \frac{m\pi x}{L} \sin n(\theta-\phi) \tag{5.5.62}$$

Thus, Eqs. (5.5.15) to (5.8.17) become

$$\left\{\rho h\omega^2 - K\left[\left(\frac{m\pi}{L}\right)^2 + \frac{1 - \mu}{2} \left(\frac{n}{a}\right)^2\right]\right\} A + \left(K \frac{1 + \mu}{2} \frac{n}{a} \frac{m\pi}{L}\right) B + \left(K \frac{\mu}{a} \frac{m\pi}{L}\right) C = 0 \tag{5.5.63}$$

$$\left(K \frac{1 + \mu}{2} \frac{m\pi}{L} \frac{n}{a}\right) A + \left\{\rho h\omega^2 - \left(K + \frac{D}{a^2}\right)\left[\frac{1 - \mu}{2}\left(\frac{m\pi}{L}\right)^2 + (\frac{n}{a})^2\right]\right\} B + \left\{- \frac{K}{a} \frac{n}{a} - \frac{D}{a} \frac{n}{a}\left[\left(\frac{m\pi}{L}\right)^2 + \left(\frac{n}{a}\right)^2\right]\right\} C = 0 \tag{5.5.64}$$

$$\left(\frac{\mu K}{a}\frac{m\pi}{L}\right)A + \left\{-\frac{K}{a}\frac{n}{a} - \frac{D}{a}\frac{n}{a}\left[\left(\frac{m\pi}{L}\right)^2 + \left(\frac{n}{a}\right)^2\right]\right\}B$$
$$+ \left\{\rho h\omega^2 - D\left[\left(\frac{m\pi}{L}\right)^2 + \left(\frac{n}{a}\right)^2\right]^2 - \frac{K}{a^2}\right\}C = 0 \tag{5.5.65}$$

or

$$\begin{bmatrix} \rho h\omega^2 - k_{11} & k_{12} & k_{13} \\ k_{21} & \rho h\omega^2 - k_{22} & k_{23} \\ k_{31} & k_{32} & \rho h\omega^2 - k_{33} \end{bmatrix} \begin{Bmatrix} A \\ B \\ C \end{Bmatrix} = 0 \tag{5.5.66}$$

where

$$k_{11} = K\left[\left(\frac{m\pi}{L}\right)^2 + \frac{1-\mu}{2}\left(\frac{n}{a}\right)^2\right] \tag{5.5.67}$$

$$k_{12} = k_{21} = K\frac{1+\mu}{2}\frac{m\pi}{L}\frac{n}{a} \tag{5.5.68}$$

$$k_{13} = k_{31} = \frac{\mu K}{a}\frac{m\pi}{L} \tag{5.5.69}$$

$$k_{22} = \left(K + \frac{D}{a^2}\right)\left[\frac{1-\mu}{2}\left(\frac{m\pi}{L}\right)^2 + \left(\frac{n}{a}\right)^2\right] \tag{5.5.70}$$

$$k_{23} = k_{32} = -\frac{K}{a}\frac{n}{a} - \frac{D}{a}\frac{n}{a}\left[\left(\frac{m\pi}{L}\right)^2 + \left(\frac{n}{a}\right)^2\right] \tag{5.5.71}$$

$$k_{33} = D\left[\left(\frac{m\pi}{L}\right)^2 + \left(\frac{n}{a}\right)^2\right]^2 + \frac{K}{a^2} \tag{5.5.72}$$

For a nontrivial solution, the determinant of Eq. (5.5.66) has to be zero. Expanding the determinant gives

$$\omega^6 + a_1\omega^4 + a_2\omega^2 + a_3 = 0 \tag{5.5.73}$$

where

$$a_1 = -\frac{1}{\rho h}(k_{11} + k_{22} + k_{33}) \tag{5.5.74}$$

$$a_2 = \frac{1}{(\rho h)^2}(k_{11}k_{33} + k_{22}k_{33} + k_{11}k_{22} - k_{23}^2 - k_{12}^2 - k_{13}^2) \tag{5.5.75}$$

$$a_3 = \frac{1}{(\rho h)^3}(k_{11}k_{23}^2 + k_{22}k_{13}^2 + k_{33}k_{12}^2 + 2k_{12}k_{23}k_{13} - k_{11}k_{22}k_{33}) \tag{5.5.76}$$

The solutions of this equation are

$$\omega_{1mn}^2 = -\frac{2}{3}\sqrt{a_1^2 - 3a_2}\cos\frac{\alpha}{3} - \frac{a_1}{3} \tag{5.5.77}$$

$$\omega_{2mn}^2 = -\frac{2}{3}\sqrt{a_1^2 - 3a_2}\cos\frac{\alpha + 2\pi}{3} - \frac{a_1}{3} \tag{5.5.78}$$

$$\omega_{3mn}^2 = -\frac{2}{3}\sqrt{a_1^2 - 3a_2}\cos\frac{\alpha + 4\pi}{3} - \frac{a_1}{3} \tag{5.5.79}$$

where

$$\alpha = \cos^{-1}\left(\frac{27a_3 + 2a_1^3 - 9a_1a_2}{2\sqrt{(a_1^2 - 3a_2)^3}}\right) \tag{5.5.80}$$

For every m, n combination we have therefore three frequencies. The lowest is associated with the mode where the transverse component dominates, while the other two are usually higher by an order of magnitude and are associated with the mode where the displacements in the tangent plane dominate. For every m, n combination we have therefore three different combinations of A, B, and C. Solving A and B in terms of C, we have

$$\begin{bmatrix} \rho h\omega_{imn}^2 - k_{11} & k_{12} \\ k_{21} & \rho h\omega_{imn}^2 - k_{22} \end{bmatrix} \begin{Bmatrix} A_i \\ B_i \end{Bmatrix} = C_i \begin{Bmatrix} k_{13} \\ k_{23} \end{Bmatrix} \tag{5.5.81}$$

where i = 1, 2, 3. Thus

$$\frac{A_i}{C_i} = -\frac{\begin{vmatrix} k_{13} & k_{12} \\ k_{23} & \rho h\omega_{imn}^2 - k_{22} \end{vmatrix}}{D} \tag{5.5.82}$$

$$\frac{B_i}{C_i} = -\frac{\begin{vmatrix} \rho h\omega_{imn}^2 - k_{11} & k_{13} \\ k_{21} & k_{23} \end{vmatrix}}{D} \tag{5.5.83}$$

$$\text{where } D = \begin{vmatrix} \rho h \omega_{imn}^2 - k_{11} & k_{12} \\ k_{21} & \rho h \omega_{imn}^2 - k_{22} \end{vmatrix} \tag{5.5.84}$$

$$\text{or} \quad \frac{A_i}{C_i} = -\frac{k_{13}(\rho h \omega_{imn}^2 - k_{22}) - k_{12}k_{23}}{(\rho h \omega_{imn}^2 - k_{11})(\rho h \omega_{imn}^2 - k_{22}) - k_{12}^2} \tag{5.5.85}$$

$$\frac{B_i}{C_i} = -\frac{k_{23}(\rho h \omega_{imn}^2 - k_{11}) - k_{21}k_{13}}{(\rho h \omega_{imn}^2 - k_{11})(\rho h \omega_{imn}^2 - k_{22}) - k_{12}^2} \tag{5.5.86}$$

Thus, in summary, the three natural modes that are associated with the three natural frequencies ω_{imn} at each m, n combination are

$$\begin{Bmatrix} U_x \\ U_\theta \\ U_3 \end{Bmatrix}_i = C_i \begin{Bmatrix} \frac{A_i}{C_i} \cos \frac{m\pi x}{L} \cos n(\theta - \phi) \\ \frac{B_i}{C_i} \sin \frac{m\pi x}{L} \sin n(\theta - \phi) \\ \sin \frac{m\pi x}{L} \cos n(\theta - \phi) \end{Bmatrix} \tag{5.5.87}$$

where the C_i are arbitrary constants.

Let us assume that we have a steel shell ($E = 20.6 \times 10^4$ N/mm^2, $\rho = 7.85 \times 10^{-9}$ Nsec^2/mm^4, $\mu = 0.3$) of thickness h = 2mn, radius a = 100 mm, and length L = 200 mm. The three frequencies ω_{imn} and the ratios A_i/C_i and B_i/C_i are plotted in Figs. 5.5.2 through 5.5.6.

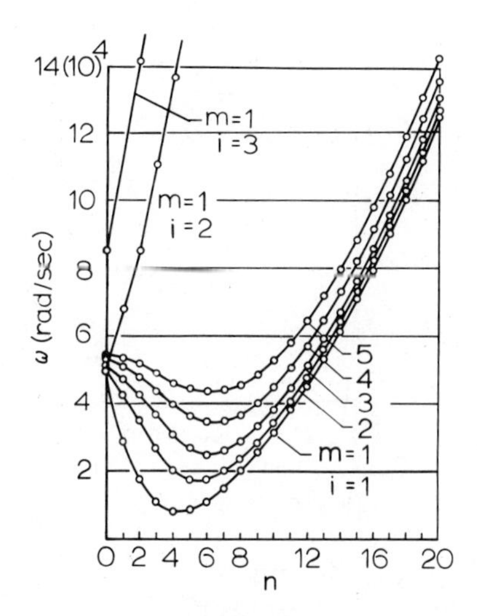

Figure 5.5.2

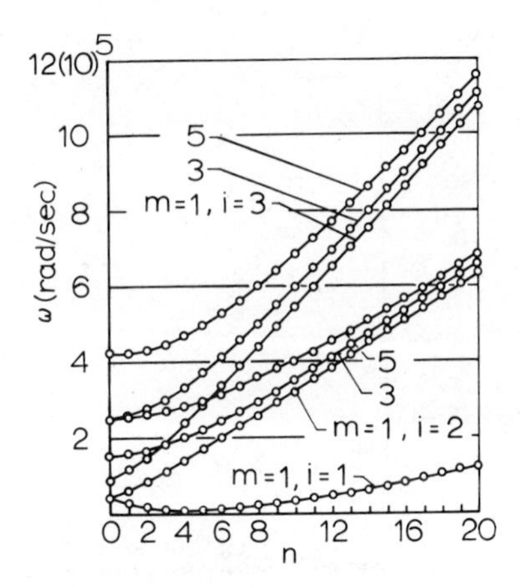

Figure 5.5.3

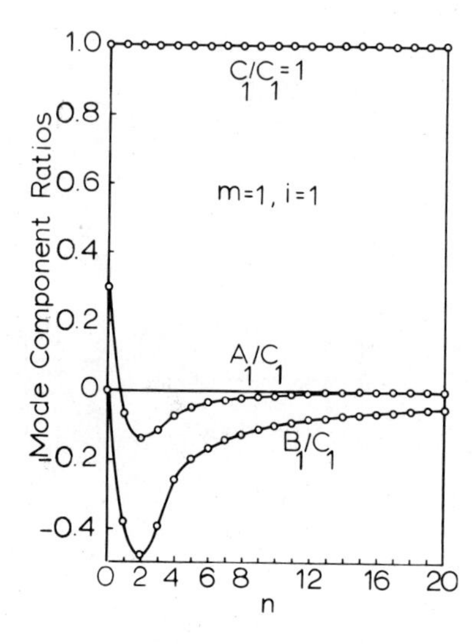

Figure 5.5.4

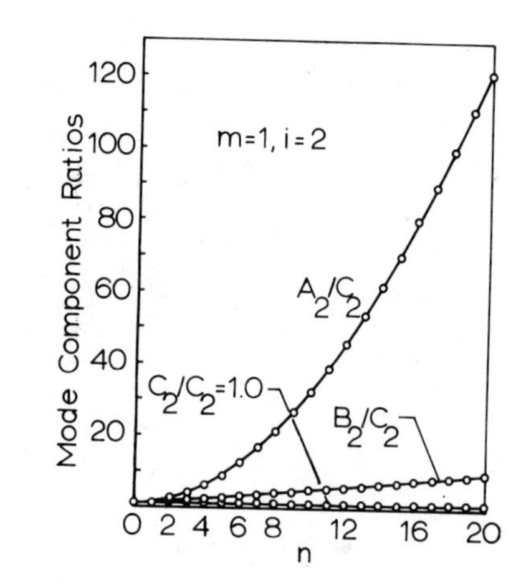

Figure 5.5.5

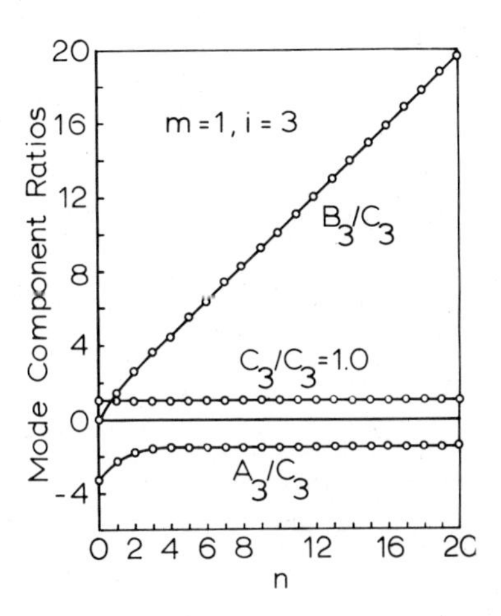

Figure 5.5.6

5.6 CIRCULAR PLATES VIBRATING TRANSVERSELY

Another category for which exact solutions are available (if series solutions can be termed *exact*) are circular plates. Circular plates are common structural elements in engineering.

The equation of motion for free vibration is

$$D\nabla^4 u_3 + \rho h \frac{\partial^2 u_3}{\partial t^2} = 0 \tag{5.6.1}$$

where

$$\nabla^2(\cdot) = \frac{\partial^2(\cdot)}{\partial r^2} + \frac{1}{r}\frac{\partial(\cdot)}{\partial r} + \frac{1}{r^2}\frac{\partial^2(\cdot)}{\partial \theta^2} \tag{5.6.2}$$

At a natural frequency

$$u_3(r,\theta,t) = U_3(r,\theta)e^{j\omega t} \tag{5.6.3}$$

If we substitute this into Eq. (5.6.1), we obtain

$$D\nabla^4 U_3 - \rho h\omega^2 U_3 = 0 \tag{5.6.4}$$

Let

$$\lambda^4 = \frac{\rho h \omega^2}{D} \tag{5.6.5}$$

Equation (5.6.4) can then be written

$$(\nabla^2 + \lambda^2)(\nabla^2 - \lambda^2)U_3 = 0 \tag{5.6.6}$$

This equation is satisfied by every solution of

$$(\nabla^2 \pm \lambda^2)U_3 = 0 \tag{5.6.7}$$

It is possible to separate variables by substituting

$$U_3(r,\theta) = R(r)\Theta(\theta) \tag{5.6.8}$$

This gives

$$r^2\left[\left(\frac{d^2R}{dr^2} + \frac{1}{r}\frac{dR}{dr}\right)\frac{1}{R} \pm \lambda^2\right] = -\frac{1}{\Theta}\frac{d^2\Theta}{d\theta^2} \tag{5.6.9}$$

This equation can only be satisfied if each expression is equal to the same constant k^2. This allows us to write

$$\frac{d^2\Theta}{d\theta^2} + k^2\Theta = 0 \tag{5.6.10}$$

and
$$\frac{d^2R}{dr^2} + \frac{1}{r}\frac{dR}{dr} + \left(\pm\lambda^2 - \frac{k^2}{r^2}\right)R = 0 \tag{5.6.11}$$

The solution of Eq. (5.6.10) is

$$\Theta = A'\cos k\theta + B\sin k\theta \tag{5.6.12}$$

or
$$\Theta = A\cos k(\theta - \phi) \tag{5.6.13}$$

where ϕ is a constant. In general, k can be a fractional number. But for plates that are closed in θ direction, Θ must be a function of period 2π. In this case, k becomes an integer.

$$k = n = 0, 1, 2, 3, \ldots \tag{5.6.14}$$

Let us now introduce a new variable

$$\xi = \begin{cases} \lambda r & \text{for } +\lambda^2 \\ j\lambda r & \text{for } -\lambda^2 \end{cases} \tag{5.6.15}$$

Equation (5.6.11) becomes

$$\frac{d^2R}{d\xi^2} + \frac{1}{\xi}\frac{dR}{d\xi} + \left(1 - \frac{k^2}{\xi 2}\right)R = 0 \qquad (5.6.16)$$

This is Bessel's equation of fractional order. The solutions are in form of series. They are classified in terms of Bessel functions. For $\xi = \lambda r$, the solution is in terms of Bessel functions of the first and second kind, $J_k(\lambda r)$ and $Y_k(\lambda r)$. For $\xi = j\lambda r$, the solution is in terms of modified Bessel functions of the first and second kind, $I_k(\lambda r)$ and $K_k(\lambda r)$.

For the special category of circular plates that are closed in the θ direction so that $k = n$, the solution R is therefore

$$R = CJ_n(\lambda r) + DI_n(\lambda r) + EY_n(\lambda r) + FK_n(\lambda r) \qquad (5.6.17)$$

Both $Y_n(\lambda r)$ and $K_n(\lambda r)$ are singular at $\lambda r = 0$. Thus, for a plate with no central hole, we set $E = F = 0$. Typical plots of the Bessel functions are shown in Figs. 5.6.1 and 5.6.2.

The general solution was first given by Kirchhof [5.1]. Numerous examples are collected in Ref. 5.2.

5.7 EXAMPLE: PLATE CLAMPED AT BOUNDARY

If the plate has no central hole

$$E = F = 0 \qquad (5.7.1)$$

The boundary conditions are, at the boundary radius $r = a$,

$$u_3(a,\theta,t) = 0 \qquad (5.7.2)$$

$$\frac{\partial u_3}{\partial r}(a,\theta,t) = 0 \qquad (5.7.3)$$

This translates into

$$R(a) = 0 \qquad (5.7.4)$$

$$\frac{dR}{dr}(a) = 0 \qquad (5.7.5)$$

Substituting Eq. (5.6.17) in these conditions gives

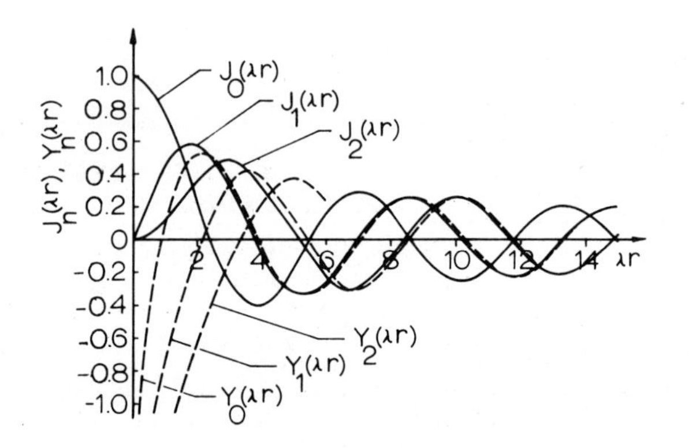

Figure 5.6.1

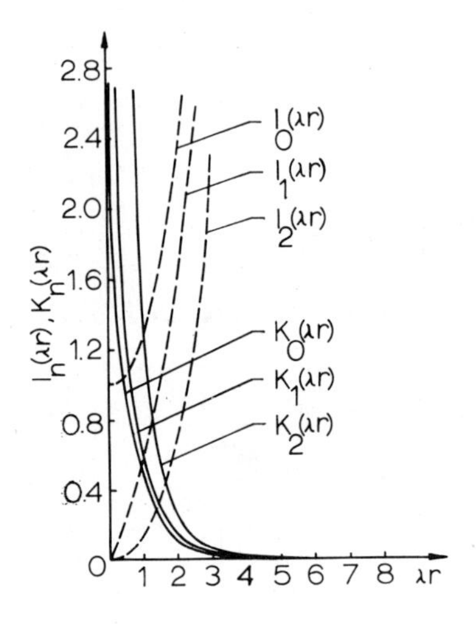

Figure 5.6.2

$$\begin{bmatrix} J_n(\lambda a) & I_n(\lambda a) \\ \frac{dJ_n}{dr}(\lambda a) & \frac{dI_n}{dr}(\lambda a) \end{bmatrix} \begin{Bmatrix} C \\ D \end{Bmatrix} = 0 \tag{5.7.6}$$

This equation is only satisfied in a meaningful way if the determinant is zero. This gives the frequency equation

$$J_n(\lambda a)\,\frac{dI_n}{dr}(\lambda a) - \frac{dJ_n}{dr}(\lambda a)\,I_n(\lambda a) = 0 \qquad (5.7.7)$$

Searching this equation for its roots λa, labeled successively $m = 0, 1, 2, \ldots$ for each $n = 0, 1, 2, \ldots$, gives the natural frequencies. Values of the roots λa are collected in Table 5.7.1

Table 5.7.1 Values for $(\lambda a)_{mn}$

m \ n	0	1	2	3
0	3.196	4.611	5.906	7.143
1	6.306	7.799	9.197	10.537
2	9.440	10.958	12.402	13.795
3	12.577	14.108	15.579	17.005

The natural frequencies are related to these roots by

$$\omega_{mn} = \frac{(\lambda a)_{mn}^2}{a^2}\sqrt{\frac{D}{\rho h}} \qquad (5.7.8)$$

Equation (5.7.7) can be simplified by using the identities

$$a\,\frac{dJ_n}{dr}(\lambda a) = nJ_n(\lambda a) - \lambda a J_{n+1}(\lambda a) \qquad (5.7.9)$$

$$a\,\frac{dI_n}{dr}(\lambda a) = nI_n(\lambda a) + \lambda a I_{n+1}(\lambda a) \qquad (5.7.10)$$

Equation (5.7.7) is then replaced by

$$J_n(\lambda a)I_{n+1}(\lambda a) + I_n(\lambda a)J_{n+1}(\lambda a) = 0 \qquad (5.7.11)$$

To find the mode shapes, we formulate from Eq. (5.7.6)

$$\frac{D}{C} = -\frac{J_n(\lambda a)}{I_n(\lambda a)} \qquad (5.7.12)$$

This gives then the mode shape expression

$$U_3(r,\theta) = A\left[J_n(\lambda r) - \frac{J_n(\lambda a)}{I_n(\lambda a)}\,I_n(\lambda r)\right]\cos n(\theta - \phi) \qquad (5.7.13)$$

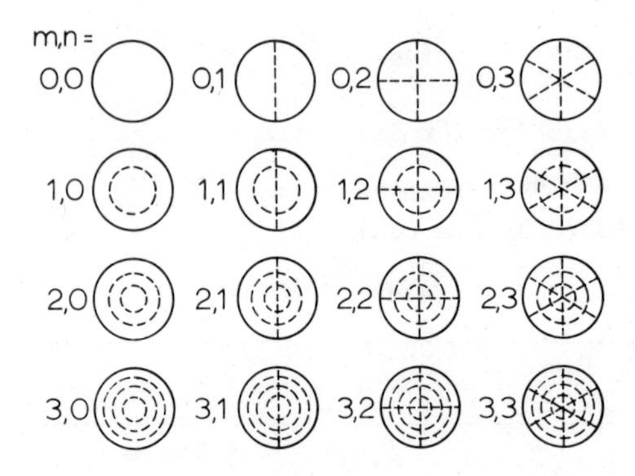

Figure 5.7.1

Setting this expression equal to zero defines the node lines. It turns out that there will be concentric circles and diametral lines. The number of concentric circles will be m and the number of diametral lines will be n. Examples are shown in Fig. 5.7.1. The values of the ratio of the nodal circles are shown in Table 5.7.2, in terms of the ratios r/a.

Table 5.7.2 Nodal radii in terms of r/a

m \ n	0	1	2	3
0	1.00	1.00	1.00	1.00
1	1.00 .38	1.00 .49	1.00 .56	1.00 .61
2	1.00 .58 .26	1.00 .64 .35	1.00 .68 .41	1.00 .71 .46
3	1.00 .69 .44 .19	1.00 .72 .50 .27	1.00 .75 .54 .33	1.00 .77 .57 .38

5.8 THE ORTHOGONALITY PROPERTY OF NATURAL MODES

Natural modes have the same property that is utilized in Fourier series formulations where sine and cosine functions are used. This property is orthogonality.

Let us start with Hamilton's principle

$$\delta \int_{t_o}^{t_1} (\Pi - K)\, dt = 0 \tag{5.8.1}$$

Because natural modes satisfy all boundary conditions, the energy put into a shell by boundary force resultants and moment resultants expressed by Eq. (2.6.8) is zero.

$$E_B = 0 \tag{5.8.2}$$

Because in the eigenvalue problem forcing is not considered

$$E_L = 0 \tag{5.8.3}$$

Therefore, as the shell vibrates in a natural mode, the potential energy is equal to the strain energy

$$\Pi = U \tag{5.8.4}$$

Hamilton's principle becomes, for this case,

$$\int_{t_o}^{t_1} \delta U\, dt - \int_{t_o}^{t_1} \delta K\, dt = 0 \tag{5.8.5}$$

where, from Eq. (2.7.15),

$$\int_{t_o}^{t_1} \delta U\, dt = \int_{t_o}^{t_1} \int_{\alpha_1} \int_{\alpha_2} \int_{\alpha_3} (\sigma_{11}\delta\varepsilon_{11} + \sigma_{22}\delta\varepsilon_{22} + \sigma_{12}\delta\varepsilon_{12} + \sigma_{13}\delta\varepsilon_{13} + \sigma_{23}\delta\varepsilon_{23})\, A_1 A_2\, d\alpha_1 d\alpha_2 d\alpha_3 dt \tag{5.8.6}$$

and where, from Eq. (2.7.8),

$$\int_{t_o}^{t_1} \delta K\, dt = -\rho h \int_{t_o}^{t_1} \int_{\alpha_1} \int_{\alpha_2} (\ddot{u}_1 \delta u_1 + \ddot{u}_2 \delta u_2 + \ddot{u}_3 \delta u_3)\, A_1 A_2\, d\alpha_1 d\alpha_2 dt \tag{5.8.7}$$

The displacements, when the shell is vibrating with mode k, are

$$u_i(\alpha_1,\alpha_2,t) = U_{ik}(\alpha_1,\alpha_2)e^{j\omega_k t} \tag{5.8.8}$$

We substitute this in Eqs. (5.8.6) and (5.8.7). Since the virtual displacements have to satisfy the boundary conditions also, but are in any other respect arbitrary, let us select mode p to represent the virtual displacement

$$\delta u_i(\alpha_1,\alpha_2,t) = U_{ip}e^{j\omega_p t} \tag{5.8.9}$$

Substituting this also gives

$$\begin{aligned}\int_{\alpha_1}\int_{\alpha_2}\int_{\alpha_3} &(\sigma_{11}^{(k)}\, \delta\varepsilon_{11}^{(p)} + \sigma_{22}^{(k)}\, \delta\varepsilon_{22}^{(p)} + \sigma_{12}^{(k)}\, \delta\varepsilon_{12}^{(p)} \\ &+ \sigma_{13}^{(k)}\, \delta\varepsilon_{13}^{(p)} + \sigma_{23}^{(k)}\, \delta\varepsilon_{23}^{(p)})\, A_1A_2d\alpha_1d\alpha_2d\alpha_3 \\ &= \omega_k^2\rho h \int_{\alpha_1}\int_{\alpha_2} (U_{1k}U_{1p} + U_{2k}U_{2p} + U_{3k}U_{3p})\, A_1A_2\, d\alpha_1d\alpha_2\end{aligned} \tag{5.8.10}$$

We note that the time integrals have canceled out. The superscripts on the stresses and variations of strain signify the modes with which they are associated.

Let us now go through the same procedure, except that we assign the mode p to the deflection description

$$u_i(\alpha_1,\alpha_2,t) = U_{ip}(\alpha_1,\alpha_2)e^{j\omega_p t} \tag{5.8.11}$$

and the mode k to the virtual displacements

$$\delta u_i = U_{ik}e^{j\omega_k t} \tag{5.8.12}$$

We obtain then

$$\begin{aligned}\int_{\alpha_1}\int_{\alpha_2}\int_{\alpha_3} &(\sigma_{11}^{(p)}\, \delta\varepsilon_{11}^{(k)} + \sigma_{22}^{(p)}\, \delta\varepsilon_{22}^{(k)} + \sigma_{12}^{(p)}\, \delta\varepsilon_{12}^{(k)} \\ &+ \sigma_{13}^{(p)}\, \delta\varepsilon_{13}^{(k)} + \sigma_{23}^{(p)}\, \delta\varepsilon_{23}^{(k)})\, A_1A_2d\alpha_1d\alpha_2d\alpha_3 \\ &= \omega_p\rho h \int_{\alpha_1}\int_{\alpha_2} (U_{1p}U_{1k} + U_{2p}U_{2k} + U_{3p}U_{3k})\, A_1A_2d\alpha_1d\alpha_2\end{aligned} \tag{5.8.13}$$

Let us now examine the first two terms of the left space integral of Eq. (5.8.10). Since

$$\varepsilon_{11}^{(p)} = \frac{1}{E}(\sigma_{11}^{(p)} - \mu\sigma_{22}^{(p)}) \qquad \varepsilon_{22}^{(p)} = \frac{1}{E}(\sigma_{22}^{(p)} - \mu\sigma_{11}^{(p)}) \tag{5.8.14}$$

we obtain

$$\sigma_{11}^{(k)}\delta\varepsilon_{11}^{(p)} + \sigma_{22}^{(k)}\delta\varepsilon_{22}^{(p)} = \frac{1}{E}\delta(\sigma_{11}^{(k)}\sigma_{11}^{(p)} + \sigma_{22}^{(k)}\sigma_{22}^{(p)} - \mu\sigma_{11}^{(k)}\sigma_{22}^{(p)} - \mu\sigma_{22}^{(k)}\sigma_{11}^{(p)}) \tag{5.8.15}$$

Similarly, if we examine the first two terms of Eq. (5.8.13), we obtain

$$\sigma_{11}^{(p)}\delta\varepsilon_{11}^{(k)} + \sigma_{22}^{(p)}\delta\varepsilon_{22}^{(k)} = \frac{1}{E}\delta(\sigma_{11}^{(p)}\sigma_{11}^{(k)} + \sigma_{22}^{(p)}\sigma_{22}^{(k)} - \mu\sigma_{11}^{(p)}\sigma_{22}^{(k)} - \mu\sigma_{22}^{(p)}\sigma_{11}^{(k)}) \tag{5.8.16}$$

Therefore, subtracting Eq. (5.8.16) from Eq. (5.8.15) gives

$$\sigma_{11}^{(k)}\delta\varepsilon_{11}^{(p)} - \sigma_{11}^{(p)}\delta\varepsilon_{11}^{(k)} + \sigma_{22}^{(k)}\delta\varepsilon_{22}^{(p)} - \sigma_{22}^{(p)}\delta\varepsilon_{22}^{(k)} = 0 \tag{5.8.17}$$

Since the shear terms subtract out to zero also, we may subtract Eq. (5.8.13) from Eq. (5.8.10) and obtain

$$(\omega_k^2 - \omega_p^2)\int_{\alpha_1}\int_{\alpha_2}(U_{1k}U_{1p} + U_{2k}U_{2p} + U_{3k}U_{3p})\,A_1A_2d\alpha_1d\alpha_2 = 0 \tag{5.8.18}$$

This equation is satisfied whenever p = k since

$$\omega_k^2 - \omega_k^2 = 0 \tag{5.8.19}$$

In this case the integral has a numerical value which we designate as N_k:

$$N_k = \int_{\alpha_1}\int_{\alpha_2} (U_{1k}^2 + U_{2k}^2 + U_{3k}^2)A_1A_2d\alpha_1 d\alpha_2 \qquad (5.8.20)$$

Whenever $k \neq p$, the only way Eq. (5.7.18) can be satisfied is when

$$\int_{\alpha_1}\int_{\alpha_2} (U_{1k}U_{1p} + U_{2k}U_{2p} + U_{3k}U_{3p})A_1A_2d\alpha_1 d\alpha_2 = 0 \qquad (5.8.21)$$

We may summarize this by using the Kronecker delta symbol

$$\int_{\alpha_1}\int_{\alpha_2} (U_{1k}U_{1p} + U_{2k}U_{2p} + U_{3p}U_{3k})A_1A_2d\alpha_1 d\alpha_2 = \delta_{pk}N_k \qquad (5.8.22)$$

where

$$\delta_{pk} = \begin{cases} 1 & p = k \\ 0 & p \neq k \end{cases} \qquad (5.8.23)$$

It is important to recognize the generality of this relationship. Any two modes of any system of uniform thickness, when multiplied with each other in the prescribed way, will integrate out to zero. This fact can, for instance, be used to check the accuracy of experimentally determined modes. But most important, it allows us to express a general solution of the forced equation in terms of an infinite series of modes.

5.9 SUPERPOSITION MODES

A standard procedure in structural vibrations is to determine natural frequencies and mode shapes experimentally. To find the natural frequencies and modes, the structural system is excited by a shaker, a magnetic driver, a periodic airblast, etc. One natural frequency after the other is identified and the characteristic shape of vibration (mode shape) at each of these natural frequencies is recorded. As long as the natural frequencies are spaced apart like in beam or rod type applications, no experimental difficulty is encountered. But in plate and shell type structures, it is possible that two or more entirely different mode shapes occur at the same frequency. These mode shapes superimpose in a ratio that is dependent on the location of the exciter. An infinite variety of shapes can thus be created. The experimenter, however, does not usually become aware of this if he goes about his task with the standard procedures which do not necessarily require him to move his excitation location, and will possibly only record a single mode shape at such a superposition frequency where he should have measured a complete set. Proof that this actually happens can be found in the experimental literature. There is extensive published material on experimental mode shapes that form incomplete sets.

The fact that modal superposition can occur in membrane plate, and shell type structures has been known for a long time. One of the earliest published discussions of it dates back to Byerly [5.3] in 1893. There, it is shown, for the examples of a rectangular membrane, how the superposition of two modes can produce an infinite variety of Chladni figures at the same natural frequency. However, what has not been shown is the procedure to extract from superposition modes information that is useful for further study of the system.

The classical superposition of modes occurs when two distinct modes are associated with the same natural frequency.

Let us look, as an example, at the simply supported square plate. This is a case which has a particularly large number of superposition mode possibilities. The natural frequencies are given by

$$\omega_{mn} = \left(\frac{\pi}{a}\right)^2 (m^2 + n^2) \sqrt{\frac{D}{\rho h}} \tag{5.9.1}$$

where m and n can be any combination of integers (m,n = 1, 2, 3, ...). The mode shapes are given by

$$U_{3mn} = A_{mn} \sin \frac{m\pi x}{a} \sin \frac{n\pi y}{a} \tag{5.9.2}$$

It is clear that superposition modes occur for those combinations of m and n for which $m^2 + n^2$ is the same. This is shown in Table 5.9.1. We see that any combination $(m,n) = (i,j)$ and (j,i) has the identical natural frequency. In one case, namely $(m,n) = (1,7)$, $(7,1)$, and $(5,5)$, we find that three distinct mode shapes are associated with the same natural frequency. Such triple occurrences happen more often as we increase the values of (m,n). In general, as $m^2 + n^2$ becomes large, the number of modes that have the identical natural frequency also becomes large.

Table 5.9.1 Numerical Value of $m^2 + n^2$

n \ m	1	2	3	4	5	6	7
1	2	5	10	17	26	37	50
2	5	8	13	20	29	40	53
3	10	13	18	25	34	45	58
4	17	20	25	32	41	52	65
5	26	29	34	41	50	61	74
6	37	40	45	52	61	72	85
7	50	53	58	65	74	85	98

Let us consider as an example the mode $(m,n) = (1,2)$. Thus,

$$U_{312} = A_{12} \sin \frac{\pi x}{a} \sin \frac{2\pi y}{a} \tag{5.9.3}$$

The associated natural frequency is

$$\omega_{12} = 5\left(\frac{\pi}{a}\right)^2 \sqrt{\frac{D}{\rho h}} \tag{5.9.4}$$

At the same natural frequency, the mode $(m,n) = (2,1)$ exists, and

$$U_{321} = A_{21} \sin \frac{2\pi x}{a} \sin \frac{\pi y}{a} \tag{5.9.5}$$

The two modes are distinctly different, as shown in Fig. 5.9.1,

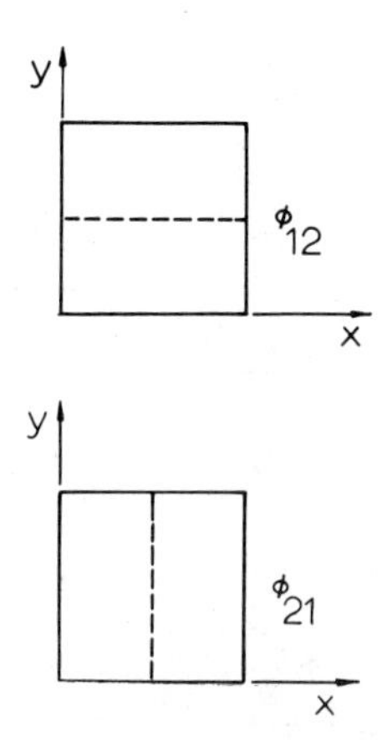

Figure 5.9.1

where the Chladni figures (node lines) are given. Node lines define locations of zero transverse displacement. Let us now suppose that the experimenter does not know in advance what mode pattern to expect. If he happens to excite the plate with a harmonically varying point force at the node line of the (m,n) = (1,2) mode, he will find the (m,n) = (2,1) mode excited. If he does not move his excitation point, he will never be aware of the existence of the (1,2) mode. The reason for this is, of course, that any transverse plate mode has an infinite transverse impedance along its node lines. This can easily be shown theoretically or verified by experiment.

If the experimenter happens to locate his excitation force at any place of the plate that is not a node line of the (1,2) or (2,1) mode, he will excite both modes. A superposition mode will be generated of the form

$$\psi = U_{312} + hU_{321} \tag{5.9.6}$$

where h is a number that depends on the exciter location.

Let us suppose his exciter location is along the line y = x, but not at $y = x = 0, \frac{a}{2}, a$. In this case, he will excite each of the two fundamental modes equally strong

$$A_{12} = A_{21} \tag{5.9.7}$$

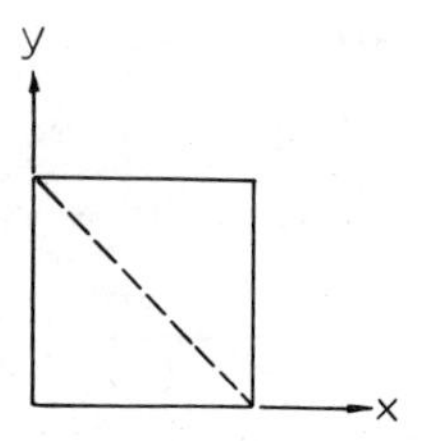

Figure 5.9.2

This gives h = 1 or

$$\psi_1 = A_{12}\left(\sin\frac{\pi x}{a}\sin\frac{2\pi y}{a} + \sin\frac{2\pi x}{a}\sin\frac{\pi y}{a}\right) \tag{5.9.8}$$

Let us find the configuration of the resulting new node line. From the requirement that $\psi_1 = 0$ along a node line, we find the new node line equation to be

$$y = a - x \tag{5.9.9}$$

This is illustrated in Fig. 5.9.2, where the new node line is shown. This mode shape is a superposition mode, and by itself, it is not able to replace the two fundamental modes in information content. At least a second mode shape has to be found. For instance, if we locate the exciter such that

$$A_{21} = 2A_{12} \tag{5.9.10}$$

we get as the superposition mode (h = 2)

$$\psi_2 = A_{12}\left(\sin\frac{\pi x}{a}\sin\frac{2\pi y}{a} + 2\sin\frac{2\pi x}{a}\sin\frac{\pi y}{a}\right) \tag{5.9.11}$$

and the resulting node line is a wavy line as shown in Fig. 5.9.3. By moving the excitation location around, a theoretically infinite number of superposition mode shapes can be found.

Note that the two superposition modes that were generated are not necessarily orthogonal to each other:

$$\int_A\int \psi_1\psi_2 dA \neq 0 \tag{5.9.12}$$

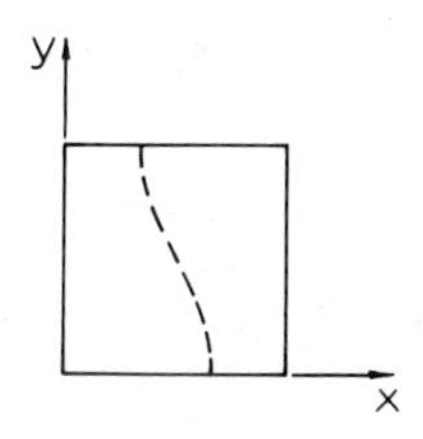

Figure 5.9.3

To prove this in general, let

$$\psi_1 = U_{3(1)} + a_1 U_{3(2)} \tag{5.9.13}$$

$$\psi_2 = U_{3(1)} + a_2 U_{3(2)} \tag{5.9.14}$$

where $U_{3(1)}$ and $U_{3(2)}$ are two basic modes that are orthogonal and have the same natural frequency. Formulation of the integral gives

$$\int_A \int (U_{3(1)}^2 + a_1 a_2 \, U_{3(2)}^2 + (a_1 + a_2) \, U_{3(1)} U_{3(2)}) dA \neq 0 \tag{5.9.15}$$

Since, by definition, the two basic modes $U_{3(1)}$ and $U_{3(2)}$ are orthogonal, the third term disappears, but the other terms will not be in general zero. Thus, it is necessary to go through an orthogonalization process to make the information useful for forced vibration prediction. The experimenter may argue that this is not his affair, as long as he produces two superposition modes for two basic modes. This is certainly true, but how does the experimenter know that there were only two basic modes and not three or more? Only by performing the orthogonalization process, before the experimental set up is removed, will he be able to be certain that he has measured enough superposition modes to give all the information that is needed. This will be discussed next.

5.10 ORTHOGONAL MODES FROM NONORTHOGONAL SUPERPOSITION MODES

In the following the Schmidt orthogonalization [5.4] procedure for vectors is adapted to the superposition mode problem.

Let us assume we have found two superposition modes, ψ_1 and

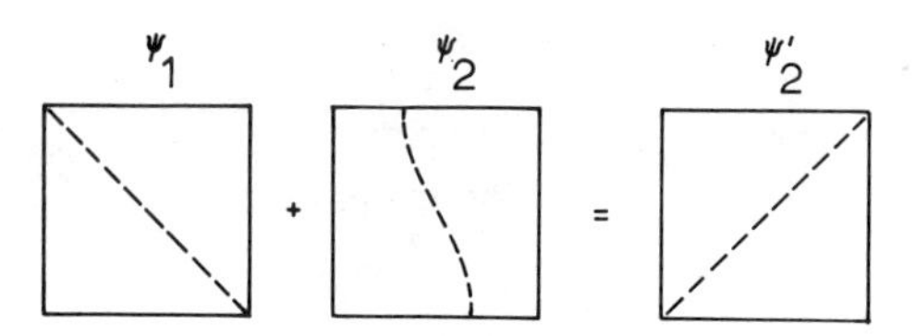

Figure 5.10.1

ψ_2, of the many possible ones. These two modes are in general not orthogonal. To make them useful experimental information, one has to go through an orthogonalization process. We select one of the modes as a base mode. Let us choose ψ_1. To obtain a second mode, ψ_2', that is orthogonal to ψ_1, we subtract from ψ_2 a scalar multiple of ψ_1 (Fig. 5.10.1):

$$\psi_2' = \psi_2 - a_1\psi_1 \tag{5.10.1}$$

The requirement is that

$$\iint_A \psi_2'\psi_1 \, dA = 0 \tag{5.10.2}$$

Multiplying Eq. (5.10.1) with ψ_1 and integrating, we obtain

$$\iint_A \psi_2'\psi_1 \, dA = \iint_A \psi_2\psi_1 \, dA - a_1 \iint_A (\psi_1)^2 \, dA \tag{5.10.3}$$

This gives

$$a_1 = \frac{\iint_A \psi_2\psi_1 \, dA}{\iint_A (\psi_1)^2 \, dA} \tag{5.10.4}$$

In the case of our example of a square plate, let us use

$$\psi_1 = A_{12} \left(\sin \frac{\pi x}{a} \sin \frac{2\pi y}{a} + \sin \frac{2\pi x}{a} \sin \frac{\pi y}{a} \right) \tag{5.10.5}$$

$$\psi_2 = A_{12} \left(\sin \frac{\pi x}{a} \sin \frac{2\pi y}{a} + 2 \sin \frac{2\pi x}{a} \sin \frac{\pi y}{a} \right) \tag{5.10.6}$$

These modes are not orthogonal. They are shown in Figures 5.9.2 and 5.9.3. Let us now use ψ_1 as the base mode and construct a mode orthogonal to it. Since

$$\iint_A (\psi_1)^2\, dA = \int_0^a\int_0^a \left(\sin\frac{\pi x}{a}\sin\frac{2\pi y}{a} + \sin\frac{2\pi x}{a}\sin\frac{\pi y}{a}\right)^2 dx\, dy$$

$$= \frac{a^2}{2} \qquad (5.10.7)$$

$$\iint_A \psi_1\psi_2\, dA = \int_0^a\int_0^a (\sin\frac{\pi x}{a}\sin\frac{2\pi y}{a}$$

$$+ \sin\frac{2\pi x}{a}\sin\frac{\pi y}{a})(\sin\frac{\pi x}{a}\sin\frac{2\pi y}{a}$$

$$+ 2\sin\frac{2\pi x}{a}\sin\frac{\pi y}{a})\, dx\, dy = \frac{3a^2}{4} \qquad (5.10.8)$$

we get $a_1 = \frac{3}{2}$ and

$$\psi_2' = A_{12}'\left(\sin\frac{\pi x}{a}\sin\frac{2\pi y}{a} - \sin\frac{2\pi x}{a}\sin\frac{\pi y}{a}\right) \qquad (5.10.9)$$

where A_{12}' is again an arbitrary constant. This mode is now indeed orthogonal to ψ_1, as a check will easily reveal. The mode, in terms of its node line, is sketched in Fig. 5.10.1.

If there are three or more basic modes that superimpose, we proceed in a similar manner, except that we now have to measure n superposition modes $\psi_1, \psi_2, \ldots, \psi_n$, where n is the number of base modes that are superimposed. We choose one of these as the base mode, let us say ψ_1. The procedure is then as before, except that we have to go through the orthogonalization process n - 1 times. Let us illustrate this for the case of superposition of three modes ψ_1, ψ_2, and ψ_3. We choose ψ_1 as the base mode. Next, we obtain a mode ψ_2' that is orthogonal to ψ_1, utilizing information from ψ_2.

$$\psi_2' = \psi_2 - a_{12}\psi_1 \qquad (5.10.10)$$

As before, we obtain, from the requirement that

$$\iint_A \psi_2'\psi_1\, dA = 0 \qquad (5.10.11)$$

the value for a_{12}

$$a_{12} = \frac{\int_A\int \psi_1\psi_2 \, dA}{\int_A\int (\psi_1)^2 \, dA} \tag{5.10.12}$$

Next, we obtain the mode ψ_3' that is orthogonal to both ψ_1 and ψ_2', utilizing ψ_2 and ψ_3 information.

$$\psi_3' = \psi_3 - a_{13}\psi_1 - a_{23}\psi_2' \tag{5.10.13}$$

This time we have two requirements, namely that ψ_3' be orthogonal to both ψ_1 and ψ_2'.

$$\int_A\int \psi_3'\psi_1 \, dA = 0 \tag{5.10.14}$$

$$\int_A\int \psi_3'\psi_2' \, dA = 0 \tag{5.10.15}$$

Multiplying Eq. (5.10.13) by ψ_1 and integrating, then repeating the process by multiplying by ψ_2' and integrating, we have

$$a_{13} = \frac{\int_A\int \psi_3\psi_1 \, dA}{\int_A\int \psi_1^2 \, dA} \tag{5.10.16}$$

$$a_{23} = \frac{\int_A\int \psi_3\psi_2' \, dA}{\int_A\int \psi_2'^2 \, dA} \tag{5.10.17}$$

Note that all integrations have to be performed numerically since experimental mode data is almost never available in functional form but rather in the form of numerical arrays.

In general, if there is a superposition of n modes, the r-th orthogonal mode is

$$\psi'_r = \psi_r - \sum_{i=1}^{r-1} a_{ir}\,\psi'_i \qquad (r = 2, 3, \ldots, n) \tag{5.10.18}$$

where

$$\psi'_1 = \psi_1 \tag{5.10.19}$$

$$a_{ir} = \frac{\int_A\int \psi_r\psi_i \, dA}{\int_A\int \psi_i^2 \, dA} \tag{5.10.20}$$

Returning to the square plate as an example, we get at an excitation frequency that is equivalent to $m^2 + n^2 = 50$ a superposition of three modes: $(m, n) = (1,7), (7,1), (5,5)$. Let us assume that we have found by experiment the following three superposition modes:

$$\psi_1 = \sin\frac{\pi x}{a}\sin\frac{7\pi y}{a} + \sin\frac{7\pi x}{a}\sin\frac{\pi y}{a} + \sin\frac{5\pi x}{a}\sin\frac{5\pi y}{a} \tag{5.10.21}$$

$$\psi_2 = 2\sin\frac{\pi x}{a}\sin\frac{7\pi y}{a} + \sin\frac{7\pi x}{a}\sin\frac{\pi y}{a} + \frac{3}{2}\sin\frac{5\pi x}{a}\sin\frac{5\pi y}{a} \tag{5.10.22}$$

$$\psi_3 = \frac{1}{2}\sin\frac{\pi x}{a}\sin\frac{7\pi y}{a} + \frac{3}{2}\sin\frac{7\pi x}{a}\sin\frac{\pi y}{a} + 2\sin\frac{5\pi x}{a}\sin\frac{5\pi y}{a} \tag{5.10.23}$$

We obtain $a_{12} = 3/2$ and therefore

$$\psi'_2 = \sin\frac{\pi x}{a}\sin\frac{7\pi y}{a} - \sin\frac{7\pi x}{a}\sin\frac{\pi y}{a} \tag{5.10.24}$$

Next, we obtain $a_{13} = 4/3$, $a_{23} = -1$, and

$$\psi'_3 = -\sin\frac{\pi x}{a}\sin\frac{7\pi y}{a} - \sin\frac{7\pi x}{a}\sin\frac{\pi y}{a} + 2\sin\frac{5\pi x}{a}\sin\frac{5\pi y}{a} \tag{5.10.25}$$

When we check for orthogonality, we find indeed that we have now

$$\int_0^a\int_0^a \psi_1\psi'_2 \, dx\, dy = 0 \tag{5.10.26}$$

$$\int_o^a \int_o^a \psi_1 \psi_3' \, dx \, dy = 0 \tag{5.10.27}$$

$$\int_o^a \int_o^a \psi_2' \psi_3' \, dx \, dy = 0 \tag{5.10.28}$$

This example also illustrates the point that it is not necessary for an orthogonalized superposition mode to be composed of all basic modes. In our case, ψ_2' is composed of only two instead of all three. The same applies for the measured modes also, since it is always possible that the exciter is located at the node line of one of the basic modes. This will not affect the method. After all, there is a minute probability that the experimenter becomes lucky and chooses his exciter location such that he excites modes that are already orthogonal and, even more improbable, that they are equal to the mathematically generated basic modes.

We have now seen how we can construct orthogonal modes from the superposition information. Let us now ask some very practical questions. First, how does the experimenter know that he or she has a superposition effect? Answer: when the experimenter produces an apparently different mode shape at the same natural frequency as he moves the exciter to a different location. Second, how does the experimenter know how many basic modes are contributing to the superposition? Answer: he or she doesn't. What one has to do is to go through the orthogonalization procedure, assuming the worst, namely many contributing modes. To this end one should obtain three or four superposition modes. One is to select the base mode ψ_1, and then take one of the measured superposition modes, ψ_2, and generate a mode ψ_2' that is orthogonal to ψ_1. Next, take the third measured superposition mode, ψ_3, and try to generate ψ_3'. If there were only two basic modes, ψ_3' will come out to be zero, subject to the limits of experimental error. If there were more than two basic modes contributing to the superposition, ψ_3' will turn out to be an orthogonal mode and the experimenter should try to generate a ψ_4' and so on.

The procedure will suggest how many superposition modes have to be measured. Deciding if one has obtained a bona fide orthogonal mode or if resulting shapes are only due to experimental error may turn out to be tricky in some cases, but with some experience it should usually be possible to tell the difference because the quasimode resulting from experimental error should have a rather random distribution of numerous node lines. Also, any contributions from nonsuperposition modes because of coupling effects (which will be discussed next) should be recognizable. As long as one remembers that mathematical procedures alone cannot replace good judgment, reasonably good experimental results should be possible.

5.11 DISTORTION OF EXPERIMENTAL MODES BECAUSE OF DAMPING

Damping is always present, but usually in such small amounts that experimental modes approximate undamped natural modes quite well inside usual limits of experimental accuracy. One exception is the phenomenon of noncrossing node lines. This will be discussed using the example of a plate.

When a plate is excited harmonically, it will respond with a vibration that consists of a superposition of all of its natural modes. This will be discussed in a later chapter. Each mode participates with different intensity. Mathematically, this can be expressed as

$$u_3(\alpha_1,\alpha_2,t) = q_1(t)\ U_{31}(\alpha_1,\alpha_2) + q_2(t)\ U_{32}(\alpha_1,\alpha_2) + \cdots \tag{5.11.1}$$

where

q_i = modal participation factor

u_3 = transverse deflection

If the system damping is negligible and if the excitation frequency is equal to the j-th natural frequency, then all ratios q_1/q_j, q_2/q_j, ..., approach zero except for q_j/q_j, which, of course, approaches unity. This means that at a natural frequency ω_j,

$$u_3 = q_j U_{3j} \tag{5.11.2}$$

This is the reason why we are able to experimentally isolate one mode after the other.

Things become different if there is damping, either of a structural or air resistance nature. If we express the damping in terms of an equivalent viscous damping coefficient, the modal participation factors are of the form

$$q_i = \frac{F_i}{\omega_i^2 \sqrt{[1 - (\frac{\omega}{\omega_i})^2]^2 + 4\xi_i^2 (\frac{\omega}{\omega_i})^2}} \tag{5.11.3}$$

where

F_i = parameter dependent on mode shape, excitation force location and distribution

ξ_i = damping factor

ω = excitation frequency

ω_i = i-th natural frequency

We can now easily see that the presence of damping will tend to make the ratios q_i/q_j approach a small amount of ε_i instead of zero. It turns out that in the case of plates, the largest ε_i will be most likely ε_1 since ω_1 is smaller than ω_2, ω_3, ... Thus, we will often see an experimental mode shape that approaches a superposition of the resonant mode plus a small amount of the first mode

$$\psi_j = U_{3j} + \varepsilon_1 U_{31} \tag{5.11.4}$$

where

ψ_j = superposition mode at the j-th natural frequency

U_{3j} = j-th natural mode

U_{31} = first natural mode

Typically, the presence of this type of superposition will tend to prevent crossing of node lines. For a square plate with the (m,n) =

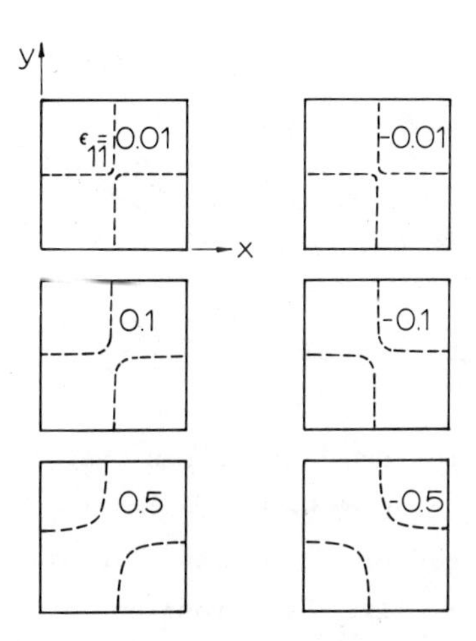

Figure 5.11.1

(2,2) mode for which no superposition modes of the first kind exist, a small amount of damping will produce a superposition mode

$$\psi_{22} = U_{322} + \varepsilon_{11} U_{311} \tag{5.11.5}$$

where

$$U_{322} = \sin \frac{2\pi x}{a} \sin \frac{2\pi y}{a} \tag{5.11.6}$$

$$U_{311} = \sin \frac{\pi x}{a} \sin \frac{\pi y}{a} \tag{5.11.7}$$

Superposition modes for various values of ε_{11} are shown in Figure 5.11.1. The larger the damping effect, the larger the distance between the non crossing lines. It is also possible, that ε_{11} is a negative number. This depends on the location of the exciter. However, all that the negative sign does is to exchange the quadrants where the node line does not cross over.

This non crossing behavior is also well known in shell structures. A typical case is shown in Figure 5.11.2. However, due to the special frequency characteristic of shells, several damping coupled modes may participate to produce the experimental mode shape.

Let us now ask the question, "What should the experimenter do

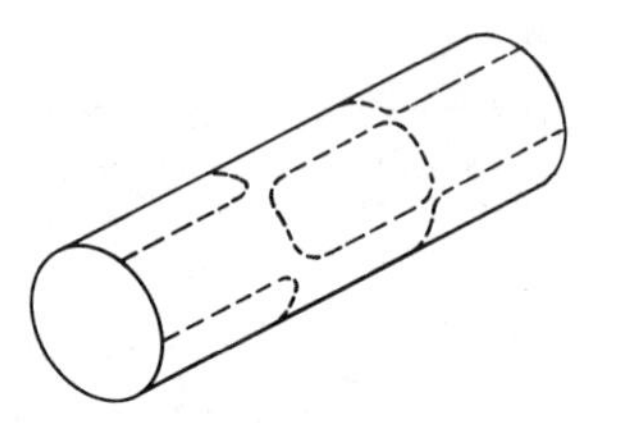

Figure 5.11.2

when he or she encounters mode shapes that look like damping superpositions?" On a first-order approximation level, it is recommended to him that he should simply indicate in his result report that the true node lines most likely do cross and offer as a choice corrected experimental data using his intuition. On a higher level of experimental fidelity, he should assume that the damping couples primarily the fundamental mode and use this idea to reconstruct the true mode. Let the superposition mode be

$$\psi_j = U_{3j} + \varepsilon_1 U_{31} \tag{5.11.8}$$

He does not know U_{3j} and E_1, but he has measured U_{31} and, of course, ψ_j. Let us multiply the equation by U_{31} and integrate over the plate area. Utilizing the fact that

$$\int_A\int U_{3j}U_{31}\, dA = 0 \quad (j \neq 1) \tag{5.11.9}$$

gives

$$\varepsilon_1 = \frac{\int_A\int \psi_j U_{31}\, dA}{\int_A\int U_{31}^2\, dA} \tag{5.11.10}$$

Thus, the true mode shape is

$$U_{3j} = \psi_j - \varepsilon_1 U_{31} \tag{5.11.11}$$

Note again that all integration will have to be done numerically because mode shape data will not be available as a function but as an array.

REFERENCES

5.1 G. R. Kirchhoff, "Über das Gleichgewicht und die Bewegung einer elastischen Scheibe," *J. Mathematik (Crelle), vol. 40,* 1850.

5.2 A. W. Leissa, *Vibration of Plates*, NASA SP-160, U.S. Government Printing Office, Washington, D.C., 1969.

5.3 W. E. Byerly, *Fourier Series*, Dover, New York, 1959 (originally published in 1893).

5.4 G. Hadley, *Linear Algebra*, Addison-Wesley, Reading, Mass., 1961.

5.5 A. W. Leissa, *Vibrations of Shells*, NASA SP-228, U.S. Government Printing Office, Washington, D.C., 1973.

5.6 W. Flügge, *Statik und Dynamik der Schalen*, Springer Verlags Berlin, 1957.

5.7 W. Soedel and M. Dhar, "Difficulties in Finding Modes Experimentally when Several Contribute to a Resonance", *J. Sound Vibration, vol. 58,* no. 1, 1978, pp. 27-38.

6

SIMPLIFIED SHELL EQUATIONS

Except for a few special cases, many of which were discussed in Chap. 5, explicit solutions are not available for Love's equations. The investigator has therefore often no choice but to use approximate solution approaches. The approximate methods will be discussed in later chapters. In this chapter we offer the alternative of using simplified versions of Love's equations.

6.1 THE MEMBRANE APPROXIMATION

A common approximation in statics of shells, but also used in the analysis of shell vibrations, is to assume that the bending stiffness in Love's quations can be neglected. Obviously, this assumption leads to disaster for transversely vibrating plates and beams but has some justification for shells and arches vibrating in shapes where the stretching of the neutral surface is a dominating contributor to the motion resistance. The approximation is also called the *extensional approximation* and Lord Rayleigh [6.1] is commonly credited with it.

Setting

$$D = 0 \tag{6.1.1}$$

implies that

$$M_{11} = M_{22} = M_{12} = Q_{13} = Q_{23} = 0 \tag{6.1.2}$$

The equations of motion are therefore

$$\frac{\partial(N_{11}A_2)}{\partial\alpha_1} + \frac{\partial(N_{12}A_1)}{\partial\alpha_2} + N_{12}\frac{\partial A_1}{\partial\alpha_2} - N_{22}\frac{\partial A_2}{\partial\alpha_1} + A_1A_2q_1 = A_1A_2\rho h\ddot{u}_1 \tag{6.1.3}$$

$$\frac{\partial(N_{12}A_2)}{\partial\alpha_1} + \frac{\partial(N_{22}A_1)}{\partial\alpha_2} + N_{12}\frac{\partial A_2}{\partial\alpha_1} - N_{11}\frac{\partial A_1}{\partial\alpha_2} + A_1A_2q_2 = A_1A_2\rho h\ddot{u}_2 \tag{6.1.4}$$

$$-A_1A_2\left(\frac{N_{11}}{R_1} + \frac{N_{22}}{R_2}\right) + A_1A_2q_3 = A_1A_2\rho h\ddot{u}_3 \tag{6.1.5}$$

Boundary conditions reduce to the type

$$N_{nn} = N^*_{nn} \quad \text{or} \quad u_n = u^*_n \tag{6.1.6}$$

and $u_3 = u^*_3$ (6.1.7)

Note that in general only two boundary conditions can now be satisfied at each edge, as compared to four for the general case.

6.2 AXISYMMETRIC EIGENVALUES OF A SPHERICAL SHELL

This example was treated first by Lamb [6.2] by reduction from the solution for a vibrating solid sphere. Here we start with the simplified Love equations (6.1.3) to (6.1.5). We are only interested in axisymmetric vibrations. All derivatives with respect to θ vanish. Since $\alpha_1 = \phi$, $\alpha_2 = \theta$, $A_1 = a$, $A_2 = a \sin\phi$, $R_1 = R_2 = a$, Eqs. (6.1.3) to (6.1.5) become

$$\frac{\partial}{\partial\phi}(N_{\phi\phi}\sin\phi) - N_{\theta\theta}\cos\phi + aq_\phi\sin\phi = a\rho h\frac{\partial^2 u_\phi}{\partial t^2}\sin\phi \tag{6.2.1}$$

$$-(N_{\phi\phi} + N_{\theta\theta}) + aq_3 = a\rho h\frac{\partial^2 u_3}{\partial t^2} \tag{6.2.2}$$

The strain-displacement relations become

$$\varepsilon^{\circ}_{\phi\phi} = \frac{1}{a}\left(\frac{\partial u_\phi}{\partial \phi} + u_3\right) \tag{6.2.3}$$

$$\varepsilon^{\circ}_{\theta\theta} = \frac{1}{a \sin \phi}\,(u_\phi \cos \phi + u_3 \sin \phi) \tag{6.2.4}$$

This gives

$$N_{\phi\phi} = \frac{K}{a}\left[\frac{\partial u_\phi}{\partial \phi} + \mu u_\phi \cot \phi + (1 + \mu)u_3\right] \tag{6.2.5}$$

$$N_{\theta\theta} = \frac{K}{a}\left[u_\phi \cot \phi + \mu \frac{\partial u_\phi}{\partial \phi} + (1 + \mu)u_3\right] \tag{6.2.6}$$

We get therefore

$$\frac{\partial^2 u_\phi}{\partial \phi^2} + \frac{\partial}{\partial \phi}(u_\phi \cot \phi) + (1 + \mu)\frac{\partial u_3}{\partial \phi} + \frac{a^2 q_\phi}{K} = \frac{a^2 \rho h}{K}\frac{\partial^2 u_\phi}{\partial t^2} \tag{6.2.7}$$

$$-\frac{\partial u_\phi}{\partial \phi} - u_\phi \cot \phi - 2u_3 + \frac{a^2}{K(1 + \mu)} q_3 = \frac{a^2 \rho h}{K(1 + \mu)}\frac{\partial^2 u_3}{\partial t^2} \tag{6.2.8}$$

To find the eigenvalues, we set $q_\phi = 0$ and $q_3 = 0$ and substitute

$$\begin{Bmatrix} u_\phi(\phi,t) \\ u_3(\phi,t) \end{Bmatrix} = \begin{Bmatrix} U_\phi(\phi) \\ U_3(\phi) \end{Bmatrix} e^{j\omega t} \tag{6.2.9}$$

We obtain

$$\frac{d^2 U_\phi}{d\phi^2} + \frac{d}{d\phi}(U_\phi \cot \phi) + (1 + \mu)\frac{dU_3}{d\phi} + (1 - \mu^2)\Omega^2 U_\phi = 0 \tag{6.2.10}$$

$$\frac{dU_\phi}{d\phi} + U_\phi \cot \phi + 2U_3 - (1 - \mu)\Omega^2 U_3 = 0 \tag{6.2.11}$$

where

$$\Omega^2 = \frac{a^2 \rho \omega^2}{E} \tag{6.2.12}$$

and solving Eq. (6.2.11) for U_3 and differentiating with respect to ϕ gives

$$\frac{dU_3}{d\phi} = \frac{1}{(1 - \mu)\Omega^2 - 2} \left[\frac{d^2U_\phi}{d\phi^2} + \frac{d}{d\phi} (U_\phi \cot \phi) \right] \tag{6.2.13}$$

Substituting this equation in Eq. (6.2.10) gives

$$\frac{d^2U_\phi}{d\phi^2} + \frac{d}{d\phi} (U_\phi \cot \phi) + \lambda(\lambda + 1)U_\phi = 0 \tag{6.2.14}$$

where

$$\lambda(\lambda + 1) = 2 + \frac{(1 + \mu)\Omega^2[3 - (1 - \mu)\Omega^2]}{1 - \Omega^2} \tag{6.2.15}$$

Let us now define a function Φ such that

$$U_\phi = \frac{d\Phi}{d\phi} \tag{6.2.16}$$

Substituting this in Eq. (6.2.14) gives

$$\frac{d}{d\phi} \left(\frac{d^2\Phi}{d\phi^2} + \cot \phi \frac{d\Phi}{d\phi} + \lambda(\lambda + 1) \right) = 0 \tag{6.2.17}$$

or, integrating

$$\frac{d^2\Phi}{d\phi^2} + \cot \phi \frac{d\Phi}{d\phi} + \lambda(\lambda + 1)\Phi = C \tag{6.2.18}$$

where C is an integration constant. This equation has a homogeneous solution and a particular solution. The latter is

$$\Phi = \frac{C}{\lambda(\lambda + 1)} \tag{6.2.19}$$

and does not contribute anything to U_ϕ according to Eq. (6.2.16). The homogeneous solution is obtained from

$$\frac{d^2\Phi}{d\phi^2} + \cot \phi \frac{d\Phi}{d\phi} + \lambda(\lambda + 1)\Phi = 0 \tag{6.2.20}$$

We recognize this equation to be Legendre's differential equation.

Its general solution is

$$\Phi = AP_\lambda(\cos\phi) + BQ_\lambda(\cos\phi) \tag{6.2.21}$$

or

$$U_\phi = \frac{d\Phi}{d\phi} = A\,\frac{dP_\lambda(\cos\phi)}{d\phi} + B\,\frac{dQ_\lambda(\cos\phi)}{d\phi} \tag{6.2.22}$$

To obtain the solution for U_3, we substitute Eq. (6.2.16) in Eq. (6.2.11):

$$-U_3[2 - (1 - \mu)\Omega^2] = \frac{d^2\Phi}{d\phi^2} + \cot\phi\,\frac{d\Phi}{d\phi} \tag{6.2.23}$$

Substituting this in Eq. (6.2.20) and utilizing Eq. (6.2.15) gives

$$U_3 = \frac{1 + (1 + \mu)\Omega^2}{1 - \Omega^2}\,\Phi \tag{6.2.24}$$

The functions $P_\lambda(\cos\phi)$ and $Q_\lambda(\cos\phi)$ are the Legendre functions of the first and second kind and of fractional order λ. The Legendre function $Q_\lambda(\cos\phi)$ is singular at $\phi = 0$. Thus, whenever we have a spherical shell that is closed at the apex $\phi = 0$, we set $B = 0$ in Eqs. (6.2.21) and (6.2.22). The Legendre function $P_\lambda(\cos\phi)$ is singular at $\phi = \pi$ unless

$$\lambda = n \quad (n = 0, 1, 2, \ldots) \tag{6.2.25}$$

In this case the $P_\lambda(\cos\phi)$ reduces to $P_n(\cos\phi)$ which are called *Legendre polynomials*. This fact provides us with a very simple solution for the special case of a closed spherical shell since Eq. (6.2.25) allows us an immediate formulation of the natural frequencies, using definition (6.2.15). We get

$$\Omega^4(1 - \mu^2) - \Omega^2[n(n + 1) + 1 + 3\mu] + [n(n + 1) - 2] = 0 \tag{6.2.26}$$

or

$$\Omega^2_{n_{1,2}} = \frac{1}{2(1 - \mu^2)}\left\{n(n + 1) + 1 + 3\mu \pm \sqrt{[n(n + 1) + 1 + 3\mu]^2 - 4(1 - \mu^2)[n(n + 1) - 2]}\right\} \tag{6.2.27}$$

The natural modes are

$$U_\phi = A \frac{dP_n(\cos\phi)}{d\phi} \tag{6.2.28}$$

$$U_3 = A \frac{1 + (1 + \mu)\Omega^2}{1 - \Omega^2} P_n(\cos\phi) \tag{6.2.29}$$

For instance, at n = 0 which is the breathing mode,

$$\Omega_{o_1}^2 = \frac{2}{1 - \mu} \tag{6.2.30}$$

$\Omega_{o_2}^2$ is negative, giving rise to an imaginary frequency and is therefore physically meaningless. Since $P_o(\cos\phi) = 1$, we get

$$U_3 = \frac{1 + (1 + \mu)\Omega_{o_1}^2}{1 - \Omega_{o_1}^2} \tag{6.2.31}$$

and $\quad U_\phi = 0 \qquad (6.2.32)$

The natural frequency in radians per second is obtained from Eq. (6.2.12) to be

$$\omega_{o_1}^2 = \frac{E}{a^2\rho} \Omega_{o_1}^2 \tag{6.2.33}$$

We recognize this case as the "breathing" mode of the sphere. The calculated frequency agrees well with experimental reality.

Natural frequencies for other n values are plotted in Fig. 6.2.1. The Ω_{n1} branch is dominated by in plane motion and agrees well with reality for all thickness to radius ratios. The Ω_{n2} branch is dominated by transverse motion. The zero value for n = 1 defines a rigid-body motion. For thickness-to-radius ratios up to approximately h/a = 0.01 the membrane approximation gives very good results, as shown in Fig. 6.2.1, where the results for the full theory are superimposed. Only when the shell starts to become a "thick" shell do we see a pronounced effect of bending on the transverse natural frequencies. This is shown for h/a = 0.1 The results

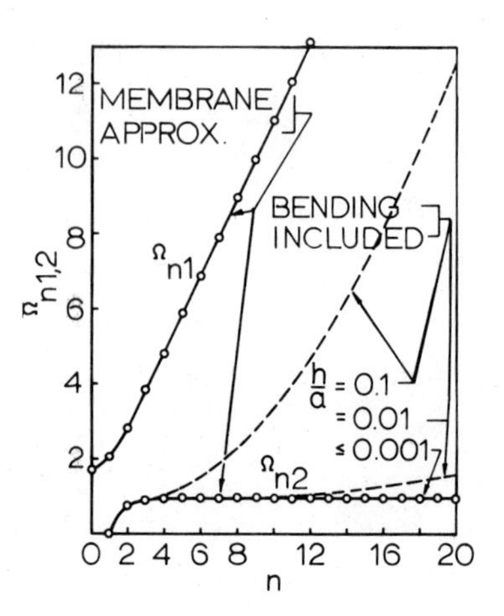

Figure 6.2.1

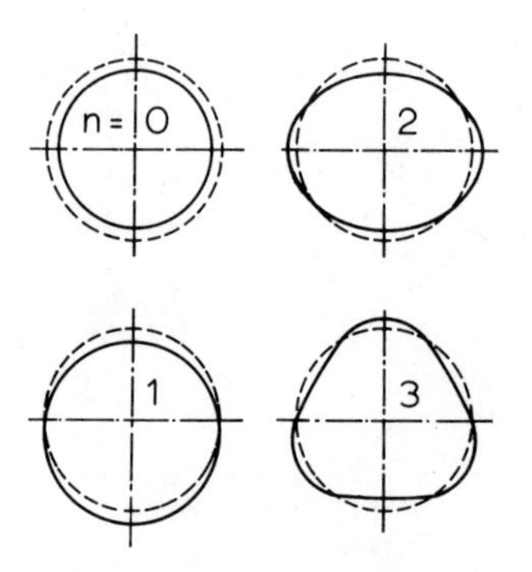

Figure 6.2.2

for the full theory where bending is considered is given by [6.3]:

$$\Omega^2_{n_{1,2}} = \frac{1}{2(1 - \mu^2)} \left(A \pm \sqrt{A^2 - 4mB} \right) \tag{6.2.34}$$

where

$$A = 3(1 + \mu) + m + \frac{1}{12}\left(\frac{h}{a}\right)^2 (m + 3)\ (m + 1 + \mu) \tag{6.2.35}$$

$$B = 1 - \mu^2 + \frac{1}{12}\left(\frac{h}{a}\right)^2 [(m + 1)^2 - \mu^2] \tag{6.2.36}$$

$$m = n(n + 1) - 2 \quad (n = 0, 1, 2, \ldots) \tag{6.2.37}$$

When h/a is set to zero, this equation reduces to Eq. (6.2.27).

An interesting fact is that if the spherical shell is very thin, all transverse frequencies above approximately $n = 4$ exist in a very narrow frequency band.

To aid in the plotting of mode shapes, the following identities are useful: $P_o(\cos\phi) = 1$, $P_1(\cos\phi) = \cos\phi$, $P_2(\cos\phi) = (3\cos 2\phi + 1)/4$, $P_3(\cos\phi) = (5\cos 2\phi + 3\cos\phi)/8$. A few transverse mode-shapes are shown in Fig. 6.2.2. For more work on spherical shell vibrations see Refs. 6.4 to 6.7.

6.3 THE BENDING APPROXIMATION

Also called the *inextensional approximation*, the bending approximation was first employed by Lord Rayleigh [6.8].

The simplification applies sometimes to shells that have developable surfaces, but mainly to any shell with transverse modes of a wavelength one order of magnitude smaller than the smallest shell surface dimension.

The assumption is that

$$\varepsilon_{11}^{\circ} = \varepsilon_{22}^{\circ} = \varepsilon_{12}^{\circ} = 0 \tag{6.3.1}$$

Thus, all membrane force resultants are zero. The equivalent effect is reached if we set $K = 0$. In both cases we obtain

$$\frac{\partial(Q_{13}A_2)}{\partial\alpha_1} + \frac{\partial(Q_{23}A_1)}{\partial\alpha_2} + A_1A_2q_3 = A_1A_2\rho h\ddot{u}_3 \tag{6.3.2}$$

where

$$Q_{13} = \frac{1}{A_1A_2}\left(\frac{\partial(M_{11}A_2)}{\partial\alpha_1} + \frac{\partial(M_{12}A_1)}{\partial\alpha_2} + M_{12}\frac{\partial A_1}{\partial\alpha_2} - M_{22}\frac{\partial A_2}{\partial\alpha_1}\right) \tag{6.3.3}$$

$$Q_{23} = \frac{1}{A_1A_2}\left(\frac{\partial(M_{12}A_2)}{\partial\alpha_1} + \frac{\partial(M_{22}A_1)}{\partial\alpha_2} + M_{12}\frac{\partial A_2}{\partial\alpha_1} - M_{11}\frac{\partial A_1}{\partial\alpha_2}\right) \tag{6.3.4}$$

Note that u_1 and u_2 are not zero but are related to the transverse displacement by two of the three membrane strain equations

$$\frac{1}{A_1}\frac{\partial u_1}{\partial \alpha_1} + \frac{u_2}{A_1A_2}\frac{\partial A_1}{\partial \alpha_2} = -\frac{u_3}{R_1} \qquad (6.3.5)$$

$$\frac{1}{A_2}\frac{\partial u_2}{\partial \alpha_2} + \frac{u_1}{A_1A_2}\frac{\partial A_2}{\partial \alpha_1} = -\frac{u_3}{R_2} \qquad (6.3.6)$$

$$\frac{A_2}{A_1}\frac{\partial}{\partial \alpha_1}\left(\frac{u_2}{A_2}\right) + \frac{A_1}{A_2}\frac{\partial}{\partial \alpha_2}\left(\frac{u_1}{A_1}\right) = 0 \qquad (6.3.7)$$

The bending strain expressions are the same as in Eqs. (2.4.22) to (2.4.24), with β_1 and β_2 given by Eqs. (2.4.7) and (2.4.8).

6.4 CIRCULAR CYLINDRICAL SHELL

In this case, $A_1 = 1$, $d\alpha_1 = dx$, $A_2 = a$, $d\alpha_2 = d\theta$, $R_1 = \infty$, $R_2 = a$. This gives the following relationships between transverse and inplane displacements.

$$\frac{\partial u_x}{\partial x} = 0 \qquad (6.4.1)$$

$$\frac{\partial u_\theta}{\partial \theta} = -u_3 \qquad (6.4.2)$$

$$\frac{\partial u_\theta}{\partial x} + \frac{1}{a}\frac{\partial u_x}{\partial \theta} = 0 \qquad (6.4.3)$$

For instance, for the simply supported shell vibrating at a natural frequency

$$u_3 = A \sin\frac{m\pi x}{L} \cos n(\theta - \phi)e^{j\omega t} \qquad (6.4.4)$$

Thus, from Eq. (6.4.2),

$$u_\theta = -\int u_3 \, d\theta + C_1 = -\frac{A}{n}\sin\frac{m\pi x}{L}\sin n(\theta - \phi)e^{j\omega t} \qquad (6.4.5)$$

Selecting (6.4.3) as our second equation,

$$u_x = -a \int \frac{\partial u_\theta}{\partial x} d\theta + C_2 \tag{6.4.6}$$

or

$$u_x = \frac{Aa}{n^2} \frac{m\pi}{L} \cos \frac{m\pi x}{L} \cos n(\theta - \phi) e^{j\omega t} \tag{6.4.7}$$

Note that the three displacement mode components are in functional character identical to the exact solution, but they are not any longer independent from each other. Thus, constants A, B, C of Sec. 5.5 are now replaced by $Ama\pi/Ln^2$, $-A/n$, and A. Substituting Eqs. (6.4.4), (6.4.5), and (6.4.7) in the bending strain expressions and these in Eqs. (6.3.2) to (6.3.4) gives

$$\omega_{mn}^2 = \frac{E}{12\rho(1 - \mu^2)} \left(\frac{h}{a}\right)^2 \frac{1}{a^2} \left[\left(\frac{m\pi a}{L}\right)^2 + n^2\right]\left[\left(\frac{m\pi a}{L}\right)^2 + n^2 - 1\right] \tag{6.4.8}$$

This equation is of special interest to the acoustical engineer since in the higher mode number range the influence of the boundary conditions disappears and any closed circular cylindrical shell will be governed by it. As a matter of fact, the -1 in the second bracket is negligible for higher m and n combinations and the equation simplifies to

$$\omega_{mn}^2 = \frac{E}{12\rho(1 - \mu^2)} \left(\frac{h}{a}\right)^2 \frac{1}{a^2} \left[\left(\frac{m\pi a}{L}\right)^2 + n^2\right]^2 \tag{6.4.9}$$

6.5 ZERO IN-PLANE DEFLECTION APPROXIMATION

For modes primarily associated with transverse motion, the contributions of u_1 and u_2 on strain are assumed to be negligible. This seems to work for very shallow shells and bending-dominated modes. The shell equation derived by Sophie Germaine [1.22] seems to have used this assumption.

The strain-displacement relationships become

$$\varepsilon_{11}^{\circ} = \frac{u_3}{R_1} \tag{6.5.1}$$

$$\varepsilon_{22}^{\circ} = \frac{u_3}{R_2} \tag{6.5.2}$$

$$\varepsilon_{12}^{\circ} = 0 \tag{6.5.3}$$

$$k_{11} = -\frac{1}{A_1}\frac{\partial}{\partial\alpha_1}\left(\frac{1}{A_1}\frac{\partial u_3}{\partial\alpha_1}\right) - \frac{1}{A_1A_2^{\,2}}\frac{\partial u_3}{\partial\alpha_2}\frac{\partial A_1}{\partial\alpha_2} \tag{6.5.4}$$

$$k_{22} = -\frac{1}{A_2}\frac{\partial}{\partial\alpha_2}\left(\frac{1}{A_2}\frac{\partial u_3}{\partial\alpha_2}\right) - \frac{1}{A_2A_1^{\,2}}\frac{\partial u_3}{\partial\alpha_1}\frac{\partial A_2}{\partial\alpha_1} \tag{6.5.5}$$

$$k_{12} = -\frac{A_2}{A_1}\frac{\partial}{\partial\alpha_1}\left(\frac{1}{A_1^{\,2}}\frac{\partial u_3}{\partial\alpha_2}\right) - \frac{A_1}{A_2}\frac{\partial}{\partial\alpha_2}\left(\frac{1}{A_2^{\,2}}\frac{\partial u_3}{\partial\alpha_1}\right) \tag{6.5.6}$$

Substituting these relationships into Love's equations gives

$$D\nabla^4 u_3 + Ku_3\left(\frac{1}{R_1^{\,2}} + \frac{1}{R_2^{\,2}} + \frac{2\mu}{R_1R_2}\right) + \rho h\frac{\partial^2 u_3}{\partial t^2} = q_3 \tag{6.5.7}$$

where

$$\nabla^2(\cdot) = \frac{1}{A_1A_2}\left[\frac{\partial}{\partial\alpha_1}\left(\frac{A_2}{A_1}\frac{\partial(\cdot)}{\partial\alpha_1}\right) + \frac{\partial}{\partial\alpha_2}\left(\frac{A_1}{A_2}\frac{\partial(\cdot)}{\partial\alpha_2}\right)\right] \tag{6.5.8}$$

Equation (6.5.7) can be used to estimate quickly effects of curvature in relatively shallow shells. However, the accuracy of prediction leaves a lot to be desired.

6.6 EXAMPLE: CURVED FAN BLADE

For some low-speed fan blades, where the centrifugal stiffening effect can be neglected, we have $R_1 = \infty$ and $R_2 = a$ (see Fig. 6.6.1). Let us use a cartesian coordinate system in the plane of projection. At a natural frequency,

$$u_3(x,y,t) = U_3(x,y)e^{j\omega t} \tag{6.6.1}$$

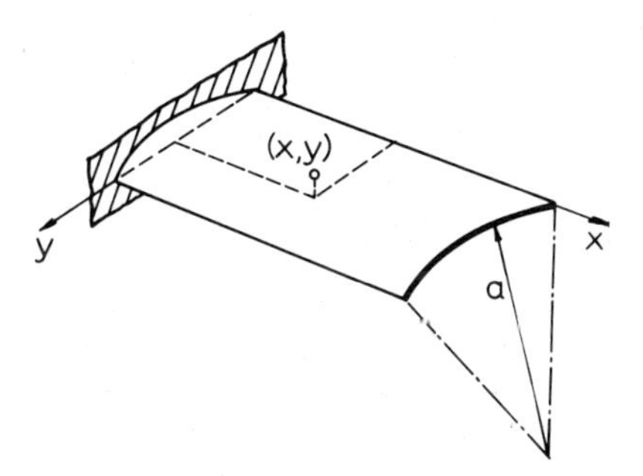

Figure 6.6.1

we get

$$D\nabla^4 U_3(x,y) + \left(\frac{K}{a^2} - \rho h\omega^2\right) U_3(x,y) = 0 \tag{6.6.2}$$

We notice therefore immediately that

$$\omega_2^2 = \omega_1^2 + \frac{K}{a^2 \rho h} \tag{6.6.3}$$

where

ω_2 = natural frequencies of the curved blade
ω_1 = natural frequencies of the flat blade

The formula will only apply approximately to the first few beam-type modes but allows quick estimates of the curvature effect.

We may now generalize this finding. According to Eq. (6.5.7), whenever the shell is so shallow that we may use plate coordinates as an approximation,

$$\omega_2^2 = \omega_1^2 + \frac{K}{\rho h}\left(\frac{1}{R_1^2} + \frac{1}{R_2^2} + \frac{2\mu}{R_1 R_2}\right) \tag{6.6.4}$$

It is implied that both curvatures are constant over the surface.

6.7 DONNELL-MUSHTARI-VLASOV EQUATIONS

Of all the presented simplifications, this one is used most widely in shell vibrations. It neither neglects bending nor membrane effects. It applies to shells that are loaded normal to their surface and concentrates on transverse deflection behavior. The approach was developed, apparently independently, by Donnell [6.9, 6.10] and Mushtari [6.11]. Donnell derived it for the circular cylindrical shell. The approach was generalized for any geometry by Vlasov [6.12]. Because Vlasov pointed out that the approach gives particularly good results for shallow shells, the equations are often referred to as shallow shell equations. This is, however, an unnecessarily severe restriction, as we will see when we develop the equations.

The first basic assumption is that contributions of in-plane deflections can be neglected in the bending strain expressions but not in the membrane strain expressions. The bending strains are therefore

$$k_{11} = -\frac{1}{A_1}\frac{\partial}{\partial\alpha_1}\left(\frac{1}{A_1}\frac{\partial u_3}{\partial\alpha_1}\right) - \frac{1}{A_1A_2^2}\cdot\frac{\partial u_3}{\partial\alpha_2}\frac{\partial A_1}{\partial\alpha_2} \tag{6.7.1}$$

$$k_{22} = -\frac{1}{A_2}\frac{\partial}{\partial\alpha_2}\left(\frac{1}{A_2}\frac{\partial u_3}{\partial\alpha_2}\right) - \frac{1}{A_2A_1^2}\frac{\partial u_3}{\partial\alpha_1}\frac{\partial A_2}{\partial\alpha_1} \tag{6.7.2}$$

$$k_{12} = -\frac{A_2}{A_1}\frac{\partial}{\partial\alpha_1}\left(\frac{1}{A_2^2}\frac{\partial u_3}{\partial\alpha_2}\right) - \frac{A_1}{A_2}\frac{\partial}{\partial\alpha_2}\left(\frac{1}{A_1^2}\frac{\partial u_3}{\partial\alpha_1}\right) \tag{6.7.3}$$

The membrane strain expressions remain the same. The next assumption is that the influence of inertia in the in-plane direction is neglected. Needless to say, the theory is restricted to normal loading. Finally; we neglect the shear terms Q_{31}/R_1 and Q_{32}/R_2.

The equations of motion are, therefore,

$$\frac{\partial(A_2N_{11})}{\partial\alpha_1} + \frac{\partial(A_1N_{12})}{\partial\alpha_2} + \frac{\partial A_1}{\partial\alpha_2}N_{12} - \frac{\partial A_2}{\partial\alpha_1}N_{22} = 0 \tag{6.7.4}$$

$$\frac{\partial(A_2N_{12})}{\partial\alpha_1} + \frac{\partial(A_1N_{22})}{\partial\alpha_2} + \frac{\partial A_2}{\partial\alpha_1}N_{12} - \frac{\partial A_1}{\partial\alpha_2}N_{11} = 0 \tag{6.7.5}$$

$$D\nabla^4 u_3 + \frac{N_{11}}{R_1} + \frac{N_{22}}{R_2} + \rho h \frac{\partial^2 u_3}{\partial t^2} = q_3 \tag{6.7.6}$$

Let us now introduce a function ϕ which we define as

$$N_{11} = \frac{1}{A_2} \frac{\partial}{\partial \alpha_2}\left(\frac{1}{A_2} \frac{\partial \phi}{\partial \alpha_2}\right) + \frac{1}{A_1{}^2 A_2} \frac{\partial A_2}{\partial \alpha_1} \frac{\partial \phi}{\partial \alpha_1} \tag{6.7.7}$$

$$N_{22} = \frac{1}{A_1} \frac{\partial}{\partial \alpha_1}\left(\frac{1}{A_1} \frac{\partial \phi}{\partial \alpha_1}\right) + \frac{1}{A_1 A_2{}^2} \frac{\partial A_1}{\partial \alpha_2} \frac{\partial \phi}{\partial \alpha_2} \tag{6.7.8}$$

$$N_{12} = -\frac{1}{A_1 A_2}\left(\frac{\partial^2 \phi}{\partial \alpha_1 \partial \alpha_2} - \frac{1}{A_1} \frac{\partial A_1}{\partial \alpha_2} \frac{\partial \phi}{\partial \alpha_1} - \frac{1}{A_2} \frac{\partial A_2}{\partial \alpha_1} \frac{\partial \phi}{\partial \alpha_2}\right) \tag{6.7.9}$$

If we substitute these definitions into Equations 6.7.4 to 6.7.6, and Equations 6.7.1 to 6.7.3 into Equation 6.7.6, we find that the first two equations are satisfied and the third equation becomes

$$D\nabla^4 u_3 + \nabla_K{}^2 \phi + \rho h \frac{\partial^2 u_3}{\partial t^2} = q_3 \tag{6.7.10}$$

where

$$\nabla_K{}^2(\cdot) = \frac{1}{A_1 A_2}\left[\frac{\partial}{\partial \alpha_1}\left[\frac{1}{R_2} \frac{A_2}{A_1} \frac{\partial(\cdot)}{\partial \alpha_1}\right] + \frac{\partial}{\partial \alpha_2}\left[\frac{1}{R_1} \frac{A_1}{A_2} \frac{\partial(\cdot)}{\partial \alpha_2}\right]\right] \tag{6.7.11}$$

This type of function was first introduced by Airy in 1863 for the two-dimensional treatment of a beam in bending [6.13] and is in general known as Airy's stress function. With it we have in effect eliminated u_1 and u_2 but still have ϕ and u_3 to contend with. To obtain a second equation, we follow the standard procedure with Airy's stress function, namely to generate the compatibility equation. The way to do this is to take the six strain displacement relationships and eliminate from them the displacement by substitutions, additions and subtraction. This is shown in detail in Refs. 6.14 and 6.15. The result is

$$\frac{k_{11}}{R_1} + \frac{k_{22}}{R_2} + \frac{1}{A_1A_2}\left\{\frac{\partial}{\partial\alpha_1}\frac{1}{A_1}\left[A_2\frac{\partial\varepsilon^\circ_{22}}{\partial\alpha_1} + \frac{\partial A_2}{\partial\alpha_1}(\varepsilon^\circ_{22} - \varepsilon^\circ_{11})\right.\right.$$

$$\left. - \frac{A_1}{2}\frac{\partial\varepsilon^\circ_{12}}{\partial\alpha_2} - \frac{\partial A_1}{\partial\alpha_2}\varepsilon^\circ_{12}\right] + \frac{\partial}{\partial\alpha_2}\frac{1}{A_2}\left[A_1\frac{\partial\varepsilon^\circ_{11}}{\partial\alpha_2}\right.$$

$$\left.\left. + \frac{\partial A_1}{\partial\alpha_2}(\varepsilon^\circ_{11} - \varepsilon^\circ_{22}) - \frac{A_2}{2}\frac{\partial\varepsilon^\circ_{12}}{\partial\alpha_1} - \frac{\partial A_2}{\partial\alpha_1}\varepsilon^\circ_{12}\right]\right\} = 0 \qquad (6.7.12)$$

We now substitute the fact that

$$\varepsilon^\circ_{11} = \frac{1}{Eh}(N_{11} - \mu N_{22}) \qquad (6.7.13)$$

$$\varepsilon^\circ_{22} = \frac{1}{Eh}(N_{22} - \mu N_{11}) \qquad (6.7.14)$$

$$\varepsilon^\circ_{12} = \frac{2(1 + \mu)}{Eh}N_{12} \qquad (6.7.15)$$

where N_{11}, N_{22}, and N_{12} are replaced by the stress function definitions of Eqs. (6.7.7) to (6.7.9). This gives us

$$Eh\nabla^2_K u_3 - \nabla^4\phi = 0 \qquad (6.7.16)$$

Thus, Eqs. (6.7.10) and (6.7.16) are the equations of motion.

There are four necessary boundary conditions at each edge, two in terms of u_3 and two in terms of ϕ. The ϕ conditions pose a problem if they are given in terms of u_1 or u_2, since we then have to solve Eqs. (6.7.7) to (6.7.9) for ϕ. However, boundary conditions where N_{11}, N_{12}, or N_{22} are specified directly are easier to handle.

6.8 NATURAL FREQUENCIES AND MODES

To obtain the eigenvalues of Eqs. (6.7.10) and (6.7.16), we substitute

$$q_3 = 0 \qquad (6.8.1)$$

$$u_3(\alpha_1,\alpha_2,t) = U_3(\alpha_1,\alpha_2)e^{j\omega t} \qquad (6.8.2)$$

$$\phi(\alpha_1,\alpha_2,t) = \Phi(\alpha_1,\alpha_2)e^{j\omega t} \tag{6.8.3}$$

and get

$$D\nabla^4 U_3 + \nabla_K^2\Phi - \rho h\omega^2 U_3 = 0 \tag{6.8.4}$$

$$Eh\nabla_K^2 U_3 - \nabla^4\Phi = 0 \tag{6.8.5}$$

Defining a function $F(\alpha_1,\alpha_2)$ such that

$$U_3 = \nabla^4 F \tag{6.8.6}$$

$$\Phi = Eh\nabla_K^2 F \tag{6.8.7}$$

we obtain, by proper substitution into Eqs. (6.8.4) and (6.8.5),

$$D\nabla^8 F + Eh\nabla_K^4 F - \rho h\omega^2\nabla^4 F = 0 \tag{6.8.8}$$

The alternative choice is to operate on Eq. (6.8.4) with ∇^4 and on Eq. 6.8.5 with ∇_K^2. Combining the two equations gives then

$$D\nabla^8 U_3 + Eh\nabla_K^4 U_3 - \rho h\omega^2\nabla^4 U_3 = 0 \tag{6.8.9}$$

This form is probably preferable.

6.9 THE CIRCULAR CYLINDRICAL SHELL

For a circular cylindrical shell, $A_1 = 1$, $\alpha_1 = x$, $A_2 = a$, $\alpha_2 = \theta$. This gives

$$\nabla^4(\cdot) = \frac{1}{a^4}\frac{\partial^4(\cdot)}{\partial\theta^4} + \frac{\partial^4(\cdot)}{\partial x^4} + \frac{2}{a^2}\frac{\partial^4(\cdot)}{\partial x^2\partial\theta^2} \tag{6.9.1}$$

$$\nabla_K^4(\cdot) = \frac{1}{a^2}\frac{\partial^4(\cdot)}{\partial x^4} \tag{6.9.2}$$

For the special category where the shell is closed in θ direction, the solution will be of the form

$$U_3(x,\theta) = U_{3n}(x)\cos n(\theta - \phi) \tag{6.9.3}$$

where ϕ is an arbitrary angle accounting for the fact that there is no preferential direction of the mode shape in circumferential direction. The equation of motion becomes

$$D\left(\frac{n^2}{a^2}-\frac{d^2}{dx^2}\right)^4 U_{3n}(x)+\frac{Eh}{a^2}\frac{d^4U_{3n}(x)}{dx^4}$$

$$-\rho h\omega^2\left(\frac{n^2}{a^2}-\frac{d^2}{dx^2}\right)^2 U_{3n}(x)=0 \qquad (6.9.4)$$

Solutions must be of the form

$$U_{3n}(x)=e^{\lambda\left(\frac{x}{L}\right)} \qquad (6.9.5)$$

This gives

$$D\left[\frac{n^2}{a^2}-\left(\frac{\lambda}{L}\right)^2\right]^4+\frac{Eh}{a^2}\left(\frac{\lambda}{L}\right)^4-\rho h\omega^2\left[\frac{n^2}{a^2}-\left(\frac{\lambda}{L}\right)^2\right]^2=0 \qquad (6.9.6)$$

This equation has the following roots $\lambda_i = \pm\eta_1, \pm j\eta_2, \pm(\eta_3+j\eta_4), \pm(\eta_3-j\eta_4)$.

The general solution must be of the form

$$U_{3n}(x)=A_1\sinh\eta_1\frac{x}{L}+A_2\cosh\eta_1\frac{x}{L}+A_3\sin\eta_2\frac{x}{L}$$

$$+A_4\cos\eta_2\frac{x}{L}+A_5e^{\eta_3\frac{x}{L}}\cos\eta_4\frac{x}{L}$$

$$+A_6e^{\eta_3\frac{x}{L}}\sin\eta_4\frac{x}{L}+A_7e^{-\eta_3\frac{x}{L}}\cos\eta_4\frac{x}{L}$$

$$+A_8e^{-\eta_3\frac{x}{L}}\sin\eta_4\frac{x}{L} \qquad (6.9.7)$$

We have to enforce four boundary conditions on each edge. However, to enforce boundary conditions involving N_{xx}, $N_{x\theta}$, u_x, or u_θ, we have to translate these conditions into a condition of the stress function ϕ. This is quite complicated and the advantage of a presumably simple theory is lost. Let us therefore follow at first a simplification introduced by Yu [6.16]. He argued that for shells and modes where

$$\frac{n^2}{a^2} >> \left(\frac{\lambda}{L}\right)^2 \tag{6.9.8}$$

we may simplify the characteristic equation to

$$D\left(\frac{n}{a}\right)^8 + \frac{Eh}{a^2}\left(\frac{\lambda}{L}\right)^4 - \rho h\omega^2\left(\frac{n}{a}\right)^4 = 0 \tag{6.9.9}$$

This gives

$$\lambda_i = \frac{nL}{a}\sqrt[4]{\frac{a^2}{Eh}\left[\rho h\omega^2 - D\left(\frac{n}{a}\right)^4\right]} \tag{6.9.10}$$

The roots are therefore $\lambda_i = \pm\eta, \pm j\eta$, where

$$\eta = \frac{nL}{a}\sqrt[4]{\frac{a^2}{Eh}\left|\rho h\omega^2 - D\left(\frac{n}{a}\right)^4\right|} \tag{6.9.11}$$

Thus, the general solution for this case is

$$U_{3n}(x) = A_1 \sin\eta\frac{x}{L} + A_2\cos\eta\frac{x}{L} + A_3\sinh\eta\frac{x}{L} + A_4\cosh\eta\frac{x}{L} \tag{6.9.12}$$

The admissable boundary conditions are

$$M_{xx} = M^*_{xx} \quad \text{or} \quad \beta_x = \beta^*_x \tag{6.9.13}$$

$$V_{x3} = V^*_{x3} \quad \text{or} \quad u_3 = u^*_3 \tag{6.9.14}$$

Equation (6.9.12) is applied to the two appropriate boundary conditions at each end. The determinant of the resulting 4 x 4 matrix equations will give the roots η_m. The natural frequencies are then obtained from Eq. (6.9.11) as

$$\omega_{mn} = \sqrt{\frac{1}{\rho h}\left[\frac{Eha^2}{L^4n^4}\eta_m^4 + D\left(\frac{n}{a}\right)^4\right]} \tag{6.9.15}$$

Note that this equation is definitely not valid for $n = 0$. It improves in accuracy as n increases. Note also that the roots η_m will be equal to the roots of the analogous beam case since Eq. (6.9.12) is of the same form as the general beam solution. This implies, of course, that the moment and shear boundary conditions of Eqs. (6.9.13) and (6.9.14) are simplified to be functions of $U_{3n}(x)$ only.

6.10 CIRCULAR DUCT CLAMPED AT BOTH ENDS

In this example we will not use the analogy to beams directly, but work with Eq. (6.9.12). The duct is shown in Fig. 6.10.1 The boundary conditions are that at $x = 0$,

$$\beta_x = 0 \tag{6.10.1}$$

$$u_3 = 0 \tag{6.10.2}$$

and at $x = L$,

$$\beta_x = 0 \tag{6.10.3}$$

$$u_3 = 0 \tag{6.10.4}$$

This translates to

$$\frac{dU_{3n}(0)}{dx} = 0 \tag{6.10.5}$$

$$U_{3n}(0) = 0 \tag{6.10.6}$$

$$\frac{dU_{3n}(L)}{dx} = 0 \tag{6.10.7}$$

$$U_{3n}(L) = 0 \tag{6.10.8}$$

Substituting Eq. (6.9.12) in these conditions gives

$$\begin{bmatrix} \eta & 0 & \eta & 0 \\ 0 & 1 & 0 & 1 \\ \eta \cos \eta & -\eta \sin \eta & \eta \cosh \eta & \eta \sinh \eta \\ \sin \eta & \cos \eta & \sinh \eta & \cosh \eta \end{bmatrix} \begin{Bmatrix} A_1 \\ A_2 \\ A_3 \\ A_4 \end{Bmatrix} = 0 \tag{6.10.9}$$

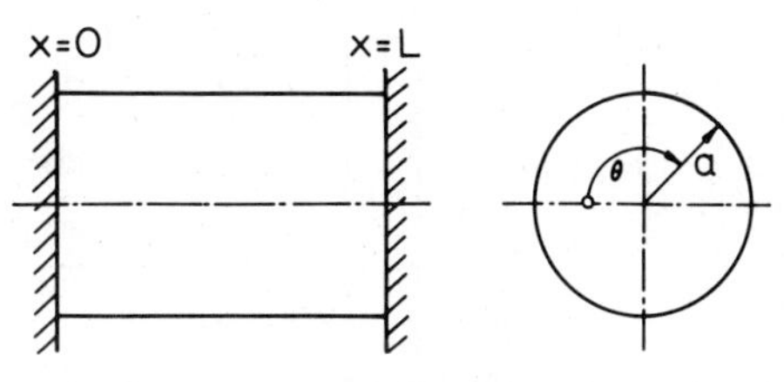

Figure 6.10.1

Setting the determinant of this equation to zero gives

$$\cos \eta \cosh \eta - 1 = 0 \tag{6.10.10}$$

The roots of this equation are $\eta_1 = 4.730$, $\eta_2 = 7.853$, $\eta_3 = 10.996$, $\eta_4 = 14.137$, etc.

Substituting the roots η_m in Eq. (6.10.9) allows us to evaluate A_2, A_3, and A_4 in terms of A_1. Utilizing Eqs. (6.9.12) and (6.9.3) gives

$$U_{3mn}(x,\theta) = \left[H\left(\cosh \eta_m \frac{x}{L} - \cos \eta_m \frac{x}{L}\right) - J\left(\sinh \eta_m \frac{x}{L} - \sin \eta_m \frac{x}{L}\right)\right] \cos n(\theta - \phi) \tag{6.10.11}$$

where

$$H = \sinh \eta_m - \sin \eta_m \tag{6.10.12}$$

$$J = \cosh \eta_m - \cos \eta_m \tag{6.10.13}$$

The solutions given here are best for pipes whose length is large as compared to their diameter. But even for short and stubby pipes, the solution agrees well with experimental evidence in the higher n range [6.17].

Natural frequencies are given by Eq. (6.9.15).

6.11 VIBRATIONS OF A FREESTANDING SMOKE STACK

Let us assume that a smoke stack can be approximated as a clamped-free cylindrical shell, as shown in Fig. 6.11.1. Initial stresses introduced by its own weight are neglected. The boundary conditions are

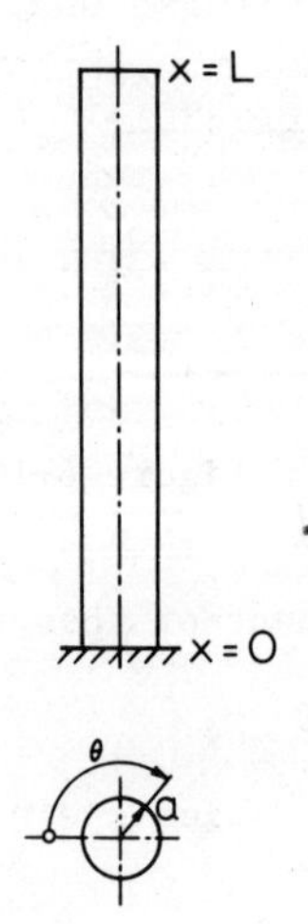

Figure 6.11.1

$$\beta_x(0) = 0 \tag{6.11.1}$$

$$u_3(0) = 0 \tag{6.11.2}$$

$$M_{xx}(L) = 0 \tag{6.11.3}$$

$$V_{3x}(L) = 0 \tag{6.11.4}$$

Let us now utilize the analogy to beams. From Ref. 6.18 we find that the characteristic equation is

$$\cos \eta \cosh \eta + 1 = 0 \tag{6.11.5}$$

The first few roots of this equation are $\eta_1 = 1.875$, $\eta_2 = 4.694$, $\eta_3 = 7.855$, $\eta_4 = 10.996$, etc. The mode shape becomes

$$U_{3mn}(x,0) = \left[F\left(\cosh \eta_m \frac{x}{L} - \cos \eta_m \frac{x}{L}\right) - G\left(\sinh \eta_m \frac{x}{L} - \sin \eta_m \frac{x}{L}\right)\right] \cos n(\theta - \phi) \tag{6.11.6}$$

where

$$F = \sinh \eta_m + \sin \eta_m \tag{6.11.7}$$

$$G = \cosh \eta_m + \cos \eta_m \tag{6.11.8}$$

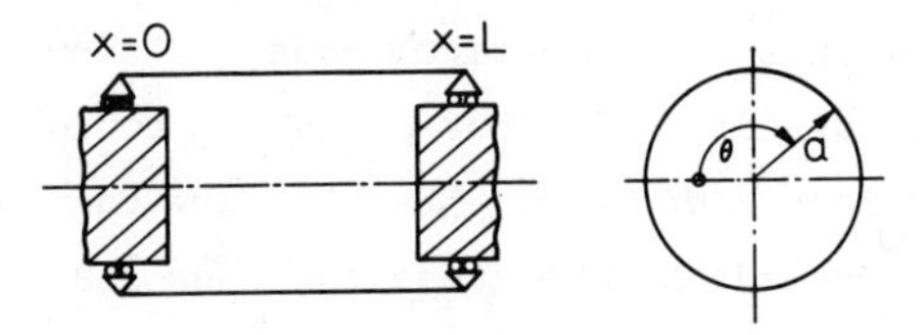

Figure 6.12.1

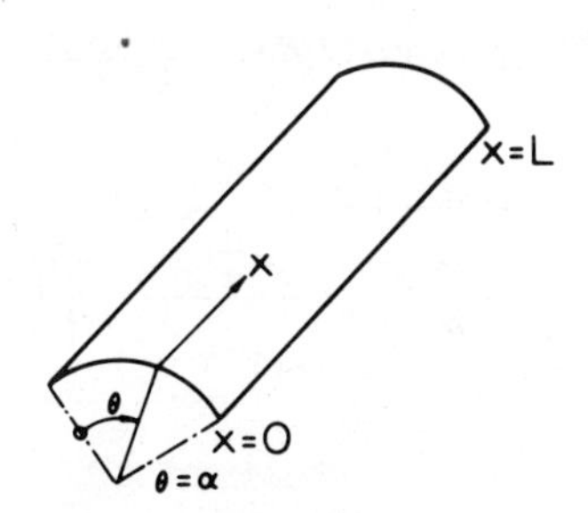

Figure 6.12.2

6.12 SPECIAL CASES OF THE SIMPLY SUPPORTED CLOSED SHELL AND CURVED PANEL

These cases are special since we do not need to submit to the approximation of Eq. (6.9.8). It is possible to guess the solution to Eq. (6.9.4) directly. We let the mode shapes be

$$U_{3mn}(x,\theta) = \sin \frac{m\pi x}{L} \cos n(\theta - \phi) \tag{6.12.1}$$

for the simply supported circular cylindrical shell shown in Fig. 6.12.1. We obtain upon substitution in Eq. (6.9.4)

$$D\left[\left(\frac{n}{a}\right)^2 + \left(\frac{m\pi}{L}\right)^2\right]^4 + \frac{Eh}{a^2}\left(\frac{m\pi}{L}\right)^4 - \rho h\omega^2\left[\left(\frac{n}{a}\right)^2 + \left(\frac{m\pi}{L}\right)^2\right]^2 = 0 \tag{6.12.2}$$

We may solve this equation directly for the natural frequencies. Writing ω_{mn} instead of ω to indicate the dependency on m and n, we get

$$\omega_{mn} = \frac{1}{a}\sqrt{\frac{\left(\frac{m\pi a}{L}\right)^4}{\left[\left(\frac{m\pi a}{L}\right)^2 + n^2\right]^2} + \frac{\left(\frac{h}{a}\right)^2}{12(1 - \mu^2)}\left[\left(\frac{m\pi a}{L}\right)^2 + n^2\right]^2}\sqrt{\frac{E}{\rho}} \tag{6.12.3}$$

Let us now treat a cylindrical panel as shown in Fig. 6.12.2. It is simply supported on all four sides and could represent an engine cover, airplane door, etc. The boundary conditions are simple support on all four edges. We guess the mode shape to be

$$U_{3mn}(x,\theta) = \sin\frac{m\pi x}{L}\sin\frac{n\pi\theta}{\alpha} \tag{6.12.4}$$

It satisfies all boundary conditions. Substituting it in Eq. (6.9.4) gives

$$\omega_{mn} = \frac{1}{a}\sqrt{\frac{(\frac{m\pi a}{L})^4}{\left[(\frac{m\pi a}{L})^2 + (\frac{n\pi}{\alpha})^2\right]^2} + \frac{(\frac{h}{a})^2}{12(1-\mu^2)}\left[\left(\frac{m\pi a}{L}\right)^2 + \left(\frac{n\pi}{\alpha}\right)^2\right]^2}\ \sqrt{\frac{E}{\rho}}$$

These two cases illustrate well the influence of the bending and membrane stiffness. The first term under the square root of Eqs. (6.12.3) and (6.12.5) is due to membrane stiffness and reduces to zero as n increases. The second term is due to the bending stiffness and increases in relative importance as n increases [6.19].

6.13 BARREL-SHAPED SHELL

This is a very common form, found in shells that seal hermetic refrigeration compressors, pump housings, etc. It is sketched in Fig. 6.13.1. If the curvature is not too pronounced, we may use cylindrical coordinates as an approximation.

$$(ds)^2 \cong (dx)^2 + a^2(d\theta)^2 \tag{6.13.1}$$

This gives $A_1 = 1$, $A_2 = a$, $\alpha_1 = x$, $\alpha_2 = \theta$. Also, we get $R_1 = R$ and $R_2 \cong a$. Thus we have

$$\nabla^2(\cdot) = \frac{1}{a^2}\ \frac{\partial^2(\cdot)}{\partial\theta^2} + \frac{\partial^2(\cdot)}{\partial x^2} \tag{6.13.2}$$

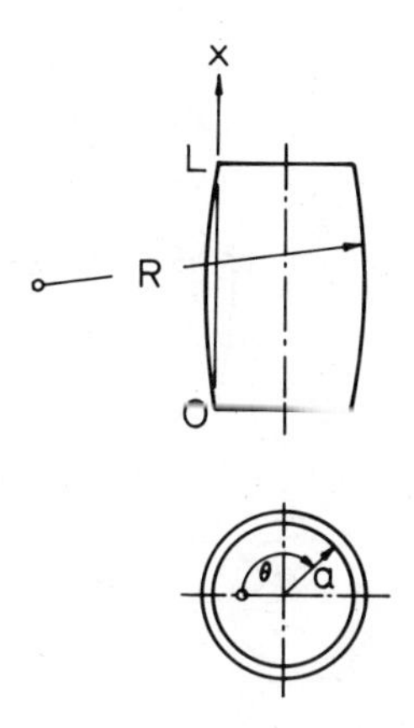

Figure 6.13.1

$$\nabla_K^2(\cdot) = \frac{1}{a}\frac{\partial^2(\cdot)}{\partial x^2} + \frac{1}{Ra^2}\frac{\partial^2(\cdot)}{\partial\theta^2} \tag{6.13.3}$$

Boundary conditions are similar to those for cylindrical shells.

For the special case of simple supports on both ends of the barrel shell, we find that the mode function

$$U_3(x,\theta) = A \sin\frac{m\pi x}{L}\cos n(\theta - \phi) \tag{6.13.4}$$

satisfies the boundary condition and Eq. (6.8.9). We obtain

$$\omega_{Bmn}^2 = \omega_{Cmn}^2 + \frac{n^2[n^2(\frac{a}{R})^2 + 2(\frac{a}{R})(\frac{m\pi a}{L})^2]}{a^2[(\frac{m\pi a}{L})^2 + n^2]^2}\frac{E}{\rho} \tag{6.13.5}$$

where

ω_{Cmn} = natural frequency of cylindrical shell of radius a given by Eq. (6.12.3)

ω_{Bmn} = natural frequency of barrel shell

Note that a negative value of R has a stiffening effect only if

$$n^2\left(\frac{a}{R}\right)^2 > 2\frac{a}{|R|}\left(\frac{m\pi a}{L}\right)^2 \tag{6.13.6}$$

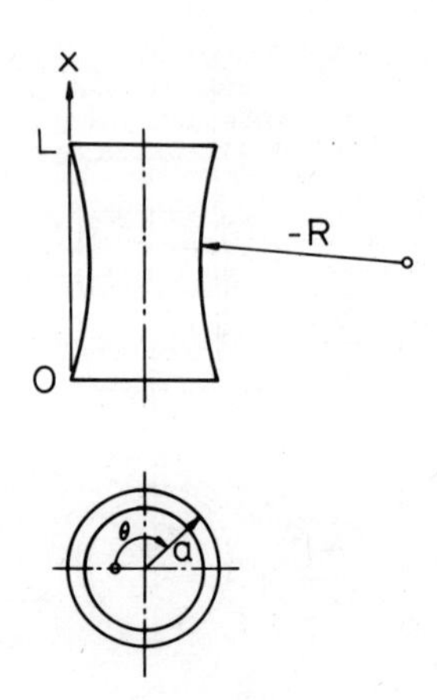

Figure 6.13.2

where $|R|$ denotes the magnitude of the radius of curvature in the x direction. Otherwise, a negative value of R will reduce the natural frequency. Barrel shells of negative R occur, for instance, as cooling towers and appear as shown in Fig. 6.13.2.

6.14 SPHERICAL CAP

If the spherical shell is shallow, it can be interpreted as a circular plate panel with spherical curvature. This allows us to formulate as approximate fundamental form (Fig. 6.14.1)

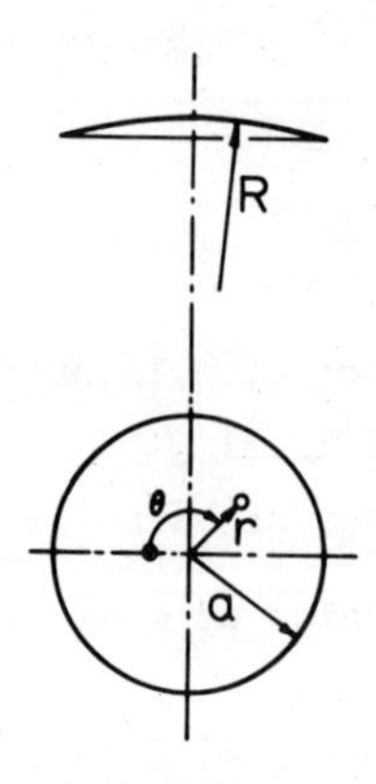

Figure 6.14.1

$$(ds)^2 \cong (dr)^2 + r^2(d\theta)^2 \qquad (6.14.1)$$

which gives $A_1 = 1$, $A_2 = r$, $\alpha_1 = r$, $\alpha_2 = \theta$. Furthermore, $R_1 = R_2 = R$. This gives

$$\nabla^2(\cdot) = \frac{1}{r}\frac{\partial}{\partial r}\left(r\frac{\partial(\cdot)}{\partial r}\right) + \frac{1}{r}\frac{\partial}{\partial \theta}\left(\frac{1}{r}\frac{\partial(\cdot)}{\partial \theta}\right) \qquad (6.14.2)$$

and

$$\nabla_k^2(\cdot) = \frac{1}{R}\nabla^2(\cdot) \qquad (6.14.3)$$

Substituting this into Eq. (6.8.9) gives

$$\nabla^4\left[D\nabla^4 + \left(\frac{Eh}{R^2} - \rho h\omega^2\right)\right]U_3 = 0 \qquad (6.14.4)$$

Solutions of

$$\nabla^4 U_3 = 0 \qquad (6.14.5)$$

and $$D\nabla^4 U_3 + \left(\frac{Eh}{R^2} - \rho h\omega^2\right)U_3 = 0 \qquad (6.14.6)$$

are solutions to the posed problem. The first equation has little physical significance. The second equation is recognized as being similar to the equation for the circular plate. The solution must be of the form of the circular plate solution

$$U_3 = [AJ_n(\lambda r) + BK_n(\lambda r) + CY_n(\lambda r) + DI_n(\lambda r)]\cos n(\theta - \phi) \qquad (6.14.7)$$

where

$$\lambda^4 = \frac{1}{D}\left(\rho h\omega^2 - \frac{Eh}{R^2}\right) \qquad (6.14.8)$$

The possible boundary conditions are identical to that of a flat plate. Therefore, all circular plate solutions apply to the spherical cap as long as the boundary conditions are the same. Therefore,

$$\omega_{smn}^2 = \omega_{pmn}^2 + \frac{E}{\rho R^2} \qquad (6.14.9)$$

where

ω_{pmn} = natural frequency of circular plate

ω_{smn} = natural frequency of spherical cap

It has been shown in Ref. 6.20 that this formula also applies to spherical caps of any shape of boundary, not only to the circular shape. Any boundary is permissible: triangular, square, etc. ω_{pmn} is the natural frequency of a plate of the same boundary shape and conditions.

REFERENCES

6.1 J. W. S. Lord Rayleigh, "Note on the free vibrations of an infinitely long shell," *Proc. Roy. Soc. (London), vol. 45*, 1889.

6.2 H. Lamb, "On the vibrations of a spherical shell," *Proc. London Math. Soc., vol. 14*, 1882.

6.3 H. Kraus, *Thin Elastic Shells*, John Wiley, New York, 1967.

6.4 A. Kalnins, Effect of bending on vibration of spherical shells, *J. Acoustical Soc. Amer. vol. 36*, 1964, pp. 74-81.

6.5 P. M. Naghdi and A. Kalnins, "On vibrations of elastic spherical shells," *J. Appl. Mech., vol. 29*, 1962, pp. 65-72.

6.6 J. P. Wilkinson and A. Kalnins, "On nonsymmetric dynamic problems of elastic spherical shells," *J. Appl. Mech. vol. 32*, 1965, pp. 525-532.

6.7 A. Kalnins and H. Kraus, "Effect of transverse shear and rotatory inertia on vibration of spherical shells," Proceedings of the 5th U.S. National Congress of Applied Mechanics, 1966.

6.8 J. W. S. Lord Rayleigh, "On the infinitesimal bending of surfaces of revolution," *Proc. London Math. Soc., vol. 13*, 1881.

6.9 L. H. Donnell, "Stability of thin walled tubes under torsion," NACA Report No. 479, 1933.

6.10 L. H. Donnell, "A discussion of thin shell theory," Proceedings of the 5th International Congress of Applied Mechanics, 1938.

6.11 K. M. Mushtari, "Certain generalizations of the theory of thin shells," *Izv. Fiz. Mat. ob-va. pri Kaz. un-te., vol. 11*, no. 8, 1938.

6.12 V. Z. Vlasov, "Basic differential equations in the general theory of elastic shells," NACA TM 1241, 1951 (translated from 1944 Russian version).

6.13 G. B. Airy, "On the strains in the interior of beams," *Phil Trans. Roy. Soc. (London), vol. 153*, 1863.

6.14 V. V. Novozhilov, *The Theory of Thin Elastic Shells*, P. Noordhoff, Groningen, The Netherlands, 1964.

6.15 W. Nowacki, *Dynamics of Elastic Systems*, John Wiley, New York, 1963.

6.16 Y. Y. Yu, "Free vibrations of thin cylindrical shells having finite lengths with freely supported and clamped edges," *J. Appl. Mech., vol. 22,* no. 4, Dec. 1955.

6.17 L. R. Koval and E. T. Cranch, "On the free vibrations of thin cylindrical shells subjected to an initial static torque," Proceedings of the 4th U.S. National Congress Applied Mechanics, 1962.

6.18 W. Flügge, *Handbook of Engineering Mechanics*, McGraw-Hill, New York, 1962.

6.19 W. Soedel, "Similitude approximations for vibrating thin shells", *J. Acoustical Soc. Amer., vol. 49*, no. 5, 1971, pp. 1535-1541.

6.20 W. Soedel, "A natural frequency analogy between spherically curved panels and flat plates," *J. Sound Vibration, vol. 29*, no. 4, 1973, pp. 457-461.

7

APPROXIMATE SOLUTION TECHNIQUES

Compared to the large number of possible shell configurations, very few exact solutions of plate and shell eigenvalue problems are possible. A representative sample was presented in the previous chapters. Included in this sample were exact solutions to the simplified equations of motion. The exact solutions are very valuable since they are the measure with which the accuracy of the approximation approaches is evaluated. They also allow an accurate and usually elegant and conclusive investigation of the various fundamental phenomena in shell vibrations. However, it is important for engineering applications to have approaches available that give numerical solutions for cases that cannot be solved exactly. To discuss these cases is the purpose of this chapter.

The approximate approaches divide roughly into two categories. In the first category, a minimization of energy approach is used. The variational integral method, the Galerkin method, and the Rayleigh-Ritz method are of this type. In the second category we find the finite difference and the finite element method. It would be wrong to distinguish these from the foregoing category as numerical methods since the Rayleigh-Ritz method becomes also a numerical method when a large number of degrees of freedom are used so that the integration cannot any longer be carried out by hand. The distinguishing mark of the second category is rather that the shell, plate, or beam surface is divided into stations or segments. However, this is where the similarity ends. The finite difference method requires knowledge of an equation of motion and is a purely mathematical solution based on this equation. The finite element method requires only knowledge of the energy expression for the segment, also called *finite elements*. The elements are then knitted together by matching them with each other at their so-called node points.

The finite element method has in recent years become the foremost method in the computer analysis of structures. However, it would be wrong to neglect the other methods because of this. There is little to be learned by a mindless application of a finite element program to every structural analysis. As the elements become more numerous, the computer program becomes the experimental structure and the finite element user is put into the situation of an experimentalist. Knowledge of solutions obtained by other methods becomes important then in classifying and interpreting the results.

7.1 APPROXIMATE SOLUTIONS BY WAY OF THE VARIATIONAL INTEGRAL

Let us start with Eq. (2.7.19). For $q_1 = q_2 = q_3 = 0$ and

$$u_i(\alpha_1,\alpha_2,t) = U_i(\alpha_1,\alpha_2)e^{j\omega t} \tag{7.1.1}$$

it becomes, with time now removed from consideration and all boundary conditions assumed to be satisfied,

$$\begin{aligned}\int_{\alpha_1}\int_{\alpha_2} \{&[L_1\{U_1,U_2,U_3\} + \rho h\omega^2\ U_1]\ \delta U_1 \\ &+ [L_2\{U_1,U_2,U_3\} + \rho h\omega^2\ U_2]\ \delta U_2 \\ &+ [L_3\{U_1,U_2,U_3\} + \rho h\omega^2\ U_3]\ \delta U_3\}\ A_1A_2\ d\alpha_1\ d\alpha_2 = 0\end{aligned} \tag{7.1.2}$$

or, in short,

$$\int_{\alpha_1}\int_{\alpha_2} \sum_{i=1}^{3} [L_i\{U_1,U_2,U_3\} + \rho h\omega^2 U_i] \; \delta U_i A_1 A_2 \; d\alpha_1 \; d\alpha_2 = 0 \tag{7.1.3}$$

where i = 1, 2, 3 and where

$$L_1\{U_1,U_2,U_3\} = \frac{1}{A_1A_2} \left(\frac{\partial(N_{11}A_2)}{\partial\alpha_1} + \frac{\partial(N_{21}A_1)}{\partial\alpha_2} + N_{12}\frac{\partial A_1}{\partial\alpha_2} - N_{22}\frac{\partial A_2}{\partial\alpha_1} \right) + \frac{Q_{13}}{R_1} \tag{7.1.4}$$

$$L_2\{U_1,U_2,U_3\} = \frac{1}{A_1A_2} \left(\frac{\partial(N_{12}A_2)}{\partial\alpha_1} + \frac{\partial(N_{22}A_1)}{\partial\alpha_2} + N_{21}\frac{\partial A_2}{\partial\alpha_1} - N_{11}\frac{\partial A_1}{\partial\alpha_2} \right) + \frac{Q_{23}}{R_2} \tag{7.1.5}$$

$$L_3\{U_1,U_2,U_3\} = \frac{1}{A_1A_2} \left(\frac{\partial(Q_{13}A_2)}{\partial\alpha_1} + \frac{\partial(Q_{13}A_1)}{\partial\alpha_2} \right) - \left(\frac{N_{11}}{R_1} + \frac{N_{22}}{R_2} \right) \tag{7.1.6}$$

We assume now functions $f_{ij}(\alpha_1,\alpha_2)$ that satisfy the boundary conditions and can represent a resonable looking mode shape. There may be one function (j = 1) or many (j = 1, 2, ..., n). In general,

$$U_1 = a_{11}f_{11} + a_{12}f_{12} + \cdots \tag{7.1.7}$$

$$U_2 = a_{21}f_{21} + a_{22}f_{22} + \cdots \tag{7.1.8}$$

$$U_3 = a_{31}f_{31} + a_{32}f_{32} + \cdots \tag{7.1.9}$$

or, in short,

$$U_i = \sum_{j=1}^{n} a_{ij}f_{ij} \tag{7.1.10}$$

and also

$$\delta U_i = \sum_{j=1}^{n} f_{ij}\,\delta a_{ij} \tag{7.1.11}$$

Note that the a_{ij} are coefficients that have to be determined.

Upon the substitution, the integral becomes

$$\int_{\alpha_1}\int_{\alpha_2} \sum_{i=1}^{3} [L_i\{\sum_{j=1}^{n} a_{1j}f_{1j}, \sum_{j=1}^{n} a_{2j}f_{2j}, \sum_{j=1}^{n} a_{3j}f_{3j}\}$$

$$+ \rho h\omega^2 \sum_{j=1}^{n} a_{ij}f_{ij}] \sum_{j=1}^{n} f_{ij}\,\delta a_{ij} A_1 A_2 \, d\alpha_1 \, d\alpha_2 = 0 \tag{7.1.12}$$

where $i = 1, 2, 3$ and $j = 1, 2, \ldots, n$. Next, we collect coefficients of δa_{ij}. Since each δa_{ij} is independent and arbitrary, the equation can only be satisfied if each coefficient is equal to zero.

$$\int_{\alpha_1}\int_{\alpha_2} [L_i\{\sum_{j=1}^{n} a_{1j}f_{1j}, \sum_{j=1}^{n} a_{2j}f_{2j}, \sum_{j=1}^{n} a_{3j}f_{3j}\}$$

$$+ \rho h\omega^2 \sum_{j=1}^{n} a_{ij}f_{ij}] f_{ij} A_1 A_2 \, d\alpha_1 \, d\alpha_2 = 0 \tag{7.1.13}$$

where $i = 1, 2, 3$ and $j = 1, 2, \ldots, n$. The approach is now to carry out the integration and to bring the result into the form

$$[A]\{a_{ij}\} = 0 \tag{7.1.14}$$

where

$$\lfloor a_{ij} \rfloor = \lfloor a_{11}, a_{12}, \ldots, a_{1n}, a_{21}, a_{22}, \ldots a_{2n}, a_{31}, a_{32}, \ldots a_{3n} \rfloor \tag{7.1.15}$$

Setting the determinant of A equal to zero

$$|A| = 0 \tag{7.1.16}$$

gives a characteristic equation for ω^2. There will be 3n roots. These are the natural frequencies. Resubstituting a natural frequency into the matrix equation allows us to solve for 3n - 1 values of a_{ij} in terms of one arbitrary a_{ij}. This establishes the mode shape that is associated with this natural frequency.

The method works well, provided the assumed functions are of sufficient variety to give good mode approximations. If n = 1 only, the mode shape is fixed by the assumption.

It can be shown that natural frequencies will always be of larger value than obtained from exact solutions, provided the mode functions satisfy all boundary conditions. This is not necessarily true if moment and shear conditions are not satisfied, which is a common approach. The rule of thumb is that the so-called primary boundary conditions, which are deflection and slope conditions, should always be satisfied, but that it is possible to ignore the so-called secondary boundary conditions which are shear and moment conditions, provided one is only interested in natural frequency predictions. If stress calculations are the objective, then ignoring moment and shear-type boundary conditions may not be permissible, depending on the circumstance and the accuracy requirement.

7.2 USE OF BEAM FUNCTIONS

A common selection of functions for rectangular plates, and cylindrical and conical shells are beam mode shapes, also called *beam functions*. The boundary conditions of the beam and the shell have to be of the same type. The argument is that, for instance, the behavior of an axial strip of cylindrical shell should be similar to that of a beam of the same type of boundary conditions.

Since the beam function has the higher mode shapes already built in, it is often sufficient to use n = 1. The matrix [A] is then a 3 x 3 matrix only.

The characteristic equation is a cubic equation in ω^2.

The advantage of using orthogonal beam functions is that the shell or plate mode shapes are orthogonal also.

For plates, the variational integral uncouples into one integral for transverse deflections,

$$\int_{\alpha_1}\int_{\alpha_2} [L_3\{\sum_{j=1}^{n} a_{3j}f_{3j}\} + \rho h\omega^2 \sum_{j=1}^{n} a_{3j}f_{3j}]\, f_{3j}A_1A_2\, d\alpha_1\, d\alpha_2 = 0 \tag{7.2.1}$$

where $L_3\{\cdot\} = -D\nabla^4(\cdot)$ (7.2.2)

and two integrals that govern in plane deflections:

$$\int_{\alpha_1}\int_{\alpha_2} [L_i\{\sum_{j=1}^{n} a_{ij}f_{ij}, \sum_{j=1}^{n} a_{2j}f_{2j}\}$$

$$+ \rho h\omega^2 \sum_{j=1}^{n} a_{ij}f_{ij}]\, f_{ij}A_1A_2\, d\alpha_1\, d\alpha_2 = 0 \tag{7.2.3}$$

where $i = 1, 2$.

Let us now examine the integral for transverse motion further. If $n = 1$, then we get

$$\int_{\alpha_1}\int_{\alpha_1} [L_3\{a_{31}f_{31}\}f_{31} + \rho h\omega^2 a_{31}^2 f_{31}^2]\, A_1A_2\, d\alpha_1\, d\alpha_2 = 0 \tag{7.2.4}$$

or

$$\omega^2 = -\frac{1}{\rho h}\, \frac{\int_{\alpha_1}\int_{\alpha_2} L_3\{f_{31}\}f_{31}A_1A_2\, d\alpha_1\, d\alpha_2}{\int_{\alpha_1}\int_{\alpha_2} f_{31}^2 A_1A_2\, d\alpha_1\, d\alpha_2} \tag{7.2.5}$$

For the case of a rectangular plate as sketched in Fig. 7.2.1, this becomes

$$\omega^2 = \frac{D}{\rho h}\, \frac{\int_0^a\int_0^b [\frac{\partial^4 f_{31}}{\partial x^4} + 2\frac{\partial^4 f_{31}}{\partial x^2 \partial y^2} + \frac{\partial^4 f_{31}}{\partial y^4}]f_{31}\, dx\, dy}{\int_0^a\int_0^b f_{31}^2\, dx\, dy} \tag{7.2.6}$$

Furthermore, if we use beam functions,

$$f_{31}(x,y) = \alpha(x)\beta(y) \tag{7.2.7}$$

where $\alpha(x)$ is the beam function in the x direction and $\beta(y)$ is the

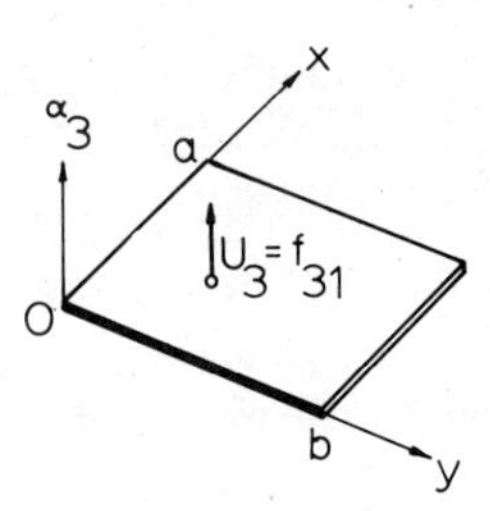

Figure 7.2.1

beam function in the y direction. Equation (7.2.6) becomes, therefore,

$$\omega^2 = \frac{D}{\rho h} \frac{\int_0^a \alpha \frac{\partial^4 \alpha}{\partial x^4} dx \int_0^b \beta^2 dy + \int_0^a \alpha^2 dx \int_0^b \beta \frac{\partial^4 \beta}{\partial y^4} dy}{\int_0^a \alpha^2 dx \int_0^b \beta^2 dy}$$

$$+ 2 \int_0^a \alpha \frac{\partial^2 \alpha}{\partial x^2} dx \int_0^b \beta \frac{\partial^2 \beta}{\partial y^2} dy \tag{7.2.8}$$

For beam functions in general,

$$\frac{\partial^4 \alpha}{\partial x^4} = \lambda_m^4 \alpha \tag{7.2.9}$$

$$\frac{\partial^4 \beta}{\partial y^4} = \lambda_n^4 \beta \tag{7.2.10}$$

where λ was defined when the beam solution was discussed in Sec. 5.2.

After substitution we obtain

$$\omega^2 = \frac{D}{\rho h} \left(\lambda_m^4 + \lambda_n^4 + 2 \frac{\int_0^a \alpha \frac{\partial^2 \alpha}{\partial x^2} dx \int_0^b \beta \frac{\partial^2 \beta}{\partial y^2} dy}{\int_0^a \alpha^2 dx \int_0^b \beta^2 dy} \right) \tag{7.2.11}$$

Let us, for instance, obtain the natural frequencies of a rectangular plate clamped on all four edges. From Sec. 5.2, the mode shape of a clamped-clamped beam is

$$\alpha(x) = C(\lambda_m x) - \frac{C(\lambda_m a)}{D(\lambda_m a)} D(\lambda_m x) \tag{7.2.12}$$

where

$$C(\lambda_m x) = \cosh \lambda_m x - \cos \lambda_m x \tag{7.2.13}$$

$$D(\lambda_m x) = \sinh \lambda_m x - \sin \lambda_m x \tag{7.2.14}$$

and where $\lambda_1 a = 4.73$, $\lambda_2 a = 7.85$, $\lambda_3 a = 11.00$, $\lambda_4 a = 14.14$, etc. Also,

$$\beta(y) = C(\lambda_n y) - \frac{C(\lambda_n b)}{D(\lambda_n b)} D(\lambda_n y) \tag{7.2.15}$$

where

$$C(\lambda_n y) = \cosh \lambda_n y - \cos \lambda_n y \tag{7.2.16}$$

$$D(\lambda_n y) = \sinh \lambda_n y - \sin \lambda_n y \tag{7.2.17}$$

and where $\lambda_1 b = 4.73$, $\lambda_2 b = 7.85$, $\lambda_3 b = 11.00$, $\lambda_4 b = 14.14$, ...

The function $f_{31}(x,y)$ now satisfies all boundary conditions for the clamped-clamped plate.

$$u_3(o,y.t) = u_3(a,y,t) = u_3(x,o,t) = u_3(x,b,t) = 0 \tag{7.2.18}$$

$$\frac{\partial u_3}{\partial x}(o,y,t) = \frac{\partial u_3}{\partial x}(a,y,t) = \frac{\partial u_3}{\partial y}(x,o,t) = \frac{\partial u_3}{\partial y}(x,b,t) = 0 \tag{7.2.19}$$

The results of Eq. (7.2.11) are provided in Table 7.2.1 in terms of $\omega a^2 \sqrt{\rho h/D}$ values.

Table 7.2.1 Natural Frequencies for Clamped Square Plate

n \ m	1	2	3
1	36.1	73.7	132.0
2	73.7	108.9	165.8
3	132.0	165.8	221.4

7.3 GALERKIN'S METHOD

Galerkin's method is an algorithmic statement of the variational approach. It is summarized for shells in Eq. (7.1.13). The usual algorithm is that the functions f_{ij} that satisfy the boundary conditions are substituted into the equations of motion and the resulting expressions are multiplied by their respective f_{ij} as so-called weighing functions. Finally, the product is integrated over the domain. This is exactly what Eq. (7.1.13) does. The Galerkin approach has become a general algorithm for solving a variety of equations and problems and its variational birthmark has disappeared.

Let us solve the example the Donnell-Mushtari-Vlasov equation for closed circular cylindrical shells directly without the Yu simplification. The Galerkin algorithm demands that we assume a function $U_{3m}(x)$ that satisfies the boundary conditions, multiply the equation with the same function as a weighing function, and integrate over the shell length. Starting with Eq. (6.9.4), this gives

$$\int_0^L \left[D\left(\frac{n^2}{a^2} - \frac{d^2}{dx^2}\right)^4 U_{3m}(x) + \frac{Eh}{a^2}\frac{d^4}{dx^4}U_{3m}(x) - \rho h\omega^2\left(\frac{n^2}{a^2} - \frac{d^2}{dx^2}\right)^2 U_{3m}(x)\right] U_{3m}(x)\, dx = 0 \qquad (7.3.1)$$

Making use of the fact that for beam functions

$$\frac{d^4}{dx^4}U_{3m}(x) = \lambda_m^4 U_{3m}(x) \qquad (7.3.2)$$

where λ_m are the roots of the beam eigenvalue problem, we get

$$\omega_{mn}^2 = \frac{\frac{Eh}{a^2}\lambda_m^4 + D[(\frac{n}{a})^8 - 4(\frac{n}{a})^6 R_m + 6(\frac{n}{a})^4\lambda_m^4 - 4(\frac{n}{a})^2\lambda_m^4 R_m + \lambda_m^8]}{\rho h[(\frac{n}{a})^4 - 2(\frac{n}{a})^2 R_m + \lambda_m^4]} \qquad (7.3.3)$$

where

$$R_m = \frac{\int_0^L \frac{d^2U_{3m}}{dx^2} U_{3m} \, dx}{\int_0^L U_{3m}^2 \, dx} \tag{7.3.4}$$

This result can now be applied to various boundary conditions. However, a further simplification is possible. Since

$$U_{3m} = U_{3m}(\lambda_m x) \tag{7.3.5}$$

we get

$$R_m = \lambda_m^2 \frac{\int_0^L \frac{d^2U_{3m}}{d(\lambda_m x)^2} U_{3m} \, dx}{\int_0^L U_{3m}^2 \, dx} \tag{7.3.6}$$

The ratio of the integrals is close to -1. For the simply supported beam function, and some other cases, it is exactly equal to -1. Thus, as an approximation, we may set

$$R_m \cong -\lambda_m^2 \tag{7.3.7}$$

This results in

$$\omega_{mn}^2 = \frac{1}{\rho h} \left\{ \frac{Eh\lambda_m^4}{a^2[(\frac{n}{a})^2 + \lambda_m^2]^2} + D\left[\left(\frac{n}{a}\right)^2 + \lambda_m^2\right]^2 \right\} \tag{7.3.8}$$

For the simply supported case, $\lambda_m = m\pi/L$, and Eq. (7.3.8) is exact. Equations (7.3.3) and (7.3.8) are compared in Fig. 7.3.1 for the clamped-clamped circular cylindrical shell for $E = 20.6 \times 10^4$ N/mm^2, $\rho = 7.85 \times 10^{-9}$ Nsec^2/mm^4, $\mu = 0.3$, $h = 2$ mm, $a = 100$ mm, $L = 200$ mm. We see that agreement is excellent. Equation (7.3.8) is very easy to use since $\lambda_m L$ values for the majority of the conceivable boundary conditions are available in handbooks [7.1]. To illustrate how this theory agrees typically with reality, results for a clamped-clamped

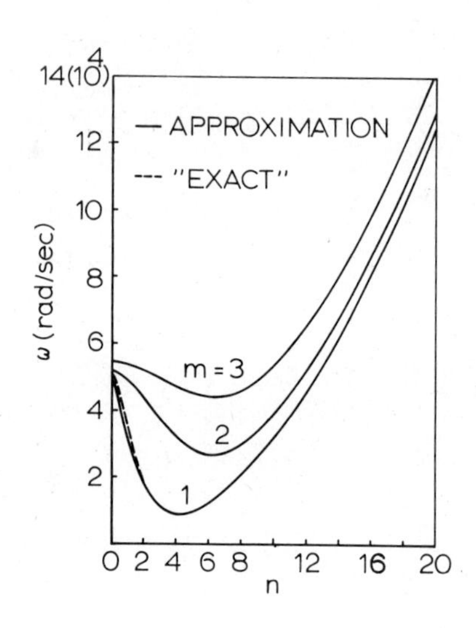

Figure 7.3.1

circular cylindrical shell are compared in Fig. 7.3.2 with experimental data collected by Koval and Cranch [7.2]. The parameters are $E = 20.6 \times 10^4$ N/mm^2, $\rho = 7.85 \times 10^{-9}$ Nsec^2/mm^4, $\mu = 0.3$, $h = 0.254$ mm, $a = 76.2$ mm, $L = 304.8$ mm. The reason that the experimental frequencies are in general lower than the theoretical frequencies

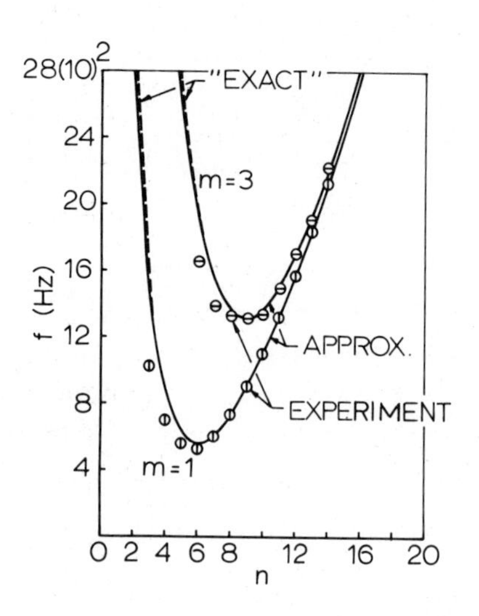

Figure 7.3.2

is that (a) the Galerkin method gives upper bound results and (b) it is virtually impossible to have a truly clamped boundary. The elasticity of the clamping device will tend to lower the experimental frequencies.

7.4 THE RAYLEIGH-RITZ METHOD

Rayleigh [7.3] used the argument that an undamped linear structure, vibrating at its natural frequency, interchanges its vibratory energy from a purely potential state at its maximum amplitude to a purely kinetic state when all vibration amplitudes are zero.

At a natural frequency, we have

$$u_i = U_i e^{j\omega t} \tag{7.4.1}$$

Substituting this in the strain energy expression of Eq. (2.6.3), we get the expression for maximum potential energy U_{max} upon taking for $e^{j\omega t}$ the maximum, namely unity:

$$U_{max} = \int_{\alpha_1}\int_{\alpha_2}\int_{\alpha_3}\left[\frac{E}{2(1-\mu^2)}\left(\varepsilon_{11}^{*2} + \varepsilon_{22}^{*2} + 2\mu\varepsilon_{11}^{*}\varepsilon_{22}^{*}\right) + \frac{G}{2}\varepsilon_{12}^{*2}\right] A_1A_2\, d\alpha_1\, d\alpha_2\, d\alpha_3 \tag{7.4.2}$$

where

$$\varepsilon_{11}^{*} = \frac{1}{A_1}\frac{\partial}{\partial\alpha_1}(U_1 + \alpha_3\beta_1^{*}) + \frac{U_2 + \alpha_3\beta_2^{*}}{A_1A_2}\frac{\partial A_1}{\partial\alpha_2} + \frac{U_3}{R_1} \tag{7.4.3}$$

$$\varepsilon_{22}^{*} = \frac{1}{A_2}\frac{\partial}{\partial\alpha_2}(U_2 + \alpha_3\beta_2^{*}) + \frac{U_1 + \alpha_3\beta_1^{*}}{A_1A_2}\frac{\partial A_2}{\partial\alpha_1} + \frac{U_3}{R_2} \tag{7.4.4}$$

$$\varepsilon_{12}^{*} = \frac{A_2}{A_1}\frac{\partial}{\partial\alpha_1}\left(\frac{U_2 + \alpha_3\beta_2^{*}}{A_2}\right) + \frac{A_1}{A_2}\frac{\partial}{\partial\alpha_2}\left(\frac{U_1 + \alpha_3\beta_1^{*}}{A_1}\right) \tag{7.4.5}$$

and where

$$\beta_1^{*} = \frac{U_1}{R_1} - \frac{1}{A_1}\frac{\partial U_3}{\partial\alpha_1} \tag{7.4.6}$$

$$\beta_2^* = \frac{U_2}{R_2} - \frac{1}{A_2}\frac{\partial U_3}{\partial \alpha_2} \tag{7.4.7}$$

Substituting Eqs. (7.4.3) to (7.4.5) in the kinetic energy expression and selecting the maximum value gives

$$K_{max} = \frac{\omega^2 \rho h}{2} \int_{\alpha_1}\int_{\alpha_2} (U_1^2 + U_2^2 + U_3^2) A_1 A_2 \, d\alpha_1 \, d\alpha_2 \tag{7.4.8}$$

Equating (7.4.8) and (7.4.2) gives

$$\omega^2 = \frac{\displaystyle\int_{\alpha_1}\int_{\alpha_2}\int_{\alpha_3} \left\{ \left[\frac{E}{(1-\mu^2)} (\varepsilon_{11}^{*2} + \varepsilon_{22}^{*2} + 2\mu\varepsilon_{11}^{*}\varepsilon_{22}^{*}) + G\varepsilon_{12}^{*2}\right\} A_1 A_2 \, d\alpha_1 \, d\alpha_2 \, d\alpha_3}{\displaystyle\rho h \int_{\alpha_1}\int_{\alpha_2} (U_1^2 + U_2^2 + U_3^2) A_1 A_2 \, d\alpha_1 \, d\alpha_2} \tag{7.4.9}$$

This formula is exact if the exact mode shape expression is substituted. However, the same formula results if we argue with Rayleigh that instead of using the unknown exact mode shape, an approximate mode shape expression with not more than one arbitrary constant can be used that satisfies the boundary conditions and resembles to a reasonable degree the actual mode shape. In this case we try to minimize the difference between the maximum potential energy and the maximum kinetic energy since only in the exact case will it be zero as required. If the assumed mode shape contains only one constant C that can be minimized,

$$\frac{d}{dC}(U_{max} - K_{max}) = 0 \tag{7.4.10}$$

results also in Eq. (7.4.9). This equation is also called *Rayleigh's quotient*. If the assumed mode shapes satisfy all boundary conditions, it can be shown that Eq. (7.4.9) results in a upper bound approximation. If ω_R is Rayleigh's frequency and ω is the exact frequency, then

$$\omega \leq \omega_R \tag{7.4.11}$$

The reason for this is that any deviation from the true mode shape is equivalent to an additional constraint, resulting in a higher value for potential energy.

It was later established by various investigators, that it is possible to relax the requirement that all boundary conditions have to be satisfied by the assumed mode shape. Most of the time it is sufficient to satisfy deflection and slope conditions and to neglect moment and shear boundary conditions, and still achieve acceptable approximations of the natural frequency.

Rayleigh's quotient can be used to investigate all natural frequencies of a plate or shell but works best for the determination of the fundamental frequency in case of plates. For higher frequencies even small deviations between the true and the assumed mode shape can easily cause large errors in the calculated results. For shells, Rayleigh's quotient is not particularly useful because of the complexity of the modes. An extension is needed that improves accuracy.

This extension was provided by Ritz [7.4]. The contribution of Ritz was to allow estimated mode shapes of more than one arbitrary constant. In the so-called Rayleigh-Ritz method we minimize with respect to each of the constants C_1, C_2, ..., C_r.

$$\begin{aligned}
&\frac{\partial}{\partial C_1}(U_{max} - K_{max}) = 0 \\
&\frac{\partial}{\partial C_2}(U_{max} - K_{max}) = 0 \\
&\cdots\cdots\cdots\cdots\cdots\cdots \\
&\frac{\partial}{\partial C_r}(U_{max} - K_{max}) = 0
\end{aligned} \tag{7.4.12}$$

If we have r constants C_r, we obtain r homogeneous equations: r - 1 equations can be solved to express r - 1 constants in terms of one arbitrarily selected constant. The requirement that the boundary conditions have to be satisfied is the same as before.

Let us illustrate all this on the example of a clamped circular plate. First, let us use the Rayleigh quotient to find the first natural frequency. We assume as fundamental mode shape

$$U_3 = C_1\left[1 - \left(\frac{r}{a}\right)^2\right]^2 \tag{7.4.13}$$

This satisfies the two boundary conditions

$$U_3(a,\theta) = 0 \tag{7.4.14}$$

$$\frac{dU_3}{dr}(a,\theta) = 0 \tag{7.4.15}$$

Setting $U_1 = U_2 = 0$, $A_1 = 1$, $A_2 = r$, $\alpha_1 = r$, $\alpha_2 = \theta$, and $d(\cdot)/d\theta = 0$, the Rayleigh quotient expression becomes

$$\omega_R^2 = \frac{D}{\rho h} \frac{\int_0^a f(r) r\, dr}{\int_0^a U_3^2 r\, dr} \tag{7.4.16}$$

where

$$f(r) = \left(\frac{\partial^2 U_3}{\partial r^2} + \frac{1}{r}\frac{\partial U_3}{\partial r}\right)^2 - 2(1 - \mu)\frac{\partial^2 U_3}{\partial r^2}\frac{1}{r}\frac{\partial U}{\partial r} \tag{7.4.17}$$

The result is

$$\omega_R = \frac{(\lambda a)^2}{a^2}\sqrt{\frac{D}{\rho h}} \tag{7.4.18}$$

where $\lambda a = 3.214$. This compares fairly well with the exact value of $\lambda a = 3.196$. (7.4.19)

Next, let us assume

$$U_3 = C_1\left[1 - \left(\frac{r}{a}\right)^2\right]^2 + C_2\left[1 - \left(\frac{r}{a}\right)^2\right]^3 \tag{7.4.20}$$

which we expect will furnish us a better approximation of the first mode and also an approximation of the second axisymmetric mode. In this case we have to use the Rayleigh-Ritz method. We get, upon substitution in Eqs. (7.4.2) and (7.4.8),

$$U_{max} - K_{max} = \pi D \frac{32}{3a^2}\left(C_1^2 + \frac{3}{2} C_1 C_2 + \frac{9}{10} C_2^2\right)$$

$$- \pi \rho h \omega^2 \frac{a^2}{10}\left(C_1^2 + \frac{5}{3} C_1 C_2 + \frac{5}{7} C_2^2\right) \tag{7.4.21}$$

To minimize this expression with respect to C_1 and C_2, we formulate

$$\frac{\partial}{\partial C_1}(U_{max} - K_{max}) = 0 \tag{7.4.22}$$

$$\frac{\partial}{\partial C_2}(U_{max} - K_{max}) = 0 \tag{7.4.23}$$

or

$$\begin{bmatrix} \frac{64}{3} - a^4\omega^2 \frac{\rho h}{5D} & \frac{64}{3} - a^4\omega^2 \frac{\rho h}{6D} \\ \frac{64}{3} - a^4\omega^2 \frac{\rho h}{6D} & \frac{96}{5} - a^4\omega^2 \frac{\rho h}{7D} \end{bmatrix} \begin{Bmatrix} C_1 \\ C_2 \end{Bmatrix} = 0 \tag{7.4.24}$$

Setting the determinant equal to zero and solving for ω gives two solutions:

$$\omega_{R1} = \frac{(\lambda a)_1^2}{a^2} \sqrt{\frac{D}{\rho h}} \tag{7.4.25}$$

$$\omega_{R_2} = \frac{(\lambda a)_2^2}{a^2} \sqrt{\frac{D}{\rho h}} \tag{7.4.26}$$

where $(\lambda a)_1 = 3.197$ and $(\lambda a)_2 = 6.565$.

This compares to the exact values of $\lambda a = 3.196$, which is very close, and $\lambda a = 6.306$, which is not an entirely unacceptable agreement as far as the second axisymmetric natural frequency is concerned.

The mode shapes are obtained with the help of Eq. (7.4.24), from which we get

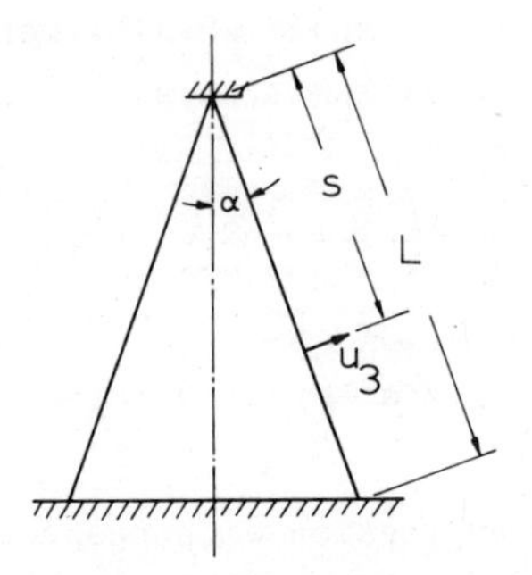

Figure 7.4.1

$$C_2 = -C_1 \frac{\frac{64}{3} - a^4\omega^2 \frac{\rho h}{5D}}{\frac{64}{3} - a^4\omega^2 \frac{\rho h}{6D}} \tag{7.4.27}$$

or

$$C_2 = -C_1 \frac{\frac{64}{3} - \frac{(\lambda a)^4}{5}}{\frac{64}{3} - \frac{(\lambda a)^4}{6}} \tag{7.4.28}$$

Substituting $(\lambda a)_1 = 3.197$ gives $C_2 = -0.112C_1$. Thus

$$U_{31} = C_1 \left\{ \left[1 - \left(\frac{r}{a}\right)^2 \right]^2 - 0.112 \left[1 - \left(\frac{r}{a}\right)^2 \right]^3 \right\} \tag{7.4.29}$$

Substituting $(\lambda a)_2 = 6.565$ gives $C_2 = -1.215C_1$, and therefore

$$U_{32} = C_1 \left\{ \left[1 - \left(\frac{r}{a}\right)^2 \right]^2 - 1.215 \left[1 - \left(\frac{r}{a}\right)^2 \right]^3 \right\} \tag{7.4.30}$$

We could keep adding terms to the series describing U_3 and get better and better agreement, describing more and more axisymmetric natural frequencies. We could also add the θ dependency into our approximate mode shape and proceed with

$$U_3 = \sum_{i=1}^{P} C_i \left[1 - \left(\frac{r}{a}\right)^2 \right]^{i+1} \cos n\theta \tag{7.4.31}$$

For a good example of the Rayleigh-Ritz application, see a

paper by Ritz [7.5] where he obtains the natural frequencies and modes of a square plate with free edges. Young [7.6] applied the method to the square plate clamped on all edges.

There are many applications of the Rayleigh-Ritz method to shells. Let us select as example the paper by Federhofer [7.7], who solved the completely clamped conical shell case by this method. As shown in Fig. 7.4.1, the cone is clamped at its base and at its apex. The assumed solution is

$$U_1 = As^2(s - L)^2 \cos n(\theta - \phi)$$

$$U_2 = Bs^2(s - L)^2 \sin n(\theta - \phi)$$

$$U_3 = Cs^2(s - L)^2 \cos n(\theta - \phi)$$

It satisfies all boundary conditions but is limited to the equivalent of the first beam mode in the axial direction.

The Rayleigh-Ritz procedure results in a cubic equation whose three roots for each value of n define the three natural frequencies. As α approaches $\pi/2$, the results approach those of a clamped plate that is also clamped at the center. As α approaches zero and $n = 1$, the result approaches that of a clamped-clamped beam of tubular cross section.

7.5 FINITE DIFFERENCES

The finite differences approach is a purely mathematical technique. The equations of motion have to be known for the structure that is to be investigated. By contrast, the finite element approach will not require knowledge of the differential equations of motion, once the element stiffness and mass matrices are defined.

The finite difference approach is based on the argument that a derivative can be approximately replaced by a difference. Let us illustrate this by the plate equation using cartesian coordinates. Since, at a natural frequency,

$$u_3 = U_3 e^{j\omega t} \tag{7.5.1}$$

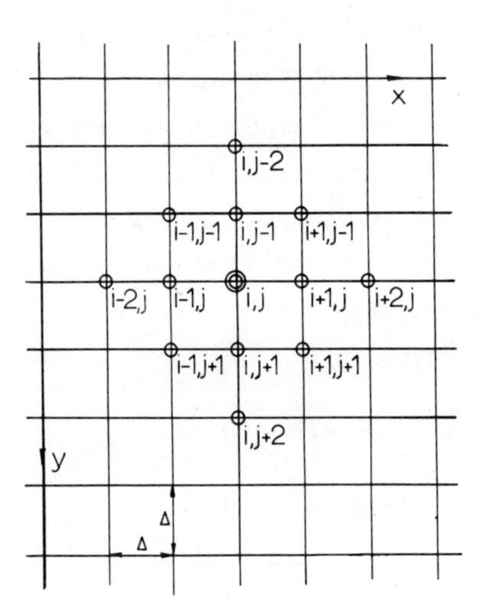

Figure 7.5.1

the plate equation becomes

$$D\nabla^4 U_3 - \rho h \omega^2 U_3 = 0 \tag{7.5.2}$$

where

$$\nabla^4 = \frac{\partial^4 U_3}{\partial x^4} + 2\frac{\partial^4 U_3}{\partial x^2 \partial y^2} + \frac{\partial^4 U_3}{\partial y^4} \tag{7.5.3}$$

The plate is divided into grids. The grids can be of unequal size, but in this discussion a square grid is used. Labeling the point at which the equation is to be evaluated the i,j point and counting forward and backward from there as shown in Fig. 7.5.1 gives, dropping the subscript 3 to make the notation easier,

$$\left(\frac{\partial U}{\partial x}\right)_{i,j} = \frac{U_{i+1,j} - U_{i-1,j}}{2\Delta} \tag{7.5.4}$$

and $$\left(\frac{\partial U}{\partial y}\right)_{i,j} = \frac{U_{i,j+1} - U_{i,j-1}}{2\Delta} \tag{7.5.5}$$

where Δ is the grid dimension. The second derivative is obtained from

$$\left(\frac{\partial^2 U}{\partial x^2}\right)_{i,j} = \frac{\left(\frac{\partial U}{\partial x}\right)_{i+1,j} - \left(\frac{\partial U}{\partial x}\right)_{i-1,j}}{2\Delta} \tag{7.5.6}$$

or, after substitution of Eq. (7.5.4), and using half steps,

$$\left(\frac{\partial^2 U}{\partial x^2}\right)_{i,j} = \frac{U_{i+1,j} - 2U_{i,j} + U_{i-1,j}}{\Delta^2} \tag{7.5.7}$$

We proceed in a similar fashion to define third, fourth, and mixed derivatives. Substitution in the equation of motion gives the equation of motion in finite difference form:

$$\begin{aligned}\frac{D}{\Delta^4}[(&U_{i+2,j} + U_{i,j+2} + U_{i-2,j} + U_{i,j-2})\\ &+ 2(U_{i+1,j+1} + U_{i-1,j+1} + U_{i-1,j-1} + U_{i+1,j-1})\\ &- 8(U_{i,j+1} + U_{i-1,j} + U_{i,j-1} + U_{i+1,j})\\ &+ 20\, U_{i,j}] - \rho h \omega^2 U_{i,j} = 0\end{aligned} \tag{7.5.8}$$

Thus, if there are N grid points, we obtain N simultaneous equations. There are more than N unknowns, since as we evaluate the equation of motion along the boundary, points outside the plate boundary will appear in the equation systems. The equations for these additional unknowns are provided by the boundary conditions. For instance, at a clamped edge, the boundary conditions

$$U_3 = 0 \tag{7.5.9}$$

$$\frac{\partial U_3}{\partial x} = 0 \tag{7.5.10}$$

become

$$U_{i,j} = 0 \tag{7.5.11}$$

$$U_{i+1,j} - U_{i-1,j} = 0 \tag{7.5.12}$$

At a free edge, the boundary conditions

$$\frac{\partial^2 U_3}{\partial x^2} + \mu \frac{\partial^2 U_3}{\partial y^2} = 0 \tag{7.5.13}$$

and $$\frac{\partial^3 U_3}{\partial x^3} + (2 - \mu) \frac{\partial^3 U_3}{\partial x \partial y^2} = 0 \tag{7.5.14}$$

become

$$U_{i+1,j} - 2U_{i,j} + U_{i-1,j} + \mu(U_{i,j+1} - 2U_{i,j} + U_{i,j-1}) = 0 \tag{7.5.15}$$

$$\begin{aligned} U_{i+2,j} &- 2U_{i+1,j} + 2U_{i-1,j} - U_{i-2,j} + (2-\mu)(U_{i+1,j+1} \\ &+ U_{i+1,j-1} - U_{i-1,j-1} - U_{i-1,j+1} + 2U_{i-1,j} \\ &- 2U_{i+1,j}) = 0 \end{aligned} \tag{7.5.16}$$

Let us illustrate this by the example of a simple supported square plate with only four interior grid points as shown in Figure 7.5.2. Because of the information requirement that the finite difference form of the plate equation imposes on us, we also have to consider the points at the boundary and dummy points outside the boundary.

From the boundary conditions

$$U_3 = 0 \tag{7.5.17}$$

for (x,y) = (0,y), (a,y), (x,0), (x,a) and

$$\frac{\partial^2 U_3}{\partial x^2} = 0 \tag{7.5.18}$$

for (x,y) = (0,y), (a,y) and

$$\frac{\partial^2 U_3}{\partial y^2} = 0 \tag{7.5.19}$$

for (x,y) = (x,0), (x,a), we obtain

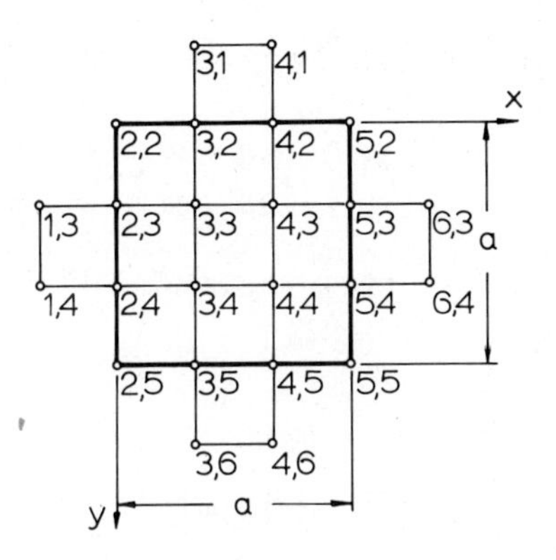

Figure 7.5.2

$$U_{2,2} = U_{3,2} = U_{4,2} = U_{5,2} = U_{5,3} = U_{5,4} = U_{5,5}$$
$$= U_{4,5} = U_{3,5} = U_{2,5} = U_{2,4} = U_{2,3} = 0 \qquad (7.5.20)$$

and
$$U_{1,3} = U_{3,1} = -U_{3,3}$$
$$U_{1,4} = U_{3,6} = -U_{3,4} \qquad (7.5.21)$$
$$U_{4,1} = U_{6,3} = -U_{4,3}$$
$$U_{6,4} = U_{4,6} = -U_{4,4}$$

Next, evaluating the finite difference form of the plate equation at the four interior points, we get, for instance, for point (3,3), making use of the boundary conditions,

$$(18 - \Delta^4 \frac{\rho h \omega^2}{D})U_{3,3} - 8U_{4,3} - 8U_{3,4} + 2U_{4,4} = 0 \qquad (7.5.22)$$

where $\Delta = a/3$. Proceeding in a similar manner for points (4,3), (3,4), and (4,4), we have, in matrix form

$$[A]\ \{U_{i,j}\} = 0 \qquad (7.5.23)$$

where

$$[A] = \begin{bmatrix} 18 - R\omega^2 & -8 & -8 & 2 \\ -8 & 18 - R\omega^2 & 2 & -8 \\ -8 & 2 & 18 - R\omega^2 & -8 \\ 2 & -8 & -8 & 18 - R\omega^2 \end{bmatrix} \quad (7.5.24)$$

$$R = \frac{\rho h \Delta^4}{D} \quad (7.5.25)$$

$$\lfloor U_{i,j} \rfloor = \lfloor U_{3,3}, U_{4,3}, U_{3,4}, U_{4,4} \rfloor \quad (7.5.26)$$

The natural frequencies are obtained by setting the determinant of the matrix to zero,

$$|A| = 0 \quad (7.5.27)$$

In this particular case we obtain approximations to the first four natural frequencies. They are

$$\omega_n = \frac{\lambda_n}{a^2}\sqrt{\frac{D}{\rho h}} \quad (7.5.28)$$

where $\lambda_1 = 18$, $\lambda_2 = \lambda_3 = 36$, $\lambda_4 = 54$. The exact values are $\lambda_1 = 2\pi^2$, $\lambda_2 = \lambda_3 = 5\pi^2$, $\lambda_4 = 8\pi^2$. We see that even with only four interior grid points, the first natural frequency is approximated well. For higher mode calculation, we need obviously a finer grid. But this is no difficulty for a computer application.

The mode shapes are obtained by substituting each natural frequency back in Eq. (7.5.23) and solving for three of the four grid points in terms of, say, $U_{3,3}$. This gives the four modes, for $U_{3,3} = 1$, as

$$\begin{aligned} [U_{i,j}]_1 &= [1, 1, 1, 1] \\ [U_{i,j}]_2 &= [1, -1, 1, -1] \\ [U_{i,j}]_3 &= [1, 1, -1, -1] \\ [U_{i,j}]_4 &= [1, -1, -1, 1] \end{aligned} \quad (7.5.29)$$

Note that for free edges, the equation of motion has to be evaluated at the edge points since these points are not motionless.

For more background on finite difference applications, see, for instance, ref. 7.8. The approach has been used mainly for plate problems. For shells, the equations of motion are of eighth order and many more grid points are involved at every equation evaluation. That makes use of finite differences more involved. For an example see ref. 7.9.

7.6 FINITE ELEMENTS

The finite element approach is a physical approach. Knowing a solution of a simple element, a shell or plate can be thought to be assembled of these elements. This assembling is done by mathematically enforced continuity conditions at the element node points (points where the elements are joined to each other or to a boundary).

To illustrate the approach, let us select as examples first a beam-bending element and second a plate-bending element. Steps that have to be taken in deriving the element properties apply essentially to all elements, even curved-shell-type elements.

There are various ways that can be used to derive the element properties. One of them is Hamilton's principle. For this purpose, we have to generate expressions for strain and kinetic energy in terms of the node displacements.

First let us obtain the expression for strain energy. From Eq. (2.6.3) it is, for a beam under transverse deflection,

$$U = \frac{EI}{2}\int_0^L \left(\frac{\partial^2 U_3}{\partial x^2}\right)^2 dx \tag{7.6.1}$$

Next, we have to assume a deflection function that allows enforcement of a transverse deflection condition and a slope condition on each end of the beam element. The four conditions require four constants. We choose

$$u_3(x,t) = a_0 + a_1 x + a_2 x^2 + a_3 x^3 \tag{7.6.2}$$

This may also be written

$$u_3(x,t) = \{A\}^T\{Z\} \tag{7.6.3}$$

where the superscript T means transpose and where

$$\{A\}^T = \lfloor a_0, a_1, a_2, a_3 \rfloor \tag{7.6.4}$$

$$\{Z\}^T = \lfloor 1, x, x^2, x^3 \rfloor \tag{7.6.5}$$

Therefore

$$\frac{\partial^2 u_3}{\partial x^2} = \{A\}^T \left\{\frac{\partial^2 Z}{\partial x^2}\right\} = \left\{\frac{\partial^2 Z}{\partial x^2}\right\}^T \{A\} \tag{7.6.6}$$

and therefore

$$\left(\frac{\partial^2 u_3}{\partial x^2}\right)^2 = \{A\}^T[D(x)]\ \{A\} \tag{7.6.7}$$

where

$$[D(x)] = \left\{\frac{\partial^2 Z}{\partial x^2}\right\} \left\{\frac{\partial^2 Z}{\partial x^2}\right\}^T \tag{7.6.8}$$

Substituting Eq. (7.6.5) in Eq. (7.6.8) gives

$$[D(x)] = \begin{bmatrix} 0 & 0 & 0 & 0 \\ 0 & 0 & 0 & 0 \\ 0 & 0 & 4 & 12x \\ 0 & 0 & 12x & 36x^2 \end{bmatrix} \tag{7.6.9}$$

The strain energy is therefore

$$U = \frac{EI}{2}\ \{A\}^T \int_0^L [D(x)]\ dx\ \{A\} \tag{7.6.10}$$

Next, let us define the so-called nodal displacements (slopes are also referred to as *displacements*). At the x = 0 end of the

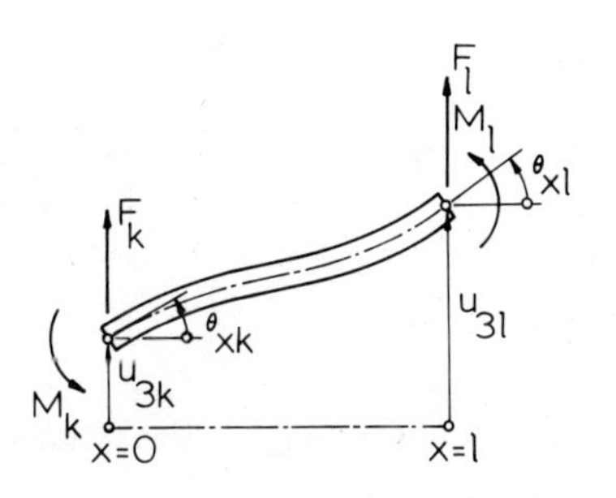

Figure 7.6.1

element designated as location k in Fig. 7.6.1 we get from Eq. (7.6.2)

$$u_3(0,t) = u_{3k} = a_0 \tag{7.6.11}$$

$$\frac{\partial u_3}{\partial x}(o,t) = \theta_{xk} = a_1 \tag{7.6.12}$$

and on the x = L end of the element designated as location ℓ we obtain

$$u_3(L,t) = u_{3\ell} = a_0 + a_1 L + a_2 L^2 + a_3 L^3 \tag{7.6.13}$$

$$\frac{\partial u_3}{\partial x}(L,t) = \theta_{x\ell} = a_1 + 2a_2 L + 3a_3 L^2 \tag{7.6.14}$$

This may be written in the form

$$\{u_3\}_i = [B]\ \{A\} \tag{7.6.15}$$

where

$$\{u_3\}_i^T = \lfloor u_{3k},\ \theta_{xk},\ u_{3\ell},\ \theta_{x\ell} \rfloor \tag{7.6.16}$$

and
$$B = \begin{bmatrix} 1 & 0 & 0 & 0 \\ 0 & 1 & 0 & 0 \\ 1 & L & L^2 & L^3 \\ 0 & 1 & 2L & 3L^2 \end{bmatrix} \tag{7.6.17}$$

Solving for {A} gives

$$\{A\} = [B]^{-1}\{u_3\}_i \tag{7.6.18}$$

Redefining

$$[B]^{-1} = [c] \tag{7.6.19}$$

allows us to write the strain energy equation as

$$U = \frac{EI}{2}\{u_3\}_i^T[c]^T\int_0^L [D(x)]\,dx\,[c]\{u_3\}_i \tag{7.6.20}$$

The variation of U in terms of variations in the node displacement is

$$\delta U = EI\,\{\delta u_3\}_i^T[c]^T\int_0^L [D(x)]\,dx\,[c]\{u_3\}_i \tag{7.6.21}$$

Next, let us obtain the kinetic energy of the beam element. It is

$$K = \frac{\rho A}{2}\int_0^L \dot{u}_3^2\,dx \tag{7.6.22}$$

From Eq. (7.6.3) we get

$$\dot{u}_3^2 = \{\dot{A}\}^T\{Z\}\{Z\}^T\{\dot{A}\} \tag{7.6.23}$$

Substituting Eqs. (7.6.18) and (7.6.19) gives

$$\dot{u}_3^2 = \{\dot{u}_3\}_i^T[c]^T[F(x)][c]\{\dot{u}_3\}_i \tag{7.6.24}$$

where

$$[F(x)] = \{Z\}\{Z\}^T = \begin{bmatrix} 1 & x & x^2 & x^3 \\ x & x^2 & x^3 & x^4 \\ x^2 & x^3 & x^4 & x^5 \\ x^3 & x^4 & x^5 & x^6 \end{bmatrix} \tag{7.6.25}$$

The kinetic energy expression is therefore

$$K = \frac{\rho A}{2} \{\dot{u}_3\}_i^T [c]^T \int_0^L [F(x)]\, dx\, [c]\{\dot{u}_3\}_i \tag{7.6.26}$$

The variation in K is

$$\delta K = \rho A\{\delta\dot{u}_3\}_i^T [c]^T \int_0^L [F(x)]\, dx\, [c]\{\dot{u}_3\}_i \tag{7.6.27}$$

We also have to consider the virtual work due to boundary forces. At x = 0, we have a shear force F_k and a bending moment M_k acting on the element. At x = L, the shear force is F_ℓ and the moment is M_ℓ. The virtual work is therefore

$$\delta W = \{F\}_i^T\{\delta u_3\}_i = \{\delta u_3\}_i^T\{F\}_i \tag{7.6.28}$$

where

$$\{F\}_i^T = \lfloor F_k, M_k, F_\ell, M_\ell \rfloor \tag{7.6.29}$$

We are now ready to apply Hamilton's principle which in this case, with boundary forces and moments, becomes

$$\int_{t_1}^{t_2} (\delta K - \delta U + \delta W)\, dt = 0 \tag{7.6.30}$$

Let us examine the kinetic energy part of the integral

$$\int_{t_1}^{t_2} \delta K\, dt = \rho A \int_{t_1}^{t_2} \{\delta\dot{u}_3\}_i^T[c]^T \int_0^L [F(x)]\, dx\, [c]\{\dot{u}_3\}_i\, dt \tag{7.6.31}$$

We have to integrate by parts in order to separate the node displacement variations from the time derivative. Since

$$\int u\, dv = uv - \int v\, du \tag{7.6.32}$$

we let

$$dv = \{\delta \dot{u}_3\}_i^T \, dt \tag{7.6.33}$$

$$u = [c]^T \int_0^L [F(x)] \, dx \, [c]\{\dot{u}_3\}_i \tag{7.6.34}$$

This gives

$$\int_{t_1}^{t_2} \delta K \, dt = \rho A \, \{\delta u_3\}_i^T \, [c]^T \int_0^L [F(x)] \, dx \, [c]\{\dot{u}_3\}_i \Bigg|_{t_1}^{t_2}$$

$$-\rho A \int_{t_1}^{t_2} \{\delta u_3\}_i^T \, [c]^T \int_0^L [F(x)] \, dx \, [c]\{\ddot{u}_3\}_i \, dt \tag{7.6.35}$$

The first term is zero because at t_1 and t_2 the variational displacements are zero by definition of Hamilton's principle. Substituting this expression and Eqs. (7.6.21) and (7.6.28) in Eq. (7.6.30) gives

$$\int_{t_1}^{t_2} \{\delta u_3\}_i^T \left[A\rho [c]^T \int_0^L [F(x)] \, dx \, [c]\{\ddot{u}_3\}_i \right.$$

$$\left. + \, EI \, [c]^T \int_0^L [D(x)] \, dx \, [c]\{u_3\}_i - \{F\}_i \right] dt = 0 \tag{7.6.36}$$

Because the variational displacements are independent and arbitrary, the equation can only be satisfied if the bracketed quantity is zero. This gives the equation of motion of the element

$$[m]\{\ddot{u}_3\}_i + [k]\{u_3\}_i = \{F\}_i \tag{7.6.37}$$

where

$$[m] = A\rho \, [c]^T \int_0^L [F(x)] \, dx \, [c] \tag{7.6.38}$$

$$[k] = EI[c]^T \int_0^L [D(x)]\ dx\ [c] \tag{7.6.39}$$

The matrix [m] is the mass matrix and the matrix [k] is the stiffness matrix. We will see that for the plate element the definition will be similar. The indicated integrations and matrix manipulations are for more complicated elements part of the finite element computer program. However, the beam case is so simple that we can easily do the integration and matrix manipulations by hand. Using Eqs. (7.6.9), (7.6.17), and (7.6.25), we obtain

$$[m] = \frac{\rho AL}{420}\begin{bmatrix} 156 & 22L & 54 & -13L \\ 22L & 4L^2 & 13L & -3L^2 \\ 54 & 13L & 156 & -22L \\ -13L & -3L^2 & -22L & 4L^2 \end{bmatrix} \tag{7.6.40}$$

$$[k] = \frac{EI}{L^3}\begin{bmatrix} 12 & 6L & -12 & 6L \\ 6L & 4L^2 & -6L & 2L^2 \\ -12 & -6L & 12 & -6L \\ 6L & 2L^2 & -6L & 4L^2 \end{bmatrix} \tag{7.6.41}$$

Now let us do the derivation of a rectangular plate element, following the identical procedure.

The strain energy for a transversely deflected plate is, from Eq. (2.6.3),

$$U = \frac{D}{2}\int_0^b \int_0^a \left\{\left(\frac{\partial^2 u_3}{\partial x^2} + \frac{\partial^2 u_3}{\partial y^2}\right)^2 - 2(1-\mu)\left[\frac{\partial^2 u_3}{\partial x^2}\frac{\partial^2 u_3}{\partial y^2} - \left(\frac{\partial^2 u_3}{\partial x\ \partial y}\right)^2\right]\right\} dx\ dy \tag{7.6.42}$$

In case of the plate element, we have to be able to enforce as a minimum continuity of deflection at each of the four corners

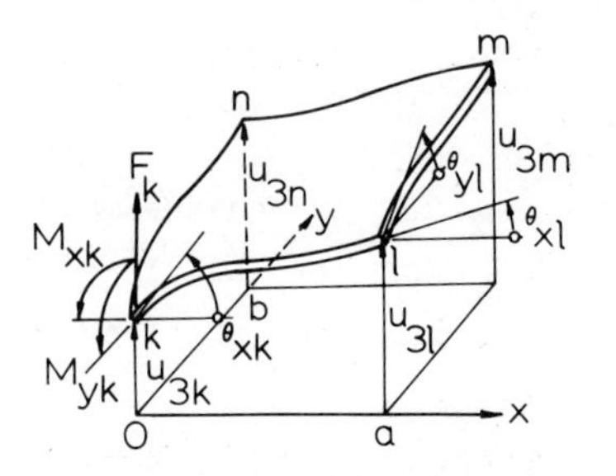

Figure 7.6.2

and continuity of slope in two orthogonal directions at each corner. It is actually better to also enforce continuity of twisting, but for simplicity's sake the easier example is used. Let us label the four corners k, ℓ, m, and n as shown in Fig. 7.6.2. Let us again use the symbol θ to designate slopes. For instance, at the corner ℓ,

$$\theta_{x\ell} = \frac{\partial u_{3\ell}}{\partial x} \tag{7.6.43}$$

$$\theta_{y\ell} = \frac{\partial u_{3\ell}}{\partial y} \tag{7.6.44}$$

To be able to enforce 12 continuity conditions, the deflection function has to have 12 constants. We choose

$$u_3(x,y,t) = a_1 + a_2x + a_3y + a_4x^2 + a_5xy + a_6y^2 + a_7x^3 + a_8x^2y + a_9xy^2 + a_{10}y^3 + a_{11}x^3y + a_{12}xy^3 \tag{7.6.45}$$

This may also be written as

$$u_3(x,y,t) = \{A\}^T\{Z\} \tag{7.6.46}$$

where

$$\{A\}^T = \lfloor a_1, a_2, \ldots, a_{12} \rfloor \tag{7.6.47}$$

$$\{Z\}^T = \lfloor 1, x, y, x^2, xy, y^2, x^3, x^2y, xy^2, y^3, x^3y, xy^3 \rfloor \tag{7.6.48}$$

The strain energy expression becomes, after substitution,

$$U = \frac{D}{2} \{A\}^T \int_0^b \int_0^a [D(x,y)] \, dx \, dy \, \{A\} \tag{7.6.49}$$

where

$$[D(x,y)] = \left\{\frac{\partial^2 z}{\partial x^2}\right\} \left\{\frac{\partial^2 z}{\partial x^2}\right\}^T + \left\{\frac{\partial^2 z}{\partial y^2}\right\} \left\{\frac{\partial^2 z}{\partial y^2}\right\}^T + \mu \left\{\frac{\partial^2 z}{\partial x^2}\right\} \left\{\frac{\partial^2 z}{\partial y^2}\right\}^T$$
$$+ \mu \left\{\frac{\partial^2 z}{\partial y^2}\right\} \left\{\frac{\partial^2 z}{\partial x^2}\right\}^T + 2(1 - \mu) \left\{\frac{\partial^2 z}{\partial x \, \partial y}\right\} \left\{\frac{\partial^2 z}{\partial x \, \partial y}\right\}^T \tag{7.6.50}$$

Next, we formulate the nodal displacements by evaluating Eq. (7.6.45) and its derivatives at the node points, enforcing the conditions that

$$\begin{aligned}
u_3(0,0,t) &= u_{3k} & \frac{\partial u_3}{\partial x}(0,0,t) &= \theta_{xk} & \frac{\partial u_3}{\partial y}(0,0,t) &= \theta_{yk} \\
u_3(a,0,t) &= u_{3\ell} & \frac{\partial u_3}{\partial x}(a,0,t) &= \theta_{x\ell} & \frac{\partial u_3}{\partial y}(a,0,t) &= \theta_{y\ell} \\
u_3(a,b,t) &= u_{3m} & \frac{\partial u_3}{\partial x}(a,b,t) &= \theta_{xm} & \frac{\partial u_3}{\partial y}(a,b,t) &= \theta_{ym} \\
u_3(0,b,t) &= u_{3n} & \frac{\partial u_3}{\partial x}(0,b,t) &= \theta_{xn} & \frac{\partial u_3}{\partial y}(0,b,t) &= \theta_{yn}
\end{aligned} \tag{7.6.51}$$

This can be written as

$$\{u_3\}_i = [B]\{A\} \tag{7.6.52}$$

where

$$\{u_3\}_i^T = \lfloor u_{3k}, \theta_{xk}, \theta_{yk}, u_{3\ell}, \theta_{x\ell}, \theta_{y\ell}, u_{3m}, \theta_{xm}, \theta_{ym}, u_{3n}, \theta_{xn}, \theta_{yn} \rfloor \tag{7.6.53}$$

and where

$$[B] = \begin{bmatrix} 1 & 0 & 0 & 0 & 0 & 0 & 0 & 0 & 0 & 0 & 0 & 0 \\ 0 & 1 & 0 & 0 & 0 & 0 & 0 & 0 & 0 & 0 & 0 & 0 \\ 0 & 0 & 1 & 0 & 0 & 0 & 0 & 0 & 0 & 0 & 0 & 0 \\ 1 & a & 0 & a^2 & 0 & 0 & a^3 & 0 & 0 & 0 & 0 & 0 \\ 0 & 1 & 0 & 2a & 0 & 0 & 3a^2 & 0 & 0 & 0 & 0 & 0 \\ 0 & 0 & 1 & 0 & a & 0 & 0 & a^2 & 0 & 0 & a^3 & 0 \\ 1 & a & b & a^2 & ab & b^2 & a^3 & a^2b & ab^2 & b^3 & a^3b & ab^3 \\ 0 & 1 & 0 & 2a & b & 0 & 3a^2 & 2ab & b^2 & 0 & 3a^2b & b^3 \\ 0 & 0 & 1 & 0 & a & 2b & 0 & a^2 & 2ab & 3b^2 & a^3 & 3ab^2 \\ 1 & 0 & b & 0 & 0 & b^2 & 0 & 0 & 0 & b^3 & 0 & 0 \\ 0 & 1 & 0 & 0 & b & 0 & 0 & 0 & b^2 & 0 & 0 & b^3 \\ 0 & 0 & 1 & 0 & 0 & 2b & 0 & 0 & 0 & 3b^2 & 0 & 0 \end{bmatrix} \tag{7.6.54}$$

Solving for {A} gives

$$\{A\} = [c]\{u_3\}_i \tag{7.6.55}$$

where

$$[c] = [B]^{-1} \tag{7.6.56}$$

The strain energy expression becomes, therefore,

$$U = \frac{D_i}{2} \{u_3\}_i^T [c]^T \int_0^b \int_0^a [D(x,y)]\, dx\, dy\, [c]\{u_3\}_i \tag{7.6.57}$$

The kinetic energy of the element is

$$K = \frac{\rho h}{2} \int_0^b \int_0^a \dot{u}_3^2\, dx\, dy \tag{7.6.58}$$

From Eq. (7.6.46) we obtain

$$\dot{u}_3^2 = \{\dot{A}\}^T\{Z\}\{Z\}^T\{\dot{A}\} \tag{7.6.59}$$

and substituting Eq. (7.6.55),

$$\dot{u}_3^2 = \{\dot{u}_3\}_i^T[c]^T[F(x,y)][c]\{\dot{u}_3\}_i \tag{7.6.60}$$

where

$$[F(x,y)] = \{Z\}\{Z\}^T \tag{7.6.61}$$

The kinetic energy expression becomes, therefore,

$$K = \frac{\rho h}{2}\{\dot{u}_3\}_i^T[c]^T\int_0^b\int_0^a [F(x,y)]\,dx\,dy\,[c]\{\dot{u}_3\}_i \tag{7.6.62}$$

The virtual work due to the nodal forces and moments is

$$\delta W = \{F\}_i^T\{\delta u_3\}_i = \{\delta u_3\}_i\{F\}_i \tag{7.6.63}$$

where

$$\{F\}_i^T = \lfloor F_k, M_{xk}, M_{yk}, F_\ell, M_{x\ell}, M_{y\ell}, F_m, M_{xm}, M_{ym}, F_n, M_{xn}, M_{yn}\rfloor \tag{7.6.64}$$

Both the potential and the kinetic energy expressions are similar to the expressions for the beam element. Thus, applying Hamilton's principle, following identical steps, we obtain the equation of motion of the element

$$[m]\{\ddot{u}_3\}_i + [k]\{u_3\}_i = \{F\}_i \tag{7.6.65}$$

where

$$[m] = \rho h\,[c]^T\int_0^b\int_0^a [F(x,y)]\,dx\,dy\,[c] \tag{7.6.66}$$

$$[k] = D[c]^T\int_0^b\int_0^a [D(x,y)]\,dx\,dy\,[c] \tag{7.6.67}$$

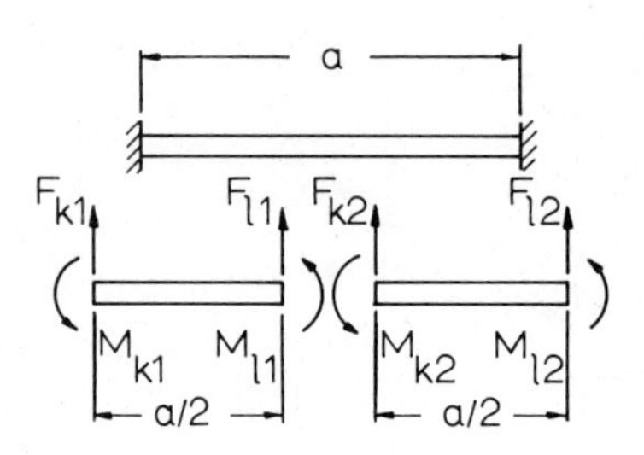

Figure 7.6.3

The stiffness matrix [k] and the mass matrix [m] are usually not given in explicit form but are generated on the computer when needed.

The remaining question is how the various elements are joined together. Let us illustrate this on the example of a clamped-clamped uniform beam that is approximated by only two beam elements of the same length $L = a/2$ as shown in Fig. 7.6.3. Using the subscript 1 for element 1 and the subscript 2 for element 2, we obtain from the continuity condition

$$u_{3\ell 1} = u_{3k2} \tag{7.6.68}$$

$$\theta_{x\ell 1} = \theta_{xk2} \tag{7.6.69}$$

The forces and moments at the junction have to add up to zero.

$$F_{\ell 1} + F_{k2} = 0 \tag{7.6.70}$$

$$M_{\ell 1} + M_{k2} = 0 \tag{7.6.71}$$

This allows us to formulate the global equation of motion from the element equations of motion by a simple addition process. We obtain

$$[m]\{\ddot{u}_3\}_i + [k]\{u_3\}_i = \{F\}_i \tag{7.6.72}$$

where the global matrix is

$$[m] = \frac{\rho AL}{420}\begin{bmatrix} 156 & 22L & 54 & -13L & 0 & 0 \\ 22L & 4L^2 & 13L & -3L^2 & 0 & 0 \\ 54 & 13L & 312 & 0 & 54 & -13L \\ -13L & -3L^2 & 0 & 8L^2 & 13L & -3L^2 \\ 0 & 0 & 54 & 13L & 156 & -22L \\ 0 & 0 & -13L & -3L^2 & -22L & 4L^2 \end{bmatrix} \tag{7.6.73}$$

and where the global stiffness matrix is

$$[k] = \frac{EI}{L^3}\begin{bmatrix} 12 & 6L & -12 & 6L & 0 & 0 \\ 6L & 4L^2 & -6L & 2L^2 & 0 & 0 \\ -12 & -6L & 24 & 0 & -12 & 6L \\ 6L & 2L^2 & 0 & 8L^2 & -6L & 2L^2 \\ 0 & 0 & -12 & -6L & 12 & -6L \\ 0 & 0 & 6L & 2L^2 & -6L & 4L^2 \end{bmatrix} \tag{7.6.74}$$

The nodal force vector becomes

$$\{F\}_i^T = \lfloor F_{k1}, M_{k1}, 0, 0, F_{\ell 2}, M_{\ell 2} \rfloor \tag{7.6.75}$$

The nodal displacement vector becomes

$$\{u_3\}_i^T = \lfloor u_{3k1}, \theta_{xk1}, u_{3k2}, \theta_{xk2}, u_{3\ell 2}, \theta_{x\ell 2} \rfloor \tag{7.6.76}$$

So far, the boundary conditions at the clamped locations have not been applied yet. If we do so, the displacement vector reduces to

$$\{u_3\}_i^T = \lfloor 0, 0, u_{3k2}, \theta_{xk2}, 0, 0 \rfloor \tag{7.6.77}$$

This means that the equation of motion reduces to

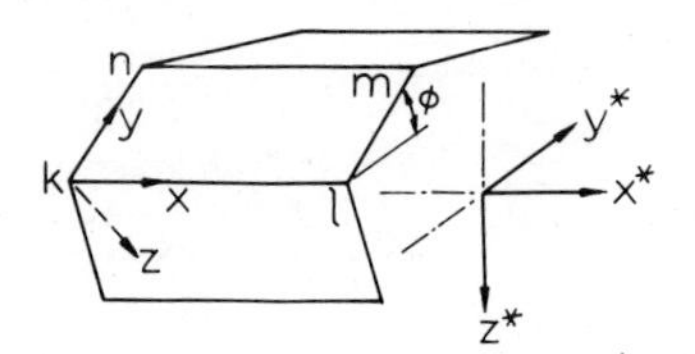

Figure 7.6.4

$$\frac{\rho AL}{420}\begin{bmatrix}312 & 0\\ 0 & 8L^2\end{bmatrix}\begin{Bmatrix}\ddot{u}_{3k2}\\ \ddot{\theta}_{xk2}\end{Bmatrix}+\frac{EI}{L^3}\begin{bmatrix}24 & 0\\ 0 & 8L^2\end{bmatrix}\begin{Bmatrix}u_{3k2}\\ \theta_{xk2}\end{Bmatrix}=\begin{Bmatrix}0\\ 0\end{Bmatrix}$$

(7.6.78)

This equation can be solved in the usual way for the first two natural frequencies and modes. The natural frequencies are given by

$$\omega_i = \frac{\lambda_i}{a^2}\sqrt{\frac{EI}{\rho A}} \tag{7.6.79}$$

where $\lambda_1 = 22.72$, $\lambda_2 = 82.0$. This compares to the exact values of $\lambda_1 = 22.3$, $\lambda_2 = 61.67$ and illustrates the need for a larger number of elements if higher modes are to be investigated.

The solution of plate problems follows the same assembly procedure.

The final point of discussion attempts to illustrate how the flat plate element can be used for shell problems. First of all, with a rectangular element we can only approximate a cylindrical shell surface. A cone or any other shell could not be approximated without leaving gaps at some node points.

A more versatile flat plate element for shell applications is the triangular element. Progressing in sophistication, elements having curvatures are next in the line of development. However, in this discussion we continue to confine ourselves to the rectangular plate element and look at the question of how it is to be modified for the cylindrical shell application.

The difficulty to be overcome is that adjacent elements are inclined to each other. This means that before the elements can be assembled, the nodal displacements and forces of the individual elements must be expressed in a common coordinate system, also called a *global coordinate system*. In this case, we pick a cartesian global coordinate system x^*, y^*, z^* as shown in Fig. 7.6.4. The global displacements are

$$\{u_3^*\}_i^T = [u_{3k}^*, \theta_{xk}^*, \theta_{yk}^*, \ldots] \tag{7.6.80}$$

Since in this special case the x axis is parallel to the x^* axis and since the angle between the y axis and the y^* axis, denoted ϕ, is equal to the angle between the z axis and the z^* axis, we get

$$\{u_3\}_i = [T]\{u_3^*\}_i \tag{7.6.81}$$

where

$$[T] = \begin{bmatrix} \cos\phi & & & & & & & & & & & \\ & 1 & & & & & & & & & & \\ & & \cos\phi & & & & & & & & & \\ & & & \cos\phi & & & & 0 & & & & \\ & & & & 1 & & & & & & & \\ & & & & & \cos\phi & & & & & & \\ & & & & & & \cos\phi & & & & & \\ & & & & & & & 1 & & & & \\ & & & & & & & & \cos\phi & & & \\ & 0 & & & & & & & & \cos\phi & & \\ & & & & & & & & & & 1 & \\ & & & & & & & & & & & \cos\phi \end{bmatrix} \tag{7.6.82}$$

A similar relationship results for the nodal forces. The equation of motion of the square plate element is therefore in global coordinates

$$[m^*]\{\ddot{u}_3^*\}_i + [k^*]\{u_3^*\}_i = \{F^*\}_i \tag{7.6.83}$$

where

$$[m^*] = [T]^T[m][T] \tag{7.6.84}$$

$$[k^*] = [T]^T[k][T] \tag{7.6.85}$$

$$\{F^*\}_i = [T]\{F\}_i \tag{7.6.86}$$

Assembly of the various elements follows the previously illustrated procedure.

There is considerable literature in finite element development and application. Of most interest are the large computer programs like NASTRAN and others that are available through government agencies, universities, or consultation companies. For a history and discussion of some of these, see Ref. 7.10.

The finite elements that were used in this chapter had both transverse deflection capabilities only. To make them more versatile, in-plane deflections can also be added. This can be done following the steps that were outlined.

7.7 SOUTHWELL'S PRINCIPLE

Southwell is credited with a formula [7.11, 7.12] which can be applied to the problem of finding natural frequencies of shells and other structures whose stiffness is controlled by several superimposed effects with benefit. Consider the eigenvalue problem

$$L\{U\} - \omega^2 N\{U\} = 0 \tag{7.7.1}$$

If it is possible to separate the operator $L\{U\}$ into

$$L\{U\} = \sum_{r=1}^{n} L_r\{U\} \tag{7.7.2}$$

and if solutions exist for each partial problem

$$L_r\{U\} - \omega_r^2 N\{U\} = 0 \tag{7.7.3}$$

we obtain for the fundamental frequency the fact that

$$\omega_s^2 \leq \omega_1^2 \tag{7.7.4}$$

where ω_s is Southwell's frequency and is given by

$$\omega_s^2 = \sum_{r=1}^{n} \omega_{1r}^2 \tag{7.7.5}$$

The symbol ω_1 designates the true fundamental frequency of the total problem. The symbol ω_{1r} designates the true fundamental frequency of the r-th partial problem.

We prove this by applying Galerkin's approach to Eq. (7.7.1). If U_1 is the true fundamental mode of the total problem, the true fundamental frequency is

$$\omega_1^2 = \frac{\int_A\int U_1 L\{U_1\}\, dA}{\int_A\int U_1 N\{U_1\}\, dA} \tag{7.7.6}$$

Let us now apply Galerkin's approach to the partial problem. If U_1 is the true mode for the total problem, we obtain the inequality

$$\omega_{1r}^2 \leq \frac{\int_A\int U_1 L_r\{U_1\}\, dA}{\int_A\int U_1 N\{U_1\}\, dA} \tag{7.7.7}$$

If it happens that U_1 is also the true mode for the partial problem, the equality applies. However, in general, it cannot be expected that U_1 is the true mode for the partial problem.

Let us now sum both sides of the equation from r = 1 to n. This gives

$$\sum_{r=1}^{n} \omega_{1r}^2 \leq \frac{\int_A\int U_1 \sum_{r=1}^{n} L_r\{U_1\}\, dA}{\int_A\int U_1 N\{U_1\}\, dA} \tag{7.7.8}$$

The left side is ω_s^2 by definition of Eq. (7.7.5) and the right side is ω_1^2 because of Eq. (7.7.2). Thus

$$\omega_s^2 \leq \omega_1^2 \tag{7.7.9}$$

and the proof is completed.

Let us illustrate this for the closed circular cylindrical shell, as described by Eq. (6.9.4). By Southwell's principle, we may formulate two partial problems:

$$D\left(\frac{n^2}{a^2} - \frac{d^2}{dx^2}\right)^4 U_{3n}(x) - \rho h\omega_1^2 \left(\frac{n^2}{a^2} - \frac{d^2}{dx^2}\right)^2 U_{3n}(x) = 0 \tag{7.7.10}$$

and
$$\frac{Eh}{a^2}\,\frac{d^4U_{3n}(x)}{dx^4} - \rho h\omega_2^2 \left(\frac{n^2}{a^2} - \frac{d^2}{dx^2}\right)^2 U_{3n}(x) = 0 \tag{7.7.11}$$

If the shell is simply supported, all boundary conditions and both equations are satisfied by

$$U_{3n}(x) = \sin\frac{m\pi x}{L} \tag{7.7.12}$$

This gives, upon substitution in Eq. (7.7.10),

$$D\left[\left(\frac{n}{a}\right)^2 + \left(\frac{m\pi}{L}\right)^2\right]^4 - \rho h\omega_1^2 \left[\left(\frac{n}{a}\right)^2 + \left(\frac{m\pi}{L}\right)^2\right]^2 = 0 \tag{7.7.13}$$

or
$$\omega_1^2 = \frac{D}{\rho h}\left[\left(\frac{n}{a}\right)^2 + \left(\frac{m\pi}{L}\right)^2\right]^2 \tag{7.7.14}$$

This equation defines the natural frequencies due to the bending effect only. Substituting Eq. (7.7.12) in Eq. (7.7.11) gives

$$\frac{Eh}{a^2}\left(\frac{m\pi}{L}\right)^4 - \rho h\omega_2^2 \left[\left(\frac{n}{a}\right)^2 + \left(\frac{m\pi}{L}\right)^2\right]^2 = 0 \tag{7.7.15}$$

or
$$\omega_2^2 = \frac{E}{\rho a^2}\,\frac{\left(\frac{m\pi}{L}\right)^4}{\left[\left(\frac{n}{a}\right)^2 + \left(\frac{m\pi}{L}\right)^2\right]^2} \tag{7.7.16}$$

This equation defines the natural frequencies due to the membrane effect only.

Thus, according to Southwell's principle, we obtain

$$\omega^2 \geq \omega_1^2 + \omega_2^2 \tag{7.7.17}$$

or
$$\omega^2 \geq \frac{1}{a^2}\left\{\frac{(\frac{m\pi a}{L})^4}{[n^2 + (\frac{m\pi a}{L})^2]^2} + \frac{1}{12(1 - \mu^2)} (\frac{h}{a})^2 [n^2 + (\frac{m\pi a}{L})^2]^2\right\}\frac{E}{\rho} \tag{7.7.18}$$

In this case, Southwell's principle gives the exact result. The reason is that for the case chosen, the fundamental mode of the total problem is equal to the fundamental mode for both partial problems. However, in general, exact results cannot be expected. Note also that we have obtained a result that can be applied for all m, n combinations, yet only for the fundamental mode is the inequality of Eq. (7.7.9) valid. For frequencies other than the lowest, nothing can be said in general about boundedness, and the quality of the prediction will have to be verified from case to case.

7.8 DUNKERLEY'S PRINCIPLE

In 1894, Dunkerley [7.13] discovered a method that would allow him to estimate the fundamental frequency of a multidegree of freedom system. In the following, a development of Dunkerley's method is shown following essentially Collatz [7.12]. Consider again the eigenvalue problem

$$L\{U\} - \omega^2 N\{U\} = 0 \tag{7.8.1}$$

If it is possible to separate the operator $N\{U\}$, which describes the mass effect, into

$$N\{U\} = \sum_{r=1}^{n} N_r\{U\} \tag{7.8.2}$$

and if we know the fundamental frequency of the partial problem, the partial problem having to be self adjoint and fully defined,

$$L\{U\} - \omega_r^2 N_r\{U\} = 0 \tag{7.8.3}$$

we obtain

$$\omega_D^2 \leq \omega_1^2 \tag{7.8.4}$$

where ω_D is Dunkerley's frequency and given by

$$\frac{1}{\omega_D{}^2} = \sum_{r=1}^{n} \frac{1}{\omega_{1r}^2} \tag{7.8.5}$$

and where ω_1 is the actual fundamental frequency of the total problem and the ω_{1r} are the fundamental frequencies of the n partial problems.

To prove Eq. (7.8.4), we use again Galerkin's method. If U_1 is the true fundamental mode, then the true fundamental frequency is

$$\omega_1^2 = \frac{\int_A\int U_1 L\{U_1\}\, dA}{\int_A\int U_1 N\{U_1\}\, dA} \tag{7.8.6}$$

Let us now apply Galerkin's approach to the partial problem. If U_1 is the true mode for the total problem, we obtain

$$\omega_{1r}^2 \leq \frac{\int_A\int U_1 L\{U_1\}\, dA}{\int_A\int U_1 N_r\{U_1\}\, dA} \tag{7.8.7}$$

or

$$\frac{1}{\omega_{1r}^2} \geq \frac{\int_A\int U_1 N_r\{U_1\}\, dA}{\int_A\int U_1 L\{U_1\}\, dA} \tag{7.8.8}$$

$$\sum_{r=1}^{n} \frac{1}{\omega_{1r}^2} \geq \frac{\int_A\int U_1 \sum_{r=1}^{n} N_r\{U_1\}\, dA}{\int_A\int U_1 L\{U_1\}\, dA} \tag{7.8.9}$$

Let us sum both sides from i = 1 to n. This gives

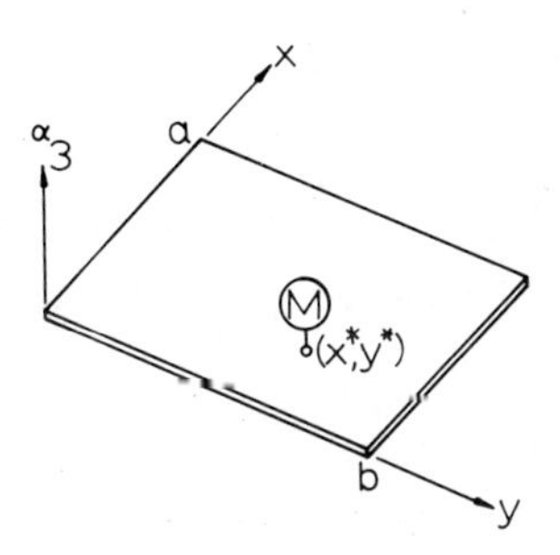

Figure 7.8.1

But, because of Eq. (7.8.2), the right side of the equation is equal to $1/\omega_1^2$. The left side is $1/\omega_D^2$ according to Eq. (7.8.5). Therefore,

$$\omega_D^2 \leq \omega_1^2 \tag{7.8.10}$$

This completes the proof. The right side is the square of the true first natural frequency. The square root of the left side is also called *Dunkerley's frequency* and given the symbol ω_D.

As example, let us consider a simply supported plate with an attached mass at location x*, y* as shown in Fig. 7.8.1. We may express the distributed mass of the plate as $\rho h + M\,\delta(x - x^*)\,\delta(y - y^*)$. The equation for free transverse motion may therefore be written as

$$D\nabla^4 u_3 + [\rho h + M\,\delta(x - x^*)\,\delta(y - y^*)]\,\frac{\partial^2 u_3}{\partial t^2} = 0 \tag{7.8.11}$$

Arguing that at a natural frequency

$$u_3(x,y,t) = U_3(x,y)e^{j\omega t} \tag{7.8.12}$$

we obtain upon substitution

$$D\nabla^4 U_3 - [\rho h + M\,\delta(x - x^*)\,\delta(y - y^*)]\omega^2 U_3 = 0 \tag{7.8.13}$$

Let us now split this problem into two partial problems

$$D\nabla^4 U_3 - \rho h\omega^2 U_3 = 0 \tag{7.8.14}$$

and $$D\nabla^4 U_3 - M\omega^2 U_3 \delta(x - x^*)\ \delta(y - y^*) = 0 \tag{7.8.15}$$

For the simply supported plate, the solution of Eq. (7.8.14) is ($r = 1$)

$$U_3 = \sin\frac{m\pi x}{a}\sin\frac{n\pi y}{b} \tag{7.8.16}$$

$$\omega_{mn1}^2 = \pi^4\left[\left(\frac{m}{a}\right)^2 + \left(\frac{n}{b}\right)^2\right]^2 \frac{D}{\rho h} \tag{7.8.17}$$

To solve the second equation, we use Galerkin's method as shown in Eq. (7.7.6), with Eq. (7.8.16) as the approximate mode expression. We obtain ($r = 2$)

$$D\pi^4\left[\left(\frac{m}{a}\right)^2 + \left(\frac{n}{b}\right)^2\right]^2 \int_0^b \int_0^a \sin^2\left(\frac{m\pi x}{a}\right)\sin^2\left(\frac{m\pi y}{b}\right) dx\,dy$$
$$- M\omega_{mn2}^2 \sin^2\left(\frac{m\pi x^*}{a}\right)\sin^2\left(\frac{n\pi y^*}{b}\right) = 0 \tag{7.8.18}$$

This gives

$$\omega_{mn2}^2 = \frac{D\pi^4\left[\left(\frac{m}{a}\right)^2 + \left(\frac{n}{b}\right)^2\right]^2 ab}{4M\sin^2\left(\frac{m\pi x}{a}\right)^*\sin^2\left(\frac{n\pi y}{b}\right)^*} \tag{7.8.19}$$

Thus, according to Dunkerley's formula, Eq. (7.8.5),

$$\omega_D^2 = \frac{1}{\dfrac{1}{\omega_{mn1}^2} + \dfrac{1}{\omega_{mn2}^2}} = \frac{\omega_{mn1}^2}{1 + \dfrac{\omega_{mn1}^2}{\omega_{mn2}^2}} \tag{7.8.20}$$

Therefore

$$\omega_D^2 = \frac{\pi^4\left[\left(\frac{m}{a}\right)^2 + \left(\frac{n}{b}\right)^2\right]^2 \frac{D}{\rho h}}{1 + 4\dfrac{M}{M_p}\sin^2\left(\frac{m\pi x}{a}\right)^*\sin^2\left(\frac{n\pi y}{b}\right)^*} \tag{7.8.21}$$

where M_p is the mass of the total plate

$$M_p = \rho hab \tag{7.8.22}$$

This result is identical to the result of Eq. (13.2.8) when curvature in Eq. (13.2.8) is set to zero. Note that we obtain frequency results for all (m, n) combinations, except that only for (m, n) = (1,1) is the inequality of Eq. (7.8.10) valid. Also, the accuracy of prediction suffers as m, n increase.

A similar, yet somewhat different result was obtained in Ref. 7.14, where the static Green's function of the simply supported plate was used to obtain ω^2_{mn2}.

REFERENCES

7.1 W. Flügge (Ed.), *Handbook of Engineering Mechanics*, McGraw-Hill, New York, 1962.

7.2 L. R. Koval and E. T. Cranch, "On the free vibrations of thin cylindrical shells subjected to an initial static torque," Proceedings of the 4th U.S. National Congress of Applied Mechanics, pp. 107-117, 1962.

7.3 J. W. S. Rayleigh, *The Theory of Sound*, Dover, New York, 1945 (originally published 1877).

7.4 W. Ritz, "Über eine neue Methode Zur Lösung gewisser Variationsprobleme der mathematischen physik," *J. Reine Angewandte Mathematik, vol. 135,* 1908.

7.5 W. Ritz, "Theorie der Transversalschwingungen einer quadratischen Platte mit freien rändern," *Ann. Physik, bd. 28,* 1909, pp. 737-786.

7.6 D. Young, "Vibration of rectangular plates by the Ritz method," *J. Appl. Mech. vol. 17*, no. 4, 1950, pp. 448-453.

7.7 K. Federhofer, "Eigenschwingungen der Kegelschale," *Ingenieur Archiv, vol. 9*, 1938, pp. 288-308.

7.8 T. Wah and L. R. Calcote, *Structural Analysis by Finite Difference Calculus*, Van Nostrand Reinhold, New York, 1970.

7.9 E. R. Kolman, "Axisymmetric configurations of oscillations of a thin conical shell," Raschety na Rasch. na Prochn. i Zhestk., Mashgiz, Moscow, 1965, pp. 49-60.

7.10 S. J. Fenves (Ed.), *Numerical and Computer Methods in Structural Mechanics*, Academic Press, New York, 1973.

7.11 R. V. Southwell, "On the free transverse vibrations of a uniform circular disk clamped at its center; and on the effects of rotation," *Proc. Roy. Soc. London* (ser. A), vol. 101, 1922, pp. 133-153.

7.12 L. Collatz, *Eigenwertprobleme und ihre numerische Behandlung*, Chelsea, New York, 1948.

7.13 S. Dunkerley, "On the whirling and vibration of shafts," *Phil. Trans. Roy. Soc. London (ser. A), vol. 185*, 1894, pp.279-360.

7.14 W. Soedel, Lower bound frequencies of plates and plate-mass systems by integration of static Green's functions, *Recent Advances in Engineering Science* (Proceedings of the Tenth Anniversary Meeting of the Society of Engineering Science), vol. 7, Scientific Publishers, Boston, 1976, pp. 157-163.

8

FORCED VIBRATIONS OF SHELLS BY MODAL EXPANSION

So far we have been concerned with the natural frequencies of shells and plates. The ultimate reason for this preoccupation is found, for the engineer, in the study of the forced response of shells. For instance, knowing the eigenvalues makes it possible to obtain the forced solution in terms of these eigenvalues. This approach is called *spectral representation* or *modal expansion* and dates back to Bernoulli's work [8.1]. It will be the major point of discussion in this chapter.

Forces will be assumed to be independent of the motion of the shell. This is an admissible approximation for most engineering shell vibration cases. For instance, a fluid impinging on a shell can be thought to be not affected by the relatively small vibration response amplitudes. Thermodynamic forces on the cylinder liner of a combustion engine can be thought to be independent of the motion of this liner. Many other examples can be listed [8.2-8.4].

8.1 THE MODAL PARTICIPATION FACTOR

A disturbance will excite the various natural modes of a shell in various amounts. The amount of participation of each mode in the total dynamic response is defined by the modal participation factor. This factor may turn out to be zero for certain modes and may approach large values for others, depending on the nature of the excitation.

In a mathematical sense, the natural modes of a shell structure represent orthogonal vectors that satisfy the boundary conditions of the structure. This vector space can be used to represent any response of the structure. In cases of finite degree of freedom systems, the vector space is of finite dimension and the number of vectors or natural modes is equal to the number of degrees of freedom. For continuous systems, such as shells, the number of degrees of freedom is infinite. This means that the general solution will be an infinite series:

$$u_i(\alpha_1,\alpha_2,t) = \sum_{k=1}^{\infty} \eta_k(t)U_{ik}(\alpha_1,\alpha_2) \tag{8.1.1}$$

where $i = 1, 2, 3$. The U_{ik} are the natural mode components in the three principal directions. The modal participation factors η_k are unknown and have to be determined in the following.

The Love equations are of the form

$$L_i(u_1,u_2,u_3) - \lambda\dot{u}_i - \rho h\ddot{u}_i = -q_i \tag{8.1.2}$$

where λ is an equivalent viscous damping factor. The viscous damping term was introduced through the forcing term, replacing the original q_i by $q_i - \lambda\dot{u}_i$. Also note that the damping factor is assumed to be the same in all three principal directions. This is not necessarily true, but since damping values are notoriously difficult to determine theoretically, and thus have more qualitative than quantitative value, and since a uniform damping factor offers computational advantages, it was decided to adopt the uniform factor here. How this factor relates to the structural damping description in the literature that uses a complex modulus will be discussed later.

The operators L_i are defined, from Love's equation, as

$$L_1(u_1,u_2,u_3) = \frac{1}{A_1A_2}\left(\frac{\partial(N_{11}A_2)}{\partial\alpha_1} + \frac{\partial(N_{21}A_1)}{\partial\alpha_2} + N_{12}\frac{\partial A_1}{\partial\alpha_2} - N_{22}\frac{\partial A_2}{\partial\alpha_1} + A_1A_2\frac{Q_{13}}{R_1}\right) \quad (8.1.3)$$

$$L_2(u_1,u_2,u_3) = \frac{1}{A_1A_2}\left(\frac{\partial(N_{12}A_2)}{\partial\alpha_1} + \frac{\partial(N_{22}A_1)}{\partial\alpha_2} + N_{21}\frac{\partial A_2}{\partial\alpha_1} - N_{11}\frac{\partial A_1}{\partial\alpha_2} + A_1A_2\frac{Q_{23}}{R_2}\right) \quad (8.1.4)$$

$$L_3(u_1,u_2,u_3) = \frac{1}{A_1A_2}\left[\frac{\partial(Q_{13}A_2)}{\partial\alpha_1} + \frac{\partial(Q_{23}A_1)}{\partial\alpha_2} - A_1A_2\left(\frac{N_{11}}{R_1} + \frac{N_{22}}{R_2}\right)\right] \quad (8.1.5)$$

Of course, any of the previously discussed simplifications can be applied. The important point is that Eq. (8.1.2) is general and will apply for all geometries and simplifications.

Substituting Eq. (8.1.1) in Eq. (8.1.2) gives

$$\sum_{k=1}^{\infty} [\eta_k L_i(U_{1k},U_{2k},U_{3k}) - \lambda\dot{\eta}_k U_{ik} - \rho h\ddot{\eta}_k U_{ik}] = -q_i \quad (8.1.6)$$

However, from our eigenvalue analysis we know that

$$L_i(U_{1k},U_{2k},U_{3k}) = -\rho h\omega_k^2 U_{ik} \quad (8.1.7)$$

Substituting this in Eq. (8.1.6) gives

$$\sum_{k=1}^{\infty} (\rho h\ddot{\eta}_k + \lambda\dot{\eta}_k + \rho h\omega_k^2\eta_k)U_{ik} = q_i \quad (8.1.8)$$

Since we know that the natural modes U_{ik} are orthogonal, we may proceed as in a Fourier analysis where we take advantage of the orthogonality of the sine and cosine functions. We multiply the equation on both sides by a mode U_{ip}, where p, in general, is either equal to k or not equal:

$$\sum_{k=1}^{\infty} (\rho h\ddot{\eta}_k + \lambda\dot{\eta}_k + \rho h\omega_k^2\eta_k)U_{ik}U_{ip} = q_iU_{ip} \tag{8.1.9}$$

In expanded form this is

$$\sum_{k=1}^{\infty} (\rho h\ddot{\eta}_k + \lambda\dot{\eta}_k + \rho h\omega_k^2\eta_k)U_{1k}U_{1p} = q_1U_{1p} \tag{8.1.10}$$

$$\sum_{k=1}^{\infty} (\rho h\ddot{\eta}_k + \lambda\dot{\eta}_k + \rho h\omega_k^2\eta_k)U_{2k}U_{2p} = q_2U_{2p} \tag{8.1.11}$$

$$\sum_{k=1}^{\infty} (\rho h\ddot{\eta}_k + \lambda\dot{\eta}_k + \rho h\omega_k^2\eta_k)U_{3k}U_{3p} = q_3U_{3p} \tag{8.1.12}$$

Adding Eqs. (8.1.10) through (8.1.12) and integrating over the shell surface gives

$$\sum_{k=1}^{\infty} (\rho h\ddot{\eta}_k + \lambda\dot{\eta}_k + \rho h\omega_k^2\eta_k)\int_{\alpha_2}\int_{\alpha_1} (U_{1k}U_{1p} + U_{2k}U_{2p}$$

$$+ U_{3k}U_{3p})A_1A_2\, d\alpha_1\, d\alpha_2 = \int_{\alpha_2}\int_{\alpha_1} (q_1U_{1p} + q_2U_{2p}$$

$$+ q_3U_{3p})A_1A_2\, d\alpha_1\, d\alpha_2 \tag{8.1.13}$$

Using the orthogonality conditions as defined by Eq. (5.8.22), we are able to remove the summation by realizing that all terms but the term for which $p = k$ vanish. We get

$$\ddot{\eta}_k + \frac{\lambda}{\rho h}\dot{\eta}_k + \omega_k^2\eta_k = F_k \tag{8.1.14}$$

where

$$F_k = \frac{1}{\rho h N_k}\int_{\alpha_2}\int_{\alpha_1} (q_1U_{1k} + q_2U_{2k} + q_3U_{3k})A_1A_2\, d\alpha_1\, d\alpha_2$$

$$N_k = \int_{\alpha_2}\int_{\alpha_1} (U_{1k}^2 + U_{2k}^2 + U_{3k}^2)A_1A_2\, d\alpha_1\, d\alpha_2 \tag{8.1.15}$$

Thus, if we take k terms of the modal expansion series as approximation to an infinite number, we have to solve the equation defining the modal participation factors k times. There is no principal difficulty connected with this. The forcing functions q_1, q_2, and q_n have to be given and the mode components U_{1k}, U_{2k}, and U_{3k} and the natural frequencies ω_k have to be known, either as direct functional or numerical theoretical solutions of the Eigenvalue problem or as experimental data in functional or numerical form. The mass density per unit shell surface ρh is obviously known also and the damping factor λ has to be given or has to be estimated.

8.2 INITIAL CONDITIONS

For the complete solution of Eq. (8.1.14), two initial conditions for each modal participation factor are required. They are the initial displacements $u_i(\alpha_1,\alpha_2,0)$ and the initial velocities $\dot{u}_i(\alpha_1,\alpha_2,0)$. They must be specified for every point of the shell. Initial velocities are in many practical cases zero, except for problems where a periodic switch of boundary conditions occurs.

Transient responses to initial conditions die down as time progresses because of damping, as will be shown. Therefore, when the steady-state solution alone is important, the initial conditions are set to zero.

When knowledge of the transient response is required and initial conditions are specified, we have to convert these initial conditions into initial conditions of the modal participation factor, which are η_k and $\dot{\eta}_k$ at t = 0.

Because of Eq. (8.1.1), we may write

$$u_i(\alpha_1,\alpha_2,0) = \sum_{k=1}^{\infty} \eta_k(0)U_{ik}(\alpha_1,\alpha_2) \tag{8.2.1}$$

$$\dot{u}_1(\alpha_1,\alpha_2,0) = \sum_{k=1}^{\infty} \dot{\eta}_k(0)U_{ik}(\alpha_1,\alpha_2) \tag{8.2.2}$$

These equations have to be solved for $\eta_k(0)$ and $\dot{\eta}_k(0)$. For instance, let us multiply Eq. (8.2.1) by $U_{ip}(\alpha_1,\alpha_2)$, where $p \neq k$ or $p = k$. We get

$$u_i(\alpha_1,\alpha_2,0)U_{ip} = \sum_{k=1}^{\infty} \eta_k(0)U_{ik}U_{ip} \tag{8.2.3}$$

In expanded form, for $i = 1$, 2, and 3, this equation becomes

$$u_1(\alpha_1,\alpha_2,0)U_{1p} = \sum_{k=1}^{\infty} \eta_k(0)U_{1k}U_{1p} \tag{8.2.4}$$

$$u_2(\alpha_1,\alpha_2,0)U_{2p} = \sum_{k=1}^{\infty} \eta_k(0)U_{2k}U_{2p} \tag{8.2.5}$$

$$u_3(\alpha_1,\alpha_2,0)U_{3p} = \sum_{k=1}^{\infty} \eta_k(0)U_{3k}U_{3p} \tag{8.2.6}$$

Summing these equations and integrating over the shell surface gives

$$\int_{\alpha_2}\int_{\alpha_1} [u_1(\alpha_1,\alpha_2,0)U_{1p} + u_2(\alpha_1,\alpha_2,0)U_{2p} + u_3(\alpha_1,\alpha_2,0)U_{3p}]A_1A_2\, d\alpha_1\, d\alpha_2 = \sum_{k=1}^{\infty} \eta_k(0)\int_{\alpha_2}\int_{\alpha_1} (U_{1k}U_{1p} + U_{2k}U_{2p} + U_{3k}U_{3p})A_1A_2\, d\alpha_1\, d\alpha_2 \tag{8.2.7}$$

Evoking the orthogonality condition of Eq. (5.8.22) eliminates the summation since the right side of the equation is zero for any p except $p = k$. We obtain

$$\eta_k(0) = \frac{1}{N_k}\int_{\alpha_2}\int_{\alpha_1} [u_1(\alpha_1,\alpha_2,0)U_{1k} + u_2(\alpha_1,\alpha_2,0)U_{2k} + u_3(\alpha_1,\alpha_2,0)U_{3k}]A_1A_2\, d\alpha_1\, d\alpha_2 \tag{8.2.8}$$

Where N_k is given by Eq. (8.1.15).

Following the same procedure, we solve Eq. (8.2.2) for the second initial condition:

$$\dot{\eta}_k(0) = \frac{1}{N_k}\int_{\alpha_2}\int_{\alpha_1} [\dot{u}_1(\alpha_1,\alpha_2,0)U_{1k} + \dot{u}_2(\alpha_1,\alpha_2,0)U_{2k}$$

$$+ \dot{u}_3(\alpha_1,\alpha_2,0)U_{3k}]A_1A_2\, d\alpha_1\, d\alpha_2 \tag{8.2.9}$$

8.3 SOLUTION OF THE MODAL PARTICIPATION FACTOR EQUATION

The modal participation factor equation is a simple oscillator equation. Thus, we may interpret the forced vibration of shells by considering the shell as composed of simple oscillators, where each oscillator consists of the shell restricted to vibrate in one of its natural modes. All these oscillators respond simultaneously, and the total shell vibration is simply the result of addition (superposition) of all the individual vibrations.

The simple oscillator equation will be solved in the following by the Laplace transformation technique. We will derive the solution for subcritical, critical, and supercritical damping, even though only the first case is of real importance in shell vibration applications.

The modal participation factor equation can be written as

$$\ddot{\eta}_k + 2\zeta_k\omega_k\dot{\eta}_k + \omega_k^2\eta_k = F_k(t) \tag{8.3.1}$$

where

$$F_k(t) = \frac{\int_{\alpha_2}\int_{\alpha_1}(q_1 U_{1k} + q_2 U_{2k} + q_3 U_{3k})A_1A_2\, d\alpha_1\, d\alpha_2}{\rho h N_k} \tag{8.3.2}$$

$$\zeta_k = \frac{\lambda}{2\rho h\omega_k} \tag{8.3.3}$$

Note that ζ_k is called the *modal damping coefficient*. It is analogous to the damping coefficient in the simple oscillator problem.

Taking the Laplace transformation of Eq. (8.3.1) allows us to solve for the modal participation factor in the Laplace domain:

$$\eta_k(s) = \frac{F_k(s) + \eta_k(0)(s + 2\zeta_k\omega_k) + \dot{\eta}_k(0)}{(s + \zeta_k\omega_k)^2 + \omega_k^2(1 - \zeta_k^2)} \tag{8.3.4}$$

The inverse transformation depends on whether the term $1 - \zeta_k^2$ is positive, zero, or negative. The positive case, when $\zeta_k < 1$, is the most common since it is very difficult to dampen shells more than that. It is called the *subcritical case*. The critical case occurs when $\zeta_k = 1$ and has no practical significance other than that it defines the damping that causes an initial modal displacement to decay in the fastest possible time without an oscillation. Supercritical damping ($\zeta_k > 1$) occurs only if a shell has such a high damping that it creeps back from an initial modal displacement without overshooting the equilibrium position.

For the subcritical case ($\zeta_k < 1$) we define a real and positive number γ_k

$$\gamma_k = \omega_k\sqrt{1 - \zeta_k^2} \tag{8.3.5}$$

The inverse Laplace transformation of Eq. (8.3.4) gives then

$$\eta_k(t) = e^{-\zeta_k\omega_k t}\left\{\eta_k(0)\cos\gamma_k t + [\eta_k(0)\zeta_k\omega_k + \dot{\eta}_k(0)]\frac{\sin\gamma_k t}{\gamma_k}\right\} + \frac{1}{\gamma_k}\int_0^t F_k(\tau)e^{-\zeta_k\omega_k(t-\tau)}\sin\gamma_k(t-\tau)\,d\tau \tag{8.3.6}$$

The solution is given in the form of the convolution integral since the forcing function $F_k(t)$ is at this point arbitrary. Once it is known the convolution integral can be evaluated. It is also possible to take the inverse Laplace transformation of Eq. (8.3.4) with a known forcing function directly.

We note that vibrations caused by initial conditions will be oscillatory but will decay exponentially with time. The convolution integral, when evaluated for a specific forcing, will divide into a transient part and possibly a steady-state part if the forcing is periodic. The transient part will decay exponentially with time.

A special case of considerable technical interest is when damping is zero. The solution reduces to

$$\eta_k(t) = \eta_k(0)\cos\omega_k t + \dot{\eta}_k(0)\frac{\sin\omega_k t}{\omega_k} + \frac{1}{\omega_k}\int_0^t F_k(\tau)\sin\omega_k(t-\tau)\,d\tau \qquad (8.3.7)$$

Since most structures are very lightly damped, Eq. (8.3.7) is often used to get an approximate response since it is much simpler to use.

Next, let us investigate the supercritical case ($\zeta_k > 1$). In this case the value of $1 - \zeta_k^2$ is negative. Defining a real and positive number ε_k.

$$\varepsilon_k = \omega_k\sqrt{\zeta_k^2 - 1} \qquad (8.3.8)$$

we obtain, taking the inverse Laplace transformation,

$$\eta_k(t) = e^{-\zeta_k\omega_k t}\left\{\eta_k(0)\cosh\varepsilon_k t + [\eta_k(0)\zeta_k\omega_k + \dot{\eta}_k(0)]\frac{\sinh\varepsilon_k t}{\varepsilon_k}\right\} + \frac{1}{\varepsilon_k}\int_0^t F_k(\tau)e^{-\zeta_k\omega_k(t-\tau)}\sinh\varepsilon_k(t-\tau)\,d\tau \qquad (8.3.9)$$

The vibrations caused by initial conditions are now nonoscillatory. However, if the forcing is periodic, an oscillatory steady-state solution will still result.

As a special case, we obtain the critical damping solution ($\zeta_k = 1$) by reduction:

$$\eta_k(t) = e^{-\omega_k t}\left\{\eta_k(0) + [\eta_k(0)\omega_k + \dot{\eta}_k(0)]t\right\} + \int_0^t F_k(\tau)e^{-\omega_k(t-\tau)}(t-\tau)\,d\tau \qquad (8.3.10)$$

8.4 REDUCED SYSTEMS

In the special case of a plate the problem simplifies to a transverse solution with

$$F_k = \frac{1}{\rho h N_k} \int_{\alpha_2} \int_{\alpha_1} q_3 U_{3k} A_1 A_2 \, d\alpha_1 \, d\alpha_2 \tag{8.4.1}$$

where

$$N_k = \int_{\alpha_2} \int_{\alpha_1} U_{3k}^2 A_1 A_2 \, d\alpha_1 \, d\alpha_2 \tag{8.4.2}$$

and an in-plane solution with

$$F_k = \frac{1}{\rho h N_k} \int_{\alpha_2} \int_{\alpha_1} (q_1 U_{1k} + q_2 U_{2k}) A_1 A_2 \, d\alpha_1 \, d\alpha_2 \tag{8.4.3}$$

where

$$N_k = \int_{\alpha_2} \int_{\alpha_1} (U_{1k}^2 + U_{2k}^2) A_1 A_2 \, d\alpha_1 \, d\alpha_2 \tag{8.4.4}$$

In the special case of a ring we get

$$F_k = \frac{1}{\rho h N_k} \int_{\alpha_2} \int_{\alpha_1} (q_1 U_{1k} + q_3 U_{3k}) A_1 A_2 \, d\alpha_1 \, d\alpha_2 \tag{8.4.5}$$

where

$$N_k = \int_{\alpha_2} \int_{\alpha_1} (U_{1k}^2 + U_{3k}^2) A_1 A_2 \, d\alpha_1 \, d\alpha_2 \tag{8.4.6}$$

For shell approximations, where the transverse modes are considered only, Eqs. (8.4.1) and (8.4.2) apply. Note that this is a good approximation since $|U_{3k}| >> |U_{1k}|, |U_{2k}|$ for transverse motion-dominated modes.

For the transversely vibrating beam

$$F_k = \frac{1}{\rho' N_k} \int_0^L q_3' U_{3k} \, dx \tag{8.4.7}$$

where

$$N_k = \int_0^L U_{3k}^2 \, dx \tag{8.4.8}$$

For the longitudinal vibration of a rod,

$$F_k = \frac{1}{\rho' N_k} \int_0^L q_1' U_{1k} \, dx \tag{8.4.9}$$

where

$$N_k = \int_0^L U_{1k}^2 \, dx \tag{8.4.10}$$

8.5 STEADY-STATE HARMONIC RESPONSE

A technically important case occurs when the load on the shell varies harmonically with time and when the onset of vibrations (the transient part) is of no interest. Using a complex notation to get the response to both sine and cosine loading, we may write the load as ($j = \sqrt{-1}$)

$$q_i(\alpha_1, \alpha_2, t) = q_i^*(\alpha_1, \alpha_2) e^{j\omega t} \tag{8.5.1}$$

We could utilize the convolution integral, but in this case it is simpler to use Eq. (8.3.1) directly. It becomes

$$\ddot{\eta}_k + 2\zeta_k \omega_k \dot{\eta}_k + \omega_k^2 \eta_k = F_k^* \, e^{j\omega t} \tag{8.5.2}$$

where

$$F_k^* = \frac{1}{\rho h N_k} \int_{\alpha_2} \int_{\alpha_1} (q_1^* U_{1k} + q_2^* U_{2k} + q_3^* U_{3k}) A_1 A_2 \, d\alpha_1 \, d\alpha_2 \tag{8.5.3}$$

At steady state, the response will be harmonic also, but lagging behind by a phase angle ϕ_k

$$\eta_k = \Lambda_k e^{j(\omega t - \phi_k)} \tag{8.5.4}$$

Substituting this gives

$$\Lambda_k e^{-j\phi_k} = \frac{F_k^*}{(\omega_k^2 - \omega^2) + 2j\zeta_k\omega_k\omega} \tag{8.5.5}$$

The magnitude of the response is, therfore,

$$\Lambda_k = \frac{F_k^*}{\omega_k^2\sqrt{[1 - (\frac{\omega}{\omega_k})^2]^2 + 4\zeta_k^2(\frac{\omega}{\omega_k})^2}} \tag{8.5.6}$$

The phase lag is

$$\phi_k = \tan^{-1}\frac{2\zeta_k(\frac{\omega}{\omega_k})}{1 - (\frac{\omega}{\omega_k})^2} \tag{8.5.7}$$

As expected, a shell will behave similar to a collection of simple oscillators. Whenever the excitation frequency coincides with one of the natural frequencies, a peak in the response curve will occur.

It has to be noted that the harmonic response solution is the same for subcritical and supercritical damping, except that for equal forcing, the response amplitudes at resonance become less and less pronounced as damping is increased until they are indistinguishable from the off resonance response.

8.6 STEP AND IMPULSE RESPONSE

Often there is a sudden onset at time $t = t_1$ of an otherwise purely static load. This can be expressed by

$$q_i(\alpha_1,\alpha_2,t) = q_i^*(\alpha_1,\alpha_2)U(t - t_1) \tag{8.6.1}$$

where

$$U(t - t_1) = \begin{cases} 0 & t < t_1 \\ 1 & t \geq t_1 \end{cases} \tag{8.6.2}$$

and where F_k^* is given by Eq. (8.5.3).

Neglecting the initial conditions by setting them equal to zero gives, for the subcritical damping case,

$$\eta_k(t) = \frac{F_k^*}{\gamma_k} \int_0^t U(\tau - t_1) e^{-\zeta_k \omega_k (t-\tau)} \sin \gamma_k (t - \tau)\, d\tau \tag{8.6.3}$$

Integrating this gives, for $t \geq t_1$,

$$\eta_k(t) = \frac{F_k^*}{\gamma_k^2} \{1 - \zeta_k^2 - \sqrt{1 - \zeta_k^2}\, e^{-\zeta_k \omega_k (t-t_1)} \cos[\gamma_k(t - t_1) - \phi_k]\} \tag{8.6.4}$$

where

$$\phi_k = \tan^{-1} \frac{\zeta_k}{\sqrt{1 - \zeta_k^2}} \tag{8.6.5}$$

We see that the step response decays exponentially until a static value of F_k^*/ω_k^2 is reached. It can also be seen that the maximum step response approaches, as damping decreases, twice the static value. From this it follows that the sudden application of a static load will produce in the limit twice the stress magnitude that a slow, careful application of the load will produce. This was first established by Krylov in 1898 during his investigations of the bursting strength of canons.

An impulse response occurs when the shell is impacted by a mass. Piston slap on the cylinder liner of a diesel engine is a representative problem of this kind. Impulse loading is a conversion of momentum process. The forcing can be expressed as

$$q_i(\alpha_1, \alpha_2, t) = M_i^*(\alpha_1, \alpha_2)\, \delta(t - t_1) \tag{8.6.6}$$

where M_i^* is a distributed momentum change per unit area, with the units newton-second per square meter (Nsec/m^2). $\delta(t - t_1)$ is the Dirac delta function which defines the occurrence of impact at time $t = t_1$. Its definitions are

$$\delta(t - t_1) = 0 \quad \text{if } t \neq t_1 \tag{8.6.7}$$

$$\int_{t=-\infty}^{t=\infty} \delta(t - t_1)\, dt = 1 \tag{8.6.8}$$

From these definitions we obtain the following integration rule:

$$\int_{t=-\infty}^{t=\infty} F(t)\, \delta(t - t_1)\, dt = F(t_1) \tag{8.6.9}$$

The unit of $\delta(t - t_1)$ is 1/sec.

The value of F_k^* is

$$F_k^* = \frac{1}{\rho h N_k} \int_{\alpha_2}\int_{\alpha_1} (M_1^* U_{1k} + M_2^* U_{2k} + M_3^* U_{3k}) A_1 A_2 \, d\alpha_1 \, d\alpha_2 \tag{8.6.10}$$

Substituting this in Eq. (8.3.6) gives, for subcritical damping and zero initial conditions,

$$\eta_k(t) = \frac{F_k^*}{\gamma_k} \int_0^t \delta(\tau - t_1) e^{-\zeta_k \omega_k (t-\tau)} \sin \gamma_k (t - \tau)\, d\tau \tag{8.6.11}$$

Applying the integration rule of Eq. (8.6.9) gives, for $t \geq t_1$,

$$\eta_k(t) = \frac{F_k^*}{\gamma_k} e^{-\zeta_k \omega_k (t-t_1)} \sin \gamma_k (t - t_1) \tag{8.6.12}$$

This result shows that no matter what the spatial distribution of the impulsive load is, the response for each mode decays exponentially to zero and the oscillation is sinusoidal at the associated natural frequency.

8.7 INFLUENCE OF LOAD DISTRIBUTION

Among the many possible load distributions there are a few that occur frequently in engineering applications. It is worth while to single these out for detailed discussion.

The most common is a spatially uniform pressure load normal to the surface

$$q_1^* = 0 \quad q_2^* = 0 \quad q_3^* = p_3 \tag{8.7.1}$$

where p_3 is the uniform pressure amplitude. In this case F_k^* of Eq. (8.5.3) becomes

$$F_k^* = \frac{P_3}{\rho h N_k} \int_{\alpha_1} \int_{\alpha_2} U_{3k} A_1 A_2 \, d\alpha_1 \, d\alpha_2 \tag{8.7.2}$$

The interesting feature is that if the shell, plate, ring, beam, etc., has modes that are symmetric or skew-symmetric about a line or lines of symmetry, none of the skew-symmetric modes are excited because they integrate out. For example, for a simply supported rectangular plate as shown in Fig. 8.7.1, we get

$$F_k^* = \frac{4p_3}{\rho h \, mn^2} (1 - \cos m\pi)(1 - \cos n\pi) \tag{8.7.3}$$

For instance, for the special case where the plate is loaded in time by a step function, the solution is

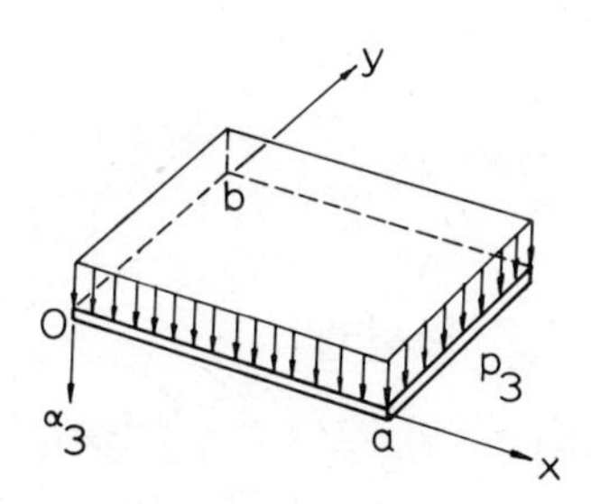

Figure 8.7.1

$$u_3(x,y,t) = \sum_{m=1}^{\infty} \sum_{n=1}^{\infty} \eta_{mn} U_{3mn} \tag{8.7.4}$$

where

$$U_{3mn} = \sin \frac{m\pi x}{a} \sin \frac{n\pi y}{b} \tag{8.7.5}$$

and where $\eta_{mn} = \eta_k$ is given by Eq. (8.6.4).

In this case only modes that are combinations of $m = 1, 3, 5, \ldots$ and $n = 1, 3, 5, \ldots$ participate in the response. Similarly, we can show that for a uniformly loaded axisymmetric circular plate only the axisymmetric modes are excited (the $n = 0$ modes).

Next, let us look at loads that are uniform along one coordinate only, let us say α_2 for the purpose of this discussion.

$$q_3^* = p_3(\alpha_1) \tag{8.7.6}$$

In this case we get

$$F_k^* = \frac{1}{\rho h N_k} \int_{\alpha_1} p_3(\alpha_1) \int_{\alpha_2} U_{3k} A_1 A_2 \, d\alpha_1 \, d\alpha_2 \tag{8.7.7}$$

For example, for the simply supported plate that has a pressure amplitude distribution as shown in Fig. 8.7.2, we have $\alpha_1 = x$, $\alpha_2 = y$ and

$$P_3 = P \frac{x}{a} \tag{8.7.8}$$

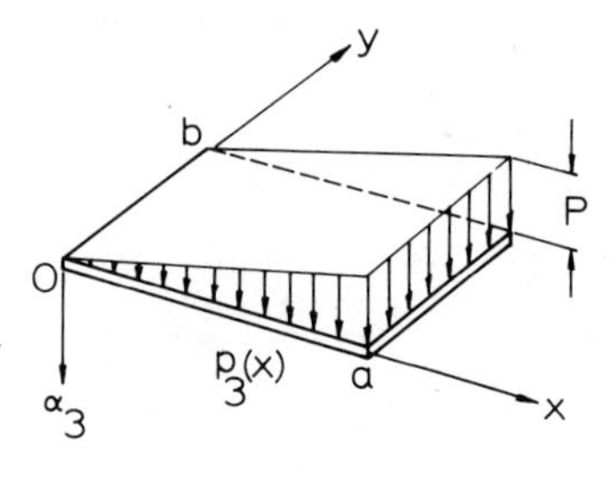

Figure 8.7.2

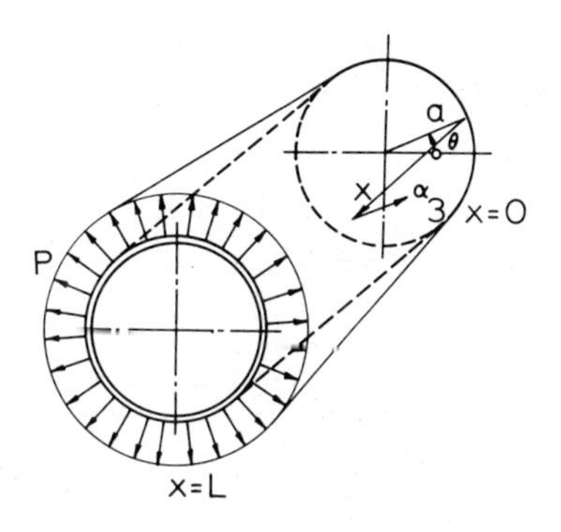

Figure 8.7.3

This gives

$$F_k^* = -\frac{4P}{\rho h\ mn\pi^2} \cos m\pi(1 - \cos n\pi) \tag{8.7.9}$$

This shows that modes that have n = 2, 4, 6 ... do not participate in the solution. But any m number will.

Another example is the cylindrical duct as shown in Fig. 8.7.3. Let the pressure distribution be axisymmetric and linearily varying with x:

$$q_3^* = P\frac{x}{L} \tag{8.7.10}$$

and the duct be a simply supported cylindrical shell of a transverse mode shape:

$$U_{3k} = \sin\frac{m\pi x}{L} \cos n(\theta - \phi) \tag{8.7.11}$$

We obtain, with $\alpha_1 = x$, $A_1 = 1$, $\alpha_2 = \theta$, $A_2 = a$, for $n = 0$

$$F_k = -\frac{2P}{\rho h m\pi} \cos m\pi \tag{8.7.12}$$

None of the modes n = 1, 2, 3, ... exist in the response. Let us take the example where the pressure distribution varies harmonically in time. The solution is then, in steady state,

$$u_3(x,\theta,t) = \sum_{m=1}^{\infty} \eta_{mo} \sin\frac{m\pi x}{L} \tag{8.7.13}$$

where $\eta_{mo} = \eta_k$ is given by Eq. (8.5.4).

It is not necessary to recognize the elimination of certain modes in advance. Including all the modes in a computer program will certainly give the correct response because the program will automatically integrate out the nonparticipating modes. The value of considering symmetry conditions in advance is more an economical one. If modes have to be generated experimentally, for instance, it will save experimental time if it is recognized that only the symmetric modes have to be excited and measured.

There is a corollary to this since we can show equally well skew-symmetric load distributions will not excite symmetric modes. Again, this is useful for the experimenter to know.

All of this assumes, of course, that the plate or shell is perfect in its symmetry and that the load does not deviate from symmetry. Since this is never exactly true for engineering applications, we find that in actual engineering systems modes other than the theoretically predicted ones will also be present, but with much reduced magnitudes. Such a tendency for modes to be present, when they should not, is especially pronounced in cylindrical ducts where the excitation pressure may be axisymmetric for all practical purposes yet other than the $n = 0$ type modes are excited. The reason here is that much less energy is required to excite to an equal amplitude modes with higher n numbers than the $n = 0$ modes because the lowest natural frequencies occur at the higher n numbers, as we have seen in Chapter 5. Thus, a slight imperfection in either pressure distribution or shell construction will be enough to bring the $n = 1, 2, \ldots$ modes into the measured response. All that the analyst can do is to allow for a small deviation of axisymmetry in his model to allow for the imperfections of manufacturing.

8.8 POINT LOADS

A type of load that is very common in engineering applications is the point load. In the immediate vicinity of the point load, some of the basic assumptions of thin shell theory are violated, for instance, that $\sigma_{33} = 0$. However, outside of the immediate vicinity of the point load, the assumptions are not affected and overall vibration responses can be calculated with excellent accuracy.

Since we do not expect to acquire exact results right under the point load anyway, keeping in mind that these results would require three-dimensional elasticity and possibly plasticity analysis and a detailed knowledge of the load application mechanism, it seems reasonable to use the Dirac delta function [8.5, 8.6] to define the point load. This function locates the load at the desired point and assures that it is of the desired magnitude, but does not define the actual mechanism and micro distribution of application.

The point load $P_i S(t)$ in newtons (N) can be expressed as a distributed load q_i in newtons per square meter (N/m^2) by

$$q_i = P_i \frac{1}{A_1 A_2} \delta(\alpha_1 - \alpha_1^*)\, \delta(\alpha_2 - \alpha_2^*) S(t) \tag{8.8.1}$$

and where α_1^* and α_2^* define the location of the point force. The function $S(t)$ represents the time dependency. For harmonic forcing, it is $e^{j\omega t}$ and for a step loading $U(t - t_1)$.

This allows us to write Eq. (8.3.1) as

$$\ddot{\eta}_k + 2\zeta_k \omega_k \dot{\eta}_k + \omega_k^2 \eta_k = F_k^* S(t) \tag{8.8.2}$$

where

$$F_k^* = \frac{1}{\rho h N_k} \int_{\alpha_2} \int_{\alpha_1} [P_1 U_{1k}(\alpha_1, \alpha_2) + P_2 U_{2k}(\alpha_1, \alpha_2) + P_3 U_{3k}(\alpha_1, \alpha_2)]\, \delta(\alpha_1 - \alpha_1^*)\, \delta(\alpha_2 - \alpha_2^*)\, d\alpha_1\, d\alpha_2 \tag{8.8.3}$$

Applying the integration rule gives

$$F_k^* = \frac{1}{\rho h N_k} [P_1 U_{1k}(\alpha_1^*, \alpha_2^*) + P_2 U_{2k}(\alpha_1^*, \alpha_2^*) + P_3 U_{3k}(\alpha_1^*, \alpha_2^*)] \tag{8.8.4}$$

This can now, for instance, be substituted in the specific modal participation factor solutions like Eq. (8.5.4) or Eq. (8.6.4). It

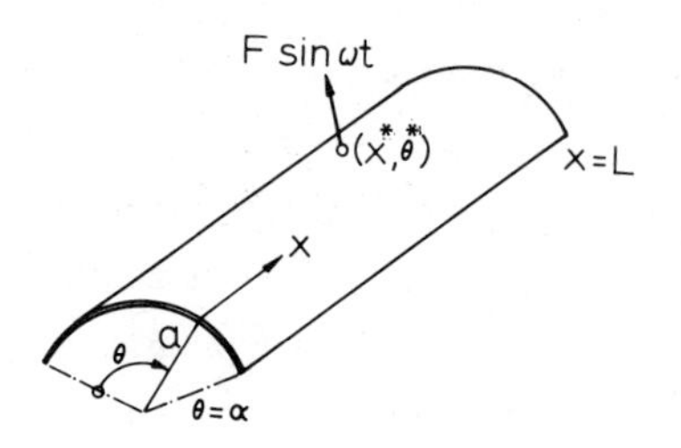

Figure 8.8.1

points out that if the point force is located on a node line of a mode component, this particular mode component will not participate in the response because $U_{ik}(\alpha_1^*,\alpha_2^*)$ is zero if (α_1^*,α_2^*) is on a node line.

Let us take as an example a harmonically varying point force acting on a circular cylindrical shell panel as shown in Fig. 8.8.1. In this case

$$P_1 = 0 \quad P_2 = 0 \quad P_3 = F \tag{8.8.5}$$

and
$$U_{3mn} = \sin\frac{m\pi x}{L}\sin\frac{n\pi\theta}{\alpha} \tag{8.8.6}$$

Thus

$$F_k^* = \frac{4F}{\rho h L\alpha}\sin\frac{m\pi x^*}{L}\sin\frac{n\pi\theta^*}{\alpha} \tag{8.8.7}$$

The value of $\eta_k = \eta_{mn}$ is given by Eq. (8.5.4) and the transverse deflection solution is

$$u_3 = \sum_{m=1}^{\infty}\sum_{n=1}^{\infty}\eta_{mn}\sin\frac{m\pi x}{L}\sin\frac{n\pi\theta}{\alpha} \tag{8.8.8}$$

An interesting special case occurs when a point load acts on a shell of revolution that is closed in the θ direction. The transverse mode components, for example, are of the general form

$$U_{3mn}(\phi,\theta) = H_{3m}(\phi)\cos n(\theta - \eta) \tag{8.8.9}$$

where η is an arbitrary angle. We have chosen η instead of the customary ϕ of Chap. 5 because ϕ is already used as a coordinate.

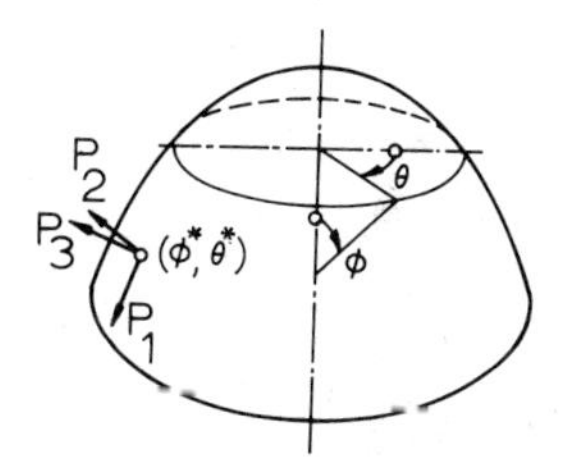

Figure 8.8.2

In order to express any shape in the θ direction, we need two orthogonal components. These we get if we let one time $\eta = 0$ and the other $\eta = \frac{\pi}{2n}$. This gives a set of modes

$$U_{3mn1}(\phi,\theta) = H_{3m}(\phi) \sin n\theta \tag{8.8.10}$$

and a second set at

$$U_{3mn2}(\phi,\theta) = H_{3m}(\phi) \cos n\theta \tag{8.8.11}$$

For a point load acting at (ϕ^*,θ^*), as shown in Fig. 8.8.2, we obtain from Eq. (8.8.4) for the first set

$$F_{mn1} = F^*_{mn1} S(t) \tag{8.8.12}$$

where

$$F^*_{mn1} = C_m(\phi^*) \sin n\theta^* \tag{8.8.13}$$

and where, for transverse loading only,

$$C_m(\phi^*) = \frac{1}{\rho h N_{mn}} P_3 H_{3m}(\phi^*) \tag{8.8.14}$$

and where, for dominating transverse mode components,

$$N_{mn} = \pi \int_{\phi} H^2_{3m}(\phi) A_\phi A_\theta \, d\phi \tag{8.8.15}$$

For the second set we get

$$F_{mn2} = F^*_{mn2} S(t) \tag{8.8.16}$$

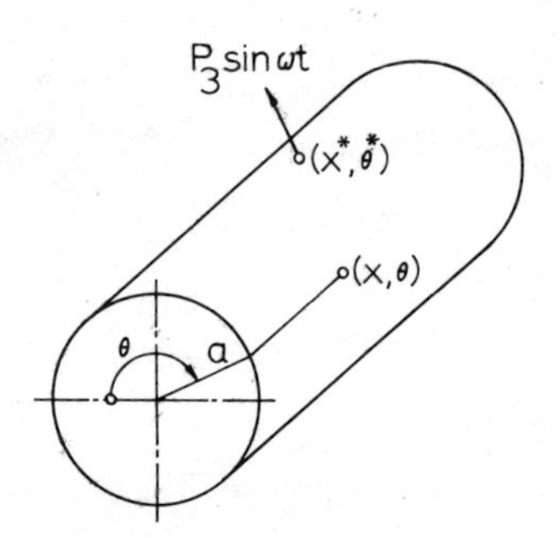

Figure 8.8.3

where

$$F^*_{mn2} = C_m(\phi^*) \cos n\theta^* \tag{8.8.17}$$

The modal participation factor solutions are therefore

$$\eta_{mn1} = T_{mn}(\phi^*,t) \sin n\theta^* \tag{8.8.18}$$

$$\eta_{mn2} = T_{mn}(\phi^*,t) \cos n\theta^* \tag{8.8.19}$$

where

$$T_{mn}(\phi^*,t) = \frac{C_m(\phi^*)}{\gamma_{mn}} \int_0^t S(\tau)e^{-\zeta_{mn}\omega_{mn}(t-\tau)} \sin \gamma_{mn}(t - \tau)\, d\tau \tag{8.8.20}$$

The total solution is, therefore, by superposition

$$u_3(\phi,\theta,t) = \sum_{m=1}^{\infty} \sum_{n=0}^{\infty} T_{mn}(\phi^*,t)\, H_{3m}(\phi)\, [\sin n\theta \sin n\theta^* + \cos n\theta \cos n\theta^*] \tag{8.8.21}$$

However, since the bracketed quantity is equal to $\cos n(\theta - \theta^*)$ we get

$$u_3(\phi,\theta,t) = \sum_{m=1}^{\infty} \sum_{n=0}^{\infty} T_{mn}(\phi^*,t)\, H_{3m}(\phi) \cos n(\theta - \theta^*) \tag{8.8.22}$$

This is an interesting result since it proves that each mode will orient itself such that its maximum deflection occurs at $\theta = \theta^*$. This example illustrates also that for closed shells of revolution the $\cos n(\theta - \eta)$ term has to be thought of as representing the two orthogonal terms $\sin n\theta$ and $\cos n\theta$.

Let us take as specific example the simply supported cylindrical shell, with loading as shown in Fig. 8.8.3, again in the transverse mode approximation. We let $A_\phi \, d\phi = dx$, $A_\theta = a$. Since

$$H_{3m}(x) = \sin \frac{m\pi x}{L} \tag{8.8.23}$$

we obtain

$$N_{mn} = \frac{\pi a L \varepsilon_n}{2} \tag{8.8.24}$$

where

$$\varepsilon_n = \begin{cases} 1 & n \neq 0 \\ 2 & n = 0 \end{cases}$$

and thus

$$C_m(x^*) = \frac{2P_3}{\rho h a L \pi \varepsilon_n} \sin \frac{m\pi x^*}{L} \tag{8.8.25}$$

Since in this example

$$S(t) = \sin \omega t \tag{8.8.26}$$

we get, for stead state,

$$T_{mn}(t,x^*) = \frac{C_m(x^*) \sin(\omega t - \phi_{mn})}{\omega_{mn}^2 \sqrt{[1 - (\frac{\omega}{\omega_{mn}})^2]^2 + 4\zeta_{mn}^2 (\frac{\omega}{\omega_{mn}})^2}} \tag{8.8.27}$$

where

$$\phi_{mn} = \tan^{-1} 2\zeta_{mn} (\frac{\omega}{\omega_{mn}}) / [1 - (\frac{\omega}{\omega_{mn}})^2] \tag{8.8.28}$$

The total solution is, therefore,

$$u_3(x,\theta,t) = \frac{2P_3}{\rho haL\pi} \sum_{m=1}^{\infty} \sum_{n=0}^{\infty} \frac{\sin \frac{m\pi x^*}{L} \sin \frac{m\pi x}{L} \cos n(\theta - \theta^*)}{\varepsilon_n \omega_{mn}^2 \sqrt{[1 - (\frac{\omega}{\omega_{mn}})^2]^2 + 4\zeta_{mn}^2 (\frac{\omega}{\omega_{mn}})^2}}$$

$$\sin(\omega t - \phi_{mn}) \tag{8.8.29}$$

8.9 LINE LOADS

Another type of loading that is relatively important in engineering is the line load. In the general discussion we will confine ourselves to line loads that occur along coordinate lines. This restriction allows us to utilize the Dirac delta function. If there is a line load along the α_1 coordinate at $\alpha_2 = \alpha_2^*$ of amplitude $Q_i^*(\alpha_1)$ in newtons per meter (N/m), as shown in Fig. 8.9.1, we may express it as

$$q_i = q_i^* \, S(t) \tag{8.9.1}$$

where

$$q_i^* = Q_i^* \frac{1}{A_2} \delta(\alpha_2 - \alpha_2^*) \tag{8.9.2}$$

Therefore, Eq. (8.5.3) becomes

$$F_k^* = \frac{1}{\rho h N_k} \int_{\alpha_1} [Q_1^*(\alpha_1) U_{1k}(\alpha_1,\alpha_2^*) + Q_2^*(\alpha_1) U_{2k}(\alpha_1,\alpha_2^*)$$

$$+ Q_3^*(\alpha_1) U_{3k}(\alpha_1,\alpha_2^*)] A_1 \, d\alpha_1 \tag{8.9.3}$$

Similarly, a line load along an α_2 coordinate can be treated.

As example, let us investigate a line load on a circular cylindrical panel as shown in Fig. 8.9.2. In this case

$$Q_1^*(\alpha_1) = 0 \qquad Q_2^*(\alpha_1) = 0 \qquad Q_3^*(\alpha_1) = Q \tag{8.9.4}$$

and $$U_{3k} = \sin \frac{m\pi x}{L} \sin \frac{n\pi\theta}{\beta} \tag{8.9.5}$$

This gives

$$N_k = \frac{\beta a L}{4} \tag{8.9.6}$$

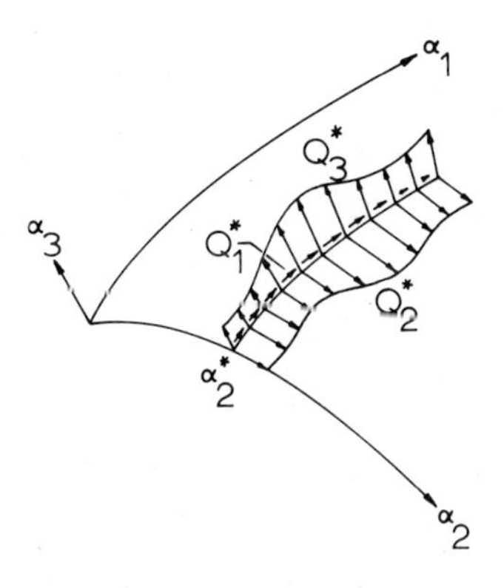

Figure 8.9.1

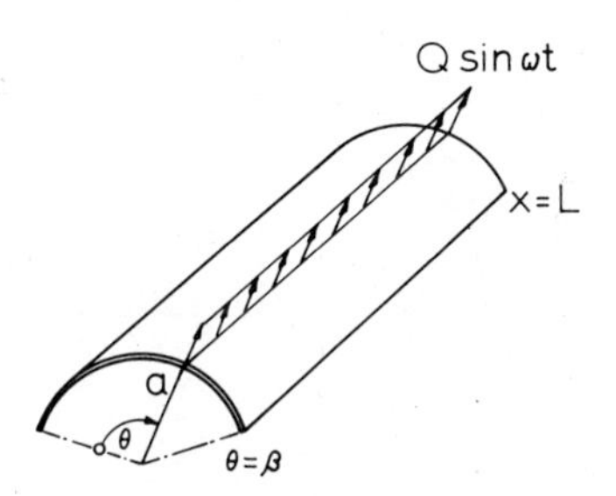

Figure 8.9.2

where we have again neglected the U_{1k} and U_{2k} contributions to N_k since they are small. This gives, therefore,

$$F_k^* = \frac{4Q}{\rho h\beta m\pi a}(1 - \cos m\pi)\sin\frac{n\pi\theta^*}{\beta} \tag{8.9.7}$$

and the rest of the solution follows.

To illustrate, why we have restricted the line load discussion to loads that are distributed along coordinate lines, let us look at the example of a simply supported rectangular plate with a transverse diagonal line load of constant magnitude as shown in Fig. 8.9.3. The load expression is in this case

$$q_3 = q_3^* S(t) \tag{8.9.8}$$

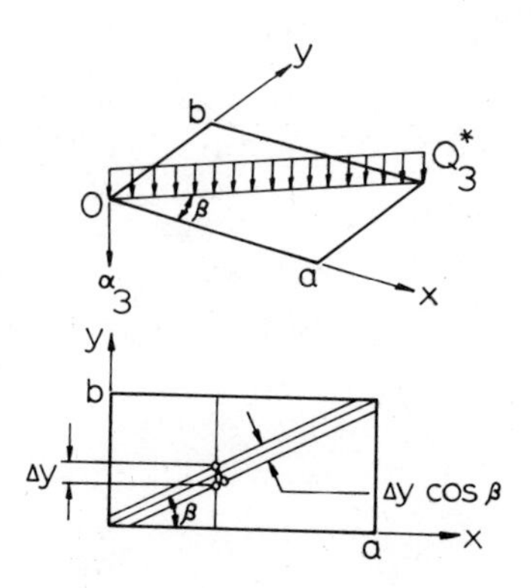

Figure 8.9.3

where

$$q_3^* = \frac{Q_3^*}{\cos\beta}\,\delta\!\left(y - \frac{b}{a}x\right) \tag{8.9.9}$$

and where

$$\cos\beta = \frac{a}{\sqrt{a^2 + b^2}} \tag{8.9.10}$$

The reason for the cos β term is that the line load Q_3^* is not distributed over a strip of width Δy, with Δy → 0, but over a narrower strip Δy cos β, with Δy → 0. This means that

$$q_3^* = \lim_{\Delta y \to 0} \frac{Q_3^*}{\Delta y \cos\beta}\left[U\!\left(y - \frac{b}{a}x\right) - U\!\left(y - \frac{b}{a}x - \Delta y\right)\right] \tag{8.9.11}$$

From this Eq. (8.9.9) results.

A more general description for loads that act along curved lines can be found in Ref. 8.7.

To finish this particular individual case, let us substitute the load description into the expression for F_k^*. This gives

$$F_k^* = \frac{Q_3^*}{\rho h N_k}\,\frac{\sqrt{a^2 + b^2}}{a}\int_0^a\int_0^b \sin\frac{m\pi x}{a}\,\sin\frac{n\pi y}{b}\,\delta\!\left(y - \frac{b}{a}x\right)dx\,dy \tag{8.9.12}$$

or

$$F_k^* = \frac{Q_3^*}{2\rho h N_k} \sqrt{a^2 + b^2}\ \delta_{mn} \tag{8.9.13}$$

where

$$\delta_{mn} = \begin{cases} 0 & \text{if} \quad m \neq n \\ 1 & \text{if} \quad m = n \end{cases} \tag{8.9.14}$$

This indicates that in this special case only the modes where $m = n$ are excited.

8.10 POINT IMPACT

Another problem of general engineering is the point impact. In contrast to the distributed impulse load treated in general in Sec. 8.6, point impact can be described by an impulse concentrated at a point, as it occurs, for example, when a shell is struck by a projectile or hammer. Piston slap in combustion engines is of this type.

The description of Eq. (8.6.6) suffices, which is

$$q_i(\alpha_1,\alpha_2,t) = M_i^*(\alpha_1,\alpha_2)\ \delta(t - t_1) \tag{8.10.1}$$

Except that we write in this case

$$M_i^*(\alpha_1,\alpha_2) = \frac{M_i}{A_1 A_2}\ \delta(\alpha_1 - \alpha_1^*)\ \delta(\alpha_2 - \alpha_2^*) \tag{8.10.2}$$

where M_1 is the momentum that is transferred to the shell by the impact. In this case Eq. (8.6.11) becomes

$$F_k^* = \frac{1}{\rho h N_k} [M_1 U_{1k}(\alpha_1^*,\alpha_2^*) + M_2 U_{2k}(\alpha_1^*,\alpha_2^*) + M_3 U_{3k}(\alpha_1^*,\alpha_2^*)] \tag{8.10.3}$$

As example, let us solve the problem where a mass m of velocity v impacts a spherical shell as shown in Fig. 8.10.1. In this case $\alpha_1 = \phi$, $A_1 = a$, $\alpha_2 = \theta$, $A_2 = a \sin \phi$, $\alpha_1^* = \phi_1^* = 0$. The mode shapes that have to be considered are given in Sec. 6.2:

$$U_{\phi n} = \frac{d}{d\phi} P_n(\cos \phi) \tag{8.10.4}$$

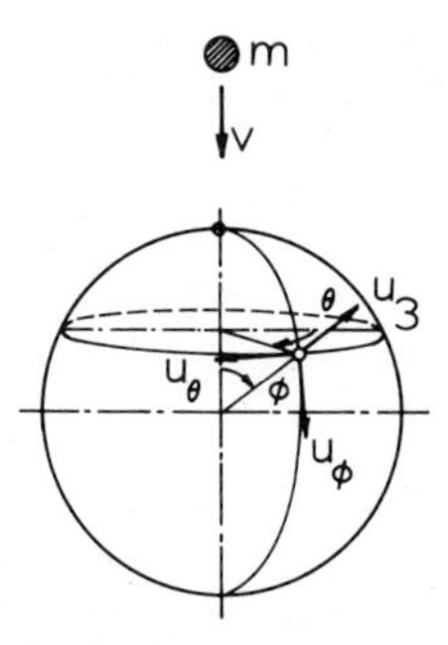

Figure 8.10.1

$$U_{3n} = \frac{1 + (1 + \mu)\Omega_n^2}{1 - \Omega_n^2} P_n(\cos \phi) \tag{8.10.5}$$

We assume that the momentum of the projectile,

$$M_3 = -mv \quad M_1 = M_2 = 0 \tag{8.10.6}$$

is completely transfered to the shell in such a brief time span that the process can be described by the Dirac delta function. Equation (8.10.3) applies, therefore, and we get (k = n)

$$F_n^* = \frac{-mv}{\rho h N_n} \frac{1 + (1 + \mu)\Omega_n^2}{1 - \Omega_n^2} P_n(1) \tag{8.10.7}$$

where

$$N_n = 2a^2\pi \int_0^\pi \left\{ \left[\frac{d}{d\phi} P_n(\cos \phi) \right]^2 + \left[\frac{1 + (1 + \mu)\Omega_n^2}{1 - \Omega_n^2} P_n(\cos \phi) \right]^2 \right\} \sin \phi \, d\phi \tag{8.10.8}$$

From Eq. (8.6.12) we obtain η_n and the total solution is, since $P_n(1) = 1$,

$$\begin{Bmatrix} u_\phi(\phi,\theta,t) \\ u_3(\phi,\theta,t) \end{Bmatrix} = - \sum_{n=0}^{\infty} \frac{1 + (1+\mu)\Omega_n^2}{1 - \Omega_n^2} \frac{mv}{\rho h \gamma_n N_n} e^{-\zeta_n \omega_n t}$$

$$\sin \gamma_n t \begin{Bmatrix} \frac{d}{d\phi} P_n(\cos\phi) \\ \frac{1 + (1+\mu)\Omega_n^2}{1 - \Omega_n^2} P_n(\cos\phi) \end{Bmatrix} \tag{8.10.9}$$

Note that the n = 1 term describes a rigid-body translation. Since $\omega_1 = \gamma_1 = \Omega_1 = 0$ for the transversely dominated mode and since

$$\lim_{\gamma_1 \to 0} \frac{\sin \gamma_1 t}{\gamma_1} = t \tag{8.10.10}$$

we obtain, at $\phi = 0$ and for n = 1,

$$u_3(0,\theta,t) = - \frac{mv}{\rho h N_1} t \tag{8.10.11}$$

Furthermore, for n = 1

$$N_1 = 2\pi a^2 \int_0^{\pi} (\sin^2\phi + \cos^2\phi) \sin\phi \, d\phi = 4\pi a^2 \tag{8.10.12}$$

This gives, taking the time derivative,

$$\dot{u}_3(0,\theta,t) = - \frac{mv}{4\pi \rho h a^2} \tag{8.10.13}$$

This is the velocity with which the spherical shell as a whole moves away from the impact. The minus sign gives the direction of the motion at $\phi = 0$, namely downward in Fig. 8.10.1.

Note that if the spherical shell would be treated as an elastic body undergoing impact by the classical impact theory, we obtain the same result as given by Eq. (8.10.13), but are unable, of course, to say anything about the resulting vibration.

REFERENCES

8.1 D. Bernoulli, *Reflections et Eclaircissements sur les Nouvelles Vibrations des Cordes*, Royal Academy, Berlin, 1755.

8.2 S. Timoshenko, *Vibration Problems in Engineering*, Van Nostrand, Princeton, N.J., 1955.

8.3 L. Meirovitch, *Analytical Methods in Vibrations*, Macmillan, London, 1967.

8.4 W. Nowacki, *Dynamics of Elastic Systems*, John Wiley, New York, 1963.

8.5 P. A. M. Dirac, "The physical interpretation of quantum mechanics," *Proc. Roy. Soc. London (ser. A), vol. 113*, 1926, p. 621.

8.6 L. Schwartz, "Theorie des distributions," Actualites Scientifiques Industrielles, no. 1091, 1950, and no. 1122, 1951.

8.7 W. Soedel and D. P. Powder, "A general dirac delta function method for calculating the vibration response of plates to loads along arbitrarily curved lines," *J. Sound Vibration, vol. 65*, no. 1, 1979, pp. 29-35.

8.8 M. M. Stanisic, "Dynamic response of a diagonal line-loaded rectangular plate," *Amer. Inst. Aeronautics Astronautics J., vol. 15*, no. 12, 1977, pp. 1804-1807.

9

THE DYNAMIC INFLUENCE (GREEN'S) FUNCTION

The dynamic influence function of a shell describes the response of each point of the shell to a unit impulse applied at some other point. For simple structures, like transversely vibrating beams, the influence function may be unidirectional. That is, the unit impulse is applied in the transverse direction and the response is in the transverse direction. Also for plates, where the in-plane response is uncoupled from the transverse response for small oscillations, a unidirectional dynamic influence function is applicable to the transverse vibration problem. However, in shell dynamics, coupling between the transverse response and the response in planes tangential to the shell surface has to be considered. Thus, a unit impulse applied transversely at a point produces a response in two principal tangential directions as well as in the transverse direction at any point of the shell. The same is true for unit impulses applied tangentially to the shell in the two principal directions. Thus, to be complete, the dynamic influence function for the general shell case has to have nine components. It can be viewed as a field of response vectors due to unit impulse vectors applied at each point of the shell.

Note that the dynamic influence function is a Green's function and is, therefore, often referred to as the dynamic Green's function of the shell. The following development follows the approach taken in Ref. 9.1.

9.1 FORMULATION OF THE INFLUENCE FUNCTION

A unit impulse, applied at location (α_1^*, α_2^*) at time t* in the direc- of α_1, may be expressed as

$$q_1(\alpha_1,\alpha_2,t) = \frac{1}{A_1A_2}\,\delta(\alpha_1 - \alpha_1^*)\,\delta(\alpha_2 - \alpha_2^*)\,\delta(t - t^*) \tag{9.1.1}$$

$$q_2(\alpha_1,\alpha_2,t) = 0 \tag{9.1.2}$$

$$q_3(\alpha_1,\alpha_2,t) = 0 \tag{9.1.3}$$

This will produce a displacement response of the shell with the three components

$$u_1(\alpha_1,\alpha_2,t) = G_{11}(\alpha_1,\alpha_2,t;\ \alpha_1^*,\alpha_2^*,t^*) \tag{9.1.4}$$

$$u_2(\alpha_1,\alpha_2,t) = G_{21}(\alpha_1,\alpha_2,t;\ \alpha_1^*,\alpha_2^*,t^*) \tag{9.1.5}$$

$$u_3(\alpha_1,\alpha_2,t) = G_{31}(\alpha_1,\alpha_2,t;\ \alpha_1^*,\alpha_2^*,t^*) \tag{9.1.6}$$

The symbol G_{ij} signifies *Green's function* and represents the response in the i direction at α_1, α_2, t to a unit impulse in the j direction at α_1^*, α_2^*, t*. Equations (5.1.1) to (5.1.3) become

$$L_1(G_{11},G_{21},G_{31}) - \lambda\dot{G}_{11} - \rho h\ddot{G}_{11} = -\frac{1}{A_1A_2}\,\delta(\alpha_1 - \alpha_1^*)\,\delta(\alpha_2 - \alpha_2^*)\,\delta(t - t^*) \tag{9.1.7}$$

$$L_2(G_{11},G_{21},G_{31}) - \lambda\dot{G}_{21} - \rho h\ddot{G}_{21} = 0 \tag{9.1.8}$$

$$L_3(G_{11},G_{21},G_{31}) - \lambda\dot{G}_{31} - \rho h\ddot{G}_{31} = 0 \tag{9.1.9}$$

Next, applying a unit impulse to the shell at location (α_1^*,α_2^*) at time t* in the α_2 direction, we obtain G_{12}, G_{22}, G_{32} from the equations

$$L_1(G_{12},G_{22},G_{32}) - \lambda\dot{G}_{12} - \rho h\ddot{G}_{12} = 0 \tag{9.1.10}$$

$$L_2(G_{12},G_{22},G_{32}) - \lambda\dot{G}_{22} - \rho h\ddot{G}_{22} =$$

$$-\frac{1}{A_1A_2}\,\delta(\alpha_1 - \alpha_1^*)\,\delta(\alpha_2 - \alpha_2^*)\,\delta(t - t^*) \tag{9.1.11}$$

$$L_3(G_{12},G_{22},G_{32}) - \lambda\dot{G}_{32} - \rho h\ddot{G}_{32} = 0 \tag{9.1.12}$$

Similarly, we may formulate the equations for G_{13}, G_{23}, G_{33}. All nine equations may be written in short notation:

$$L_i(G_{1j},G_{2j},G_{3j}) - \lambda\dot{G}_{ij} - \rho h\ddot{G}_{ij} =$$

$$-\frac{\delta_{ij}}{A_1A_2}\,\delta(\alpha_1 - \alpha_1^*)\,\delta(\alpha_2 - \alpha_2^*)\,\delta(t - t^*) \tag{9.1.13}$$

where i,j = 1, 2, 3 and where

$$\delta_{ij} = \begin{cases} 1 & i=j \\ 0 & i\neq j \end{cases} \tag{9.1.14}$$

Assuming that it is always possible to obtain the natural frequencies and modes of any shell, plate, etc., we may use modal expansion analysis to find the G_{ij} components. For instance, the solution of Eqs. (9.1.7) to (9.1.9) is

$$G_{i1}(\alpha_1,\alpha_2,t;\alpha_1^*,\alpha_2^*,t^*) = \sum_{k=1}^{\infty} \eta_{1k}(t;\alpha_1^*,\alpha_2^*,t^*)\,U_{ik}(\alpha_1,\alpha_2) \tag{9.1.15}$$

Similarly we formulate G_{i2} and G_{i3}. In general, the solution to Eq. (9.1.13) is

$$G_{ij}(\alpha_1,\alpha_2,t;\alpha_1^*,\alpha_2^*,t^*) = \sum_{k=1}^{\infty} \eta_{jk}(t;\alpha_1^*,\alpha_2^*,t^*)\,U_{ik}(\alpha_1,\alpha_2) \tag{9.1.16}$$

where i,j = 1, 2, 3. The term η_{jk} is the modal participation factor of the k-th mode due to a unit impulse in the j direction.

Note that all influence function components will automatically satisfy all boundary conditions that the natural modes satisfy.

Substituting Eq. (9.1.16) in Eq. (9.1.13) and proceeding in the usual way gives

$$\ddot{\eta}_{jk} + 2\zeta_k\omega_k\dot{\eta}_{jk} + \omega_k^2\eta_{jk} = F_{jk}(t;\alpha_1^*,\alpha_2^*,t^*) \tag{9.1.17}$$

where

$$F_{jk}(t;\alpha_1^*,\alpha_2^*,t^*) = \frac{1}{\rho h N_k} U_{jk}(\alpha_1^*,\alpha_2^*)\ \delta(t - t^*) \tag{9.1.18}$$

and where N_k is given by Eq. (8.1.15). Confining ourselves to the subcritical damping case in the following, we solve for η_{jk} and obtain, for $t \geq t^*$,

$$\eta_{jk} = \frac{U(t - t^*)}{\rho h N_k \gamma_k} U_{jk}(\alpha_1^*,\alpha_2^*)e^{-\zeta_k\omega_k(t-t^*)} \sin \gamma_k(t - t^*) \tag{9.1.19}$$

Note that $\eta_{jk} = 0$ when $t < t^*$. To keep track of this causality the unit step function $U(t - t^*)$ was introduced. The dynamic influence function is obtained by substituting Eq. (9.1.19) in Eq. (9.1.16).

$$G_{ij}(\alpha_1,\alpha_2,t;\alpha_1^*,\alpha_2^*,t^*) = \frac{1}{\rho h}\sum_{k=1}^{\infty} \frac{U_{ik}(\alpha_1,\alpha_2)U_{jk}(\alpha_1^*,\alpha_2^*)S(t - t^*)}{N_k} \tag{9.1.20}$$

where, for subcritical damping,

$$S(t-t^*) = \frac{1}{\gamma_k} e^{-\zeta_k\omega_k(t-t^*)} \sin \gamma_k(t - t^*)U(t - t^*) \tag{9.1.21}$$

Equation (9.1.20) is also valid for the other damping cases. For critical damping it is

$$S(t - t^*) = (t - t^*)e^{-\omega_k(t-t^*)} U(t - t^*) \tag{9.1.22}$$

For supercritical damping it is

$$S(t - t^*) = \frac{1}{\varepsilon_k} e^{-\zeta_k\omega_k(t-t^*)} \sin h\ \varepsilon_k(t - t^*) \tag{9.1.23}$$

Note that in Eq. (9.1.20) we may interchange α_1^* and α_2^* with α_1 and α_2 and prove that

$$G_{ij}(\alpha_1,\alpha_2,t;\alpha_1^*,\alpha_2^*,t^*) = G_{ji}(\alpha_1^*,\alpha_2^*,t^*;\alpha_1,\alpha_2,t) \tag{9.1.24}$$

This follows of course also from the Maxwell reciprocity theorem.

9.2 SOLUTION TO GENERAL FORCING USING THE DYNAMIC INFLUENCE FUNCTION

From physical reasoning, the response in the i direction has to be equal to the summation in space and time of all loads in the α_1 direction multiplied by G_{i1}, plus all loads in the α_2 direction multiplied by G_{i2}, and plus all loads in the normal direction multiplied by G_{i3}. This superposition reasoning leads immediately to the solution integral

$$u_i(\alpha_1,\alpha_2,t) = \int_0^t \int_{\alpha_2} \int_{\alpha_1} \sum_{j=1}^{3} G_{ij}(\alpha_1,\alpha_2,t;\alpha_1^*,\alpha_2^*,t^*)q_j(\alpha_1^*,\alpha_2^*,t^*) A_1^*A_2^*\, d\alpha_1^*\, d\alpha_2^*\, dt^* \tag{9.2.1}$$

Let us prove that this is true. Let us substitute Eq. (9.1.20) in Eq. (9.2.1). This gives

$$u_i(\alpha_1,\alpha_2,t) = \sum_{k=1}^{\infty} \frac{U_{ik}(\alpha_1,\alpha_2)}{\rho h N_k} \int_0^t \int_{\alpha_1} \int_{\alpha_2} \sum_{j=1}^{3} U_{jk}(\alpha_1^*,\alpha_2^*)q_j(\alpha_1^*,\alpha_2^*,t^*) S(t - t^*)A_1^*A_2^*\, d\alpha_1^*\, d\alpha_2^*\, dt^* \tag{9.2.2}$$

This expression is identical to the general modal expansion solution for zero initial conditions. From Eqs. (8.1.1), (8.3.2), and (8.3.6) we get

$$u_i(\alpha_1,\alpha_2,t) = \sum_{k=1}^{\infty} \frac{U_{ik}(\alpha_1,\alpha_2)}{\rho h N_k} \int_0^t \int_{\alpha_1} \int_{\alpha_2} \sum_{j=1}^{3} U_{jk}(\alpha_1,\alpha_2)qj(\alpha_1,\alpha_2,\tau) S(t - \tau)A_1A_2\, d\alpha_1\, d\alpha_2\, d\tau \tag{9.2.3}$$

where $S(t - \tau)$ is identical to $S(t - t^*)$ except that t^* is replaced by τ. However, we recognize that the integration variables are interchangeable. We may set $\alpha_1 = \alpha_1^*$, $\alpha_2 = \alpha_2^*$, $\tau = t^*$ without changing the result of the integration. This proves that Eqs. (9.2.2) and (9.2.3) are identical and that Eq. (9.2.1) is the general solution in terms of the dynamic influence function.

9.3 REDUCED SYSTEMS

The total definition of the dynamic influence function for a shell requires in general nine components. For reduced systems we do not need as many because of the uncoupling of the governing equations. For the in-plane vibration of a plate we need only

$$[G_{ij}] = \begin{bmatrix} G_{11} & G_{12} & 0 \\ G_{21} & G_{22} & 0 \\ 0 & 0 & 0 \end{bmatrix} \tag{9.3.1}$$

and for the transverse vibration of a plate

$$[G_{ij}] = \begin{bmatrix} 0 & 0 & 0 \\ 0 & 0 & 0 \\ 0 & 0 & G_{33} \end{bmatrix} \tag{9.3.2}$$

The components G_{13}, G_{23}, G_{31}, G_{32} do not exist because an excitation in the α_1 or α_2 direction does not produce a response in the transverse direction and vice versa, at least according to the linear approximation.

This means that for in-plane vibration

$$G_{ij} = \frac{1}{\rho h} \sum_{k=1}^{\infty} \frac{1}{N_{12k}} U_{jk}(\alpha_1^*, \alpha_2^*) U_{ik}(\alpha_1, \alpha_2) S(t - t^*) \tag{9.3.3}$$

where i, j = 1, 2 and

$$N_{12k} = \int_{\alpha_2} \int_{\alpha_1} (U_{1k}^2 + U_{2k}^2) A_1 A_2 \, d\alpha_1 \, d\alpha_2 \tag{9.3.4}$$

The solution for a general load is

$$u_i = \int_0^t \int_{\alpha_2} \int_{\alpha_1} [G_{i1} \, q_1(\alpha_1^*, \alpha_2^*, t^*) + G_{i2} \, q_2(\alpha_1^*, \alpha_2^*, t^*)] \, A_1^* A_2^* \, d\alpha_1^* \, d\alpha_2^* \, dt^* \tag{9.3.5}$$

For transverse vibrations

$$G_{33} = \frac{1}{\rho h} \sum_{k=1}^{\infty} \frac{1}{N_{3k}} U_{3k}(\alpha_1^*, \alpha_2^*) U_{3k}(\alpha_1, \alpha_2) S(t - t^*) \tag{9.3.6}$$

where

$$N_{3k} = \int_{\alpha_2} \int_{\alpha_1} U_{3k}^2 A_1 A_2 \, d\alpha_1 \, d\alpha_2 \tag{9.3.7}$$

The solution for general transverse loads is

$$u_3 = \int_0^t \int_{\alpha_2} \int_{\alpha_1} G_{33} q_3(\alpha_1^*, \alpha_2^*, t^*) A_1^* A_2^* \, d\alpha_2^* \, d\alpha_1^* \, dt^* \tag{9.3.8}$$

Another interesting case is the ring where we have

$$[G_{ij}] = \begin{bmatrix} G_{11} & 0 & G_{13} \\ 0 & 0 & 0 \\ G_{31} & 0 & G_{33} \end{bmatrix} \tag{9.3.9}$$

where

$$G_{ij} = \frac{1}{\rho h} \sum_{k=1}^{\infty} \frac{1}{N_k} U_{jk}(\theta^*) U_{ik}(\theta) S(t - t^*) \tag{9.3.10}$$

$$N_k = ba \int_0^{\alpha} (U_{1k}^2 + U_{3k}^2) \, d\theta \tag{9.3.11}$$

and where b is the width of the ring, a is the radius, and α defines the size of the segment. The general solution is

$$u_i = ba \int_0^t \int_0^{\alpha} [G_{i1} q_1(\theta^*, t^*) + G_{i3} q_3(\theta^*, t^*)] \, d\theta^* \, dt^* \tag{9.3.12}$$

9.4 DYNAMIC INFLUENCE FUNCTION FOR THE SIMPLY SUPPORTED SHELL

For the case treated in Chap. 5 that has simply supported ends and no axial end restraints the natural modes are ($\alpha_1 = x$, $\alpha_2 = \theta$):

$$U_{1k}(x,\theta) = A_{mnp} \cos \frac{m\pi x}{L} \cos n(\theta - \phi) \tag{9.4.1}$$

$$U_{2k}(x,\theta) = B_{mnp} \sin \frac{m\pi x}{L} \sin n(\theta - \phi) \tag{9.4.2}$$

$$U_{3k}(x,\theta) = C_{mnp} \sin \frac{m\pi x}{L} \cos n(\theta - \phi) \tag{9.4.3}$$

where $m = 1, 2, \ldots$; $n = 0, 1, 2, \ldots$ and $p = 1, 2, 3$. The index k implies again a certain combination of m, n and p. We remember from Sec. 5.5 that for any m, n combination there are three natural frequencies and mode combinations.

The angle ϕ is again arbitrary and indicates the non preferential nature of the modes of free vibration of the closed axisymmetric shell. For the sake of identifying orthogonal modes that can be used to define the response as function of θ, one set of modes may be formed by letting $\phi = 0$ and a second set by letting $n\phi = \pi/2$.

The $n = 0$ modes are axisymmetric. For $n = 0$ and $\phi = 0$, only the mode components that have axial and transverse motion exist, namely U_{1k} and U_{3k}. For $n = 0$ and $n\phi = \pi/2$, this is reversed and only U_{2k} exists.

The value of N_k becomes, for $n \neq 0$,

$$N_k = N_{mnp} = \frac{aL\pi}{2} (A^2_{mnp} + B^2_{mnp} + C^2_{mnp}) \tag{9.4.4}$$

For $n = 0$ we get

$$N_k = N_{mop} = aL\pi(A^2_{mop} + B^2_{mop} + C^2_{mop}) \tag{9.4.5}$$

Components of the influence function may be found simply by substituting in Eq. (9.1.20). For example, the component describing the transverse response to a transverse impulse is

$$G_{33}(x,\theta,t;\ x^*,\theta^*,t^*) = \frac{1}{\rho h} \sum_{m=1}^{\infty} \sum_{n=0}^{\infty} \sum_{p=1}^{3} \sum_{n\phi=0,\ \frac{\pi}{2}} \frac{1}{N_{mnp}} C^2_{mnp} \sin \frac{m\pi x}{L} \sin \frac{m\pi x^*}{L} \cos n(\theta - \phi) \cos n(\theta^* - \phi) S(t - t^*) \tag{9.4.6}$$

The angle ϕ and its associated summation may be eliminated from Eq. (9.4.6) and from the other components of this influence function by noting that each component contains the sum of two products of sine or cosine functions when the summation over ϕ is written out. For example, G_{33} contains the term

$$\cos n\theta \cos n\theta^* + \cos (n\theta - \pi/2) \cos (n\theta^* - \pi/2)$$

This term may be reduced by trigonometric identities to $\cos n(\theta - \theta^*)$. For G_{32}, for instance, the corresponding term and its reduction is

$$\cos n\theta \sin n\theta^* + \cos(n\theta - \frac{\pi}{2}) \sin(n\theta^* - \frac{\pi}{2}) = \sin n(\theta^* - \theta) \tag{9.4.7}$$

On the other hand, for G_{23}, the corresponding term is

$$\sin n\theta \cos n\theta^* + \sin(n\theta - \frac{\pi}{2}) \cos(n\theta^* - \frac{\pi}{2}) = -\sin n(\theta^* - \theta) \tag{9.4.8}$$

The three-directional dynamic influence function for a thin cylindrical shell with simply supported ends is then

$$G_{ij} = \frac{1}{\rho h} \sum_{m=1}^{\infty} \sum_{n=0}^{\infty} \sum_{p=1}^{3} \frac{1}{N_{mnp}} A_{ij} \; S(t - t^*) \tag{9.4.9}$$

where

$$\begin{aligned}
A_{11} &= A_{mnp}^2 \cos \frac{m\pi x}{L} \cos \frac{m\pi x^*}{L} \cos n(\theta - \theta^*) \\
A_{12} &= A_{mnp} B_{mnp} \cos \frac{m\pi x}{L} \sin \frac{m\pi x^*}{L} \sin n(\theta^* - \theta) \\
A_{13} &= A_{mnp} C_{mnp} \cos \frac{m\pi x}{L} \sin \frac{m\pi x^*}{L} \cos n(\theta - \theta^*) \\
A_{21} &= B_{mnp} A_{mnp} \sin \frac{m\pi x}{L} \cos \frac{m\pi x^*}{L} \sin n(\theta - \theta^*) \\
A_{22} &= B_{mnp}^2 \sin \frac{m\pi x}{L} \sin \frac{m\pi x^*}{L} \cos n(\theta - \theta^*) \qquad (9.4.10) \\
A_{23} &= B_{mnp} C_{mnp} \sin \frac{m\pi x}{L} \sin \frac{m\pi x^*}{L} \sin n(\theta - \theta^*) \\
A_{31} &= C_{mnp} A_{mnp} \sin \frac{m\pi x}{L} \cos \frac{m\pi x^*}{L} \cos n(\theta - \theta^*) \\
A_{32} &= C_{mnp} B_{mnp} \sin \frac{m\pi x}{L} \sin \frac{m\pi x^*}{L} \sin(\theta^* - \theta) \\
A_{33} &= C_{mnp}^2 \sin \frac{m\pi x}{L} \sin \frac{m\pi x^*}{L} \cos n(\theta - \theta^*)
\end{aligned}$$

and where, for subcritical damping, for instance, $S(t - t^*)$ is given by Eq. (9.1.21). The solution to any other type of loading is now given by the integral of Eq. (9.2.1). Note that

$$A_{ij}(x,\theta;\ x^*,\theta^*) = A_{ji}(x^*,\theta^*;\ x,\theta) \tag{9.4.11}$$

This verifies Eq. (9.1.24).

9.5 DYNAMIC INFLUENCE FUNCTION OF THE CLOSED CIRCULAR RING

If we assume that the ring deforms only in its plane, the modes obtained in Sec. 5.3 apply. They are ($\alpha_1 = \Theta$, $A_1 = a$, $k = n$)

$$U_{1k}(\theta) = V_{np} \sin n(\theta - \phi) \tag{9.5.1}$$

$$U_{3k}(\theta) = W_{np} \cos n(\theta - \phi) \tag{9.5.2}$$

where $n = 0, 1, 2, \ldots$ and $p = 1, 2$, and where k implies a combination of n and p. As seen in Sec. 5.3, for any n number there exist two separate natural frequencies and mode combinations. In one case the motion is primarily circumferential and in the other case primarily transversal.

According to Eq. (9.3.11), we obtain

$$N_k = N_{np} = \begin{cases} ab\pi(V_{np}^2 + W_{np}^2) & n \neq 0 \\ 2ab\pi(V_{op}^2 + W_{op}^2) & n = 0 \end{cases} \tag{9.5.3}$$

The four components of the dynamic influence function become $(i,j = 1,3)$

$$G_{ij} = \frac{1}{\rho h} \sum_{n=0}^{\infty} \sum_{p=1}^{2} \frac{1}{N_{np}} A_{ij}\ S(t - t^*) \tag{9.5.4}$$

where

$$A_{11} = V_{np}^2 \cos n(\theta - \theta^*)$$

$$A_{13} = V_{np} W_{np} \sin n(\theta - \theta^*) \tag{9.5.5}$$

$$A_{31} = V_{np}W_{np} \sin n(\theta^* - \theta)$$

$$A_{33} = W_{np}^2 \cos n(\theta - \theta^*)$$

This dynamic influence function automatically includes the rigid rotating mode which is described by $n = 0$ and $n\phi = \frac{\pi}{2}$ and also the rigid translating mode which is described by $n = 1$ and $n\phi = 0$.

9.6 TRAVELING POINT LOAD ON SIMPLY SUPPORTED CYLINDRICAL SHELL

In general, the advantage of the dynamic influence function in the analysis of structures is that once the function is known, the structure is defined and an infinite variety of loading combinations can be treated by a relatively simply integration process.

One application for which the dynamic influence function is particularly useful is the traveling load. Early investigations of traveling loads were made by Krylov and Timoshenko [9.2], who were concerned with the response of railroad bridges to traveling locomotives. The problem surfaced in Russia before World War I when trains started to travel with high speeds over bridges that were designed for static loads only. Kryloff and Timoshenko used modal expansion without using an influence function approach, but the formulation of such a problem in terms of the dynamic influence function is particularly simple. One of the first to use this technique was Cottis [9.3], who calculated the response of a shell to a traveling pressure wave.

Let us as example assume that a force of constant magnitude travels along a $\theta = \psi$ line in the positive x direction with constant velocity v, touching the shell surface at $(x,\theta) = (0,\psi)$ at $t = 0$ and leaving it at $(x,\theta) = (L,\psi)$ at $t = L/v$. This is sketched in Fig. 9.6.1. Physically this is a very rough approximation of the action of a piston on a cylinder liner.

Since this example falls into the category of loads moving along a coordinate line, we may express the load as

$$q_1 = q_2 = 0 \tag{9.6.1}$$

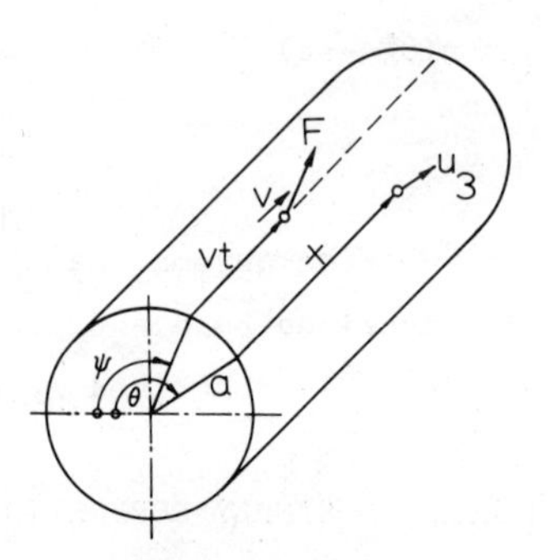

Figure 9.6.1

$$q_3(x,\theta,t) = \frac{F}{a}\,\delta(\theta - \psi)\,\delta(x - vt)[1 - U(vt - L)] \tag{9.6.2}$$

since the instantaneous position is x=vt in this case. The solution is given by Eq. (9.2.1), with G_{ij} for the simply supported shell given by Eq. (9.4.9). This results in, for zero initial conditions,

$$u_i(x,\theta,t) = \frac{F}{\rho h}\sum_{k=1}^{\infty}\int_0^t\int_0^L\int_0^{2\pi}\frac{1}{N_k}A_{i3}S(t - t^*)\,\delta(\theta^* - \psi)$$

$$\delta(x^* - vt^*)[1 - U(vt^* - L)\,d\theta^*\,dx^*\,dt^* \tag{9.6.3}$$

This gives

$$\begin{Bmatrix} u_1 \\ u_2 \\ u_3 \end{Bmatrix} = \frac{F}{\rho h}\sum_{k=1}^{\infty}\frac{1}{\gamma_k N_k}\begin{Bmatrix} A_kC_k\cos\frac{m\pi x}{L} \\ B_kC_k\sin\frac{m\pi x}{L} \\ C_k^2\sin\frac{m\pi x}{L}\end{Bmatrix}\int_0^{2\pi}\begin{Bmatrix}\cos n(\theta - \theta^*) \\ \sin n(\theta - \theta^*) \\ \cos n(\theta - \theta^*)\end{Bmatrix}\delta(\theta^*-\psi)\,d\theta^*$$

$$\int_0^t [1 - U(vt^* - L)]e^{-\zeta_k\omega_k(t-t^*)}\sin\gamma_k(t - t^*)$$

$$\int_0^L \sin\frac{m\pi x^*}{L}\,\delta(x^* - vt^*)\,dx^*\,dt^* \tag{9.6.4}$$

Evaluating the integrals one by one results in

$$\int_{0}^{2\pi} \begin{Bmatrix} \cos n(\theta - \theta^*) \\ \sin n(\theta - \theta^*) \\ \cos n(\theta - \theta^*) \end{Bmatrix} \delta(\theta^* - \psi)\, d\theta^* = \begin{Bmatrix} \cos n(\theta - \psi) \\ \sin n(\theta - \psi) \\ \cos n(\theta - \psi) \end{Bmatrix} \tag{9.6.5}$$

$$\int_{0}^{L} \sin \frac{m\pi x^*}{L}\, \delta(x^* - vt^*)\, dx^* = \sin \frac{m\pi vt^*}{L} \tag{9.6.6}$$

Let us now solve the time integral. If $0 \leq vt < L$, the gate function is $[1 - U(vt - L)] = 1$ and the integral becomes

$$\int_{t^*=0}^{t} F(t^*,t)[1 - U(vt^* - L)]\, dt^* = \int_{t^*=0}^{t} F(t^*,t)\, dt^* \tag{9.6.7}$$

If $vt \geq L$, the gate function is zero and the integral becomes

$$\int_{t^*=0}^{t} F(t^*,t)[1 - U(vt^* - L)]\, dt^* = \int_{t^*=0}^{L/v} F(t^*,t)\, dt^* \tag{9.6.8}$$

Thus, we get, for the time the load is on the shell,

$$\begin{aligned} I_k(t) &= \int_{0}^{t} e^{-a_k(t-t^*)} \sin \gamma_k(t - t^*) \sin \alpha t^*\, dt^* \\ &= \frac{1}{2[a_k^4 + 2a_k^2(\alpha^2 + \gamma_k^2) + (\alpha^2 - \gamma_k^2)^2]} \Big\{ [e^{-a_k t}(\xi_k \sin \gamma_k t \\ &\quad + a_k \cos \gamma_k t) - a_k \cos \alpha t - \xi_k \sin \alpha t](a_k^2 + \eta_k^2) \\ &\quad + [e^{-a_k t}(\eta_k \sin \gamma_k t - a_k \cos \gamma_k t) + a_k \cos \alpha t \\ &\quad + \eta_k \sin \alpha t](a_k^2 + \xi_k^2) \Big\} \end{aligned} \tag{9.6.9}$$

where

$$\alpha = \frac{m\pi v}{L} \tag{9.6.10}$$

$$a_k = \zeta_k \omega_k \tag{9.6.11}$$

$$\xi_k = \alpha - \gamma_k \tag{9.6.12}$$

$$\eta_k = \alpha + \gamma_k \tag{9.6.13}$$

During the time the load is traveling on the shell ($0 \leq vt < L$), the response of the shell is

$$\begin{Bmatrix} u_1 \\ u_2 \\ u_3 \end{Bmatrix} = \frac{F}{\rho h} \sum_{k=1}^{\infty} \frac{1}{\gamma_k N_k} \begin{Bmatrix} A_k C_k \cos \frac{m\pi x}{L} \cos n(\theta - \psi) \\ B_k C_k \sin \frac{m\pi x}{L} \sin n(\theta - \psi) \\ C_k^2 \sin \frac{m\pi x}{L} \cos n(\theta - \psi) \end{Bmatrix} I_k(t) \tag{9.6.14}$$

When the load has left the shell ($L \leq vt$), $I_k(t)$ is replaced by $I_k(L/v,t)$

$$I_k(L/v,t) = \int_{t^*=0}^{L/v} e^{-a_k(t-t^*)} \sin \gamma_k(t - t^*) \sin \alpha t^* \, dt^* \tag{9.6.15}$$

This integral will give a decaying motion of the shell and will not be evaluated here, since it is of no particular interest.

What is of more interest is what happens as the load is traversing the shell. We see from the denominator of $I_k(t)$ that we will have a resonance whenever

$$\alpha = \gamma_k \tag{9.6.16}$$

which means that all traversing velocities, setting $\gamma_k \approx \omega_k$ for small damping,

$$v = \frac{L\omega_k}{m\pi} \tag{9.6.17}$$

have to be avoided.

Another way of looking at it is to recognize that the time it takes for the moving load to traverse the shell is

$$T = \frac{L}{v} \tag{9.6.18}$$

and that α can be looked upon as an excitation frequency. Thus, the period of excitation is

$$T_\alpha = \frac{2}{m} T \tag{9.6.19}$$

This means that if it takes the load T seconds to traverse the shell, the periods of possible resonance are 2T, T, 2/3T, 2/4T, etc., provided any of the resonance periods

$$T_k = \frac{2\pi}{\omega_k} \tag{9.6.20}$$

agree with these values.

9.7 POINT LOAD TRAVELING AROUND A CLOSED CIRCULAR CYLINDRICAL SHELL IN CIRCUMFERENTIAL DIRECTION

This case will demonstrate another interesting resonance phenomenon. The traveling load is described, in this case, by

$$q_1 = q_2 = 0 \tag{9.7.1}$$

$$q_3(x,\theta,t) = \frac{F}{a} \delta(x - \xi) \delta(\theta - \Omega t) \tag{9.7.2}$$

where Ω is the angular velocity of load travel. The load travels continuously around the shell as shown in Fig. 9.7.1 For zero initial conditions, we obtain

$$u_i(x,\theta,t) = \frac{F}{\rho h} \sum_{k=1}^{\infty} \int_0^t \int_0^L \int_0^{2\pi} \frac{1}{N_{mnp}} A_{i3} S(t - t^*) \delta(x^* - \xi) \delta(\theta^* - \Omega t^*) \, d\theta^* \, dx^* \tag{9.7.3}$$

This gives, with A_{13}, A_{23}, and A_{33} given by Eq. (9.4.10),

$$\begin{Bmatrix} u_1 \\ u_2 \\ u_3 \end{Bmatrix} = \frac{F}{\rho h} \sum_{k=1}^{\infty} \frac{1}{\gamma_k N_k} \begin{Bmatrix} A_k C_k \cos \frac{m\pi x}{L} \\ B_k C_k \sin \frac{m\pi x}{L} \\ C_k^2 \sin \frac{m\pi x}{L} \end{Bmatrix} \sin \frac{m\pi\xi}{L} \int_0^t e^{-a_k(t-t^*)} \sin \gamma_k (t - t^*) \begin{Bmatrix} \cos n(\theta - \Omega t^*) \\ \sin n(\theta - \Omega t^*) \\ \cos n(\theta - \Omega t^*) \end{Bmatrix} dt^* \tag{9.7.4}$$

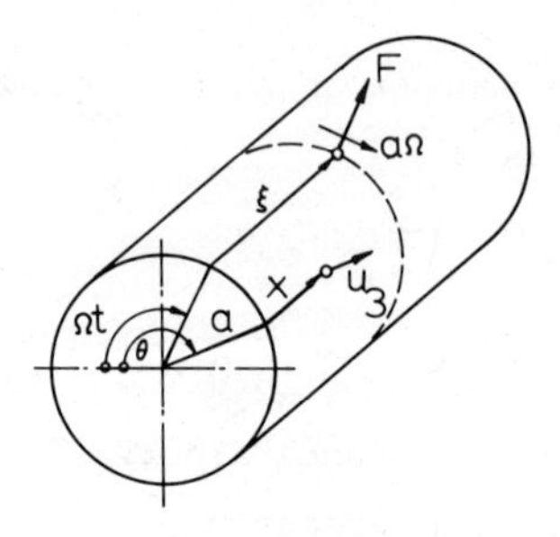

Figure 9.7.1

Let us single out, for further discussion, the transverse response u_3. The integral becomes

$$J_k(t) = \int_0^t e^{-a_k(t-t^*)} \sin \gamma_k(t - t^*) \cos n(\theta - \Omega t^*)\, dt^*$$

$$= Y_k \cos[n(\theta - \Omega t) - \phi_k] + e^{-a_k t} T_k(t) \quad (9.7.5)$$

where

$$Y_k = \frac{\sqrt{a_k^2(\xi_k^2 - \eta_k^2)^2 + [\eta_k(a_k^2 + \xi_k^2) - \xi_k(a_k^2 + \eta_k^2)]^2}}{2[a_k^4 + 2a_k^2(n^2\Omega^2 + \gamma_k^2) + (n^2\Omega^2 - \gamma_k^2)^2]} \quad (9.7.6)$$

$$\phi_k = \tan^{-1} \frac{a_k(\xi_k^2 - \eta_k^2)}{\eta_k(a_k^2 + \xi_k^2) - \xi_k(a_k^2 + \eta_k^2)} \quad (9.7.7)$$

$$T_k(t) = \frac{a_k \sin(n\theta - \gamma_k t) + \xi_k \cos(n\theta - \gamma_k t)}{2(a_k^2 + \xi_k^2)}$$

$$- \frac{a_k \sin(n\theta + \gamma_k t) + \eta_k \cos(n\theta + \gamma_k t)}{2(a_k^2 + \eta_k^2)} \quad (9.7.8)$$

and where

$$\xi_k = n\Omega - \gamma_k \quad (9.7.9)$$

$$\eta_k = n\Omega + \gamma_k \tag{9.7.10}$$

$$a_k = \zeta_k\omega_k \tag{9.7.11}$$

The transient term, due to the sudden onset of travel at t = 0, disappears with time. Of primary interest is the steady-state part of the solution. Examining the denominator of the response amplitude Y_k, we observe a resonance condition which exists whenever

$$\Omega = \frac{\gamma_k}{n} \tag{9.7.12}$$

For small damping, $\gamma_k \approx \omega_k$. Therefore, the first critical speed Ω_c occurs when

$$\Omega_c = \left(\frac{\omega_k}{n}\right)_{min} \tag{9.7.13}$$

This is illustrated on hand of the example shell of Sec. 5.5 for which natural frequencies were obtained. The lowest set of ω_k occurs when m = 1. Thus, we plot ω_{1n}/n as function of n in Fig. 9.7.2 and obtain, as the minimum value, Ω_c = 1800 rad/sec at n = 5. This

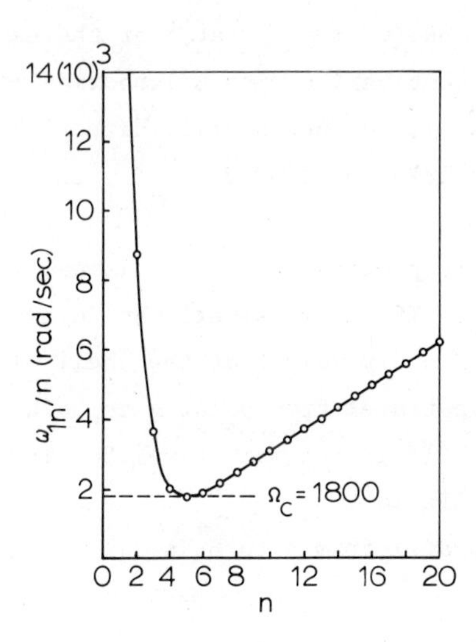

Figure 9.7.2

is the rotational speed at which the first resonance occurs. Below that speed, we do not have any resonance. Above this speed, other resonances will occur.

Another interesting result is indicated by Eq. (9.7.5). An observer that travels along side the traveling load will, in steady state, see standing wave forms only and will not see an oscillation. These waves are quasi-static with respect to the load, but an observer that is located on the shell surface without traveling along with the load will experience oscillations.

These results can be shown to be true for all closed shells of revolution when loads travel in circumferential direction. These effects were pointed out to exist in Ref. 9.4 for an automobile tire rolling on a smooth surface. The tire was treated as a shell of revolution, nonhomogeneous and nonisentropic. The contact region pressure resultant was in this case the traveling load. Critical rolling speeds for typical passenger car tires were predicted to occur at about 140 km/h, which agreed well with experimental observation.

Note also that when the critical speed is reached, the mode that dominates the response is not the simplest mode (by simplest the $n = 0$ or $n = 1$ mode is meant) but, for the example treated here, the $n = 5$ mode, in combination with neighboring modes.

The wave crest is not necessarily directly under the traveling load but may lag or lead the load depending on the damping and the modal participation.

A physical interpretation of the critical speed formula is given by Fig. 9.7.3. If one considers, as an example, the $n = 3$ mode, one can sense intuitively that the shell will go into resonance if the load can travel from point A to point B (the distance AB is the wavelength λ) in the same time that it takes the mode to go through one oscillation.

The time of travel from A to B is

$$T = \frac{\lambda}{\Omega a} \qquad (9.7.14)$$

However, the wavelength is given by

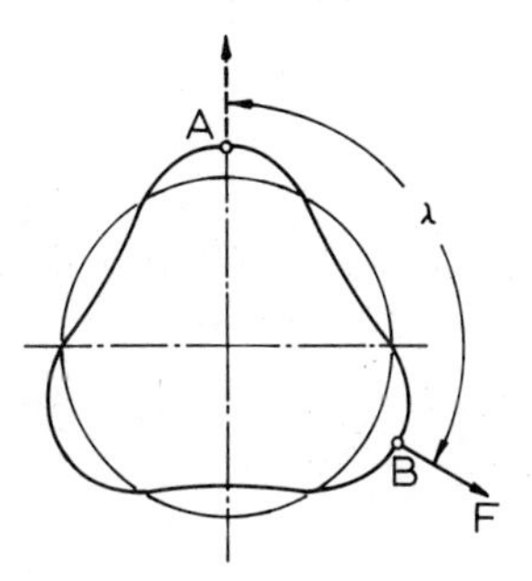

Figure 9.7.3

$$\lambda = \frac{2\pi a}{n} \tag{9.7.15}$$

Thus

$$T = \frac{2\pi}{n\Omega} \tag{9.7.16}$$

Equating this with the period of oscillation

$$T = \frac{2\pi}{\omega_k} \tag{9.7.17}$$

gives as resonance condition

$$\Omega = \frac{\omega_k}{n} \tag{9.7.18}$$

REFERENCES

9.1 I. D. Wilken and W. Soedel, "Three directional dynamic Green's function for thin shells by modal expansion," 82nd National Meeting of the Acoustical Society of America, Denver, Colorado, Oct. 1971.

9.2 S. P. Timoshenko, *Vibration of Bridges*, McGraw-Hill, New York, 1953, pp. 463-481.

9.3 M. G. Cottis, "Green's function technique in the dynamics of a finite cylindrical shell," *J. Acoustical Soc. Amer., vol. 37*, 1965, pp. 31-42.

9.4 W. Soedel, "On the dynamic response of rolling tires according to thin shell approximations," *J. Sound Vibration, vol. 41*, no. 2, 1975, pp. 233-246.

9.5 M. D. Greenberg, "Application of Green's functions in science and engineering," Prentice-Hall, Englewood Cliffs, N.J., 1971.

10

MOMENT LOADING

Let us now consider the response of shells to the excitation by moments. The formulation presented so far allows only treatment of cases where the excitation can be expressed in terms of pressure-type loading q_1, q_2, and q_3. By use of the Dirac delta function, we were able to treat also line and point loads. However, in engineering practice, we often meet problems where the excitation is a moment. For instance, consider rotating machinery with an unbalance whose plane is parallel to the surface of the shell on which the machinery is mounted. Obviously, in addition to the in-plane forces, we have also a force couple. Any kind of connection to a shell that is acted on by forces not transverse to the shell surface will produce a moment.

One approach to this problem is to consider, for instance, two transverse point forces equal in magnitude but opposed in direction, a small distance apart. They form a moment. As we let the distance approach zero in the limit, we have created a true point moment. For instance, Ref. 10.1 indicates that this approach has been used with great success in special cases, but it is difficult to formulate the general case without rather complicated notation.

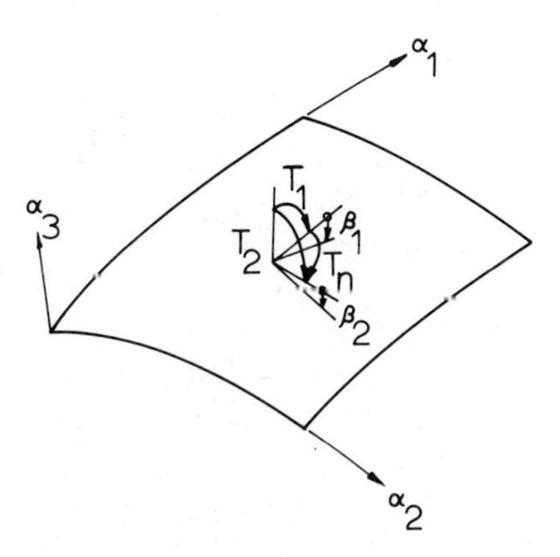

Figure 10.1.1

In this chapter, an approach is followed that is taken from Ref. 10.2. The idea of a distributed moment is used (moment per unit area). Line and point moments, the usual engineering cases, are then formulated using the Dirac delta function.

10.1 FORMULATION OF SHELL EQUATIONS THAT INCLUDE MOMENT LOADING

Let us consider three distributed moment components, T_1 in the α_1 direction, T_2 in the α_2 direction, and a twisting moment about the normal, as shown in Fig. 10.1.1. The units are newton-meters per square meter ($N \cdot m/m^2$).

To consider moment loading in addition to the classical pressure loading in the three orthogonal directions, one has to add the energy due to the applied moments. It is

$$E_m = \int_{\alpha_1}\int_{\alpha_2} (T_1\beta_1 + T_2\beta_2 + T_n\beta_n)A_1A_2 \, d\alpha_1 \, d\alpha_2 \tag{10.1.1}$$

where β_1 and β_2 are defined by Eqs. (2.4.7) and (2.4.8) and where

$$\beta_n = \frac{1}{2A_1A_2}\left(\frac{\partial(A_2u_2)}{\partial\alpha_1} - \frac{\partial(A_1u_1)}{\partial\alpha_2}\right) \tag{10.1.2}$$

This is the expression for a twisting angle. Let us derive it for

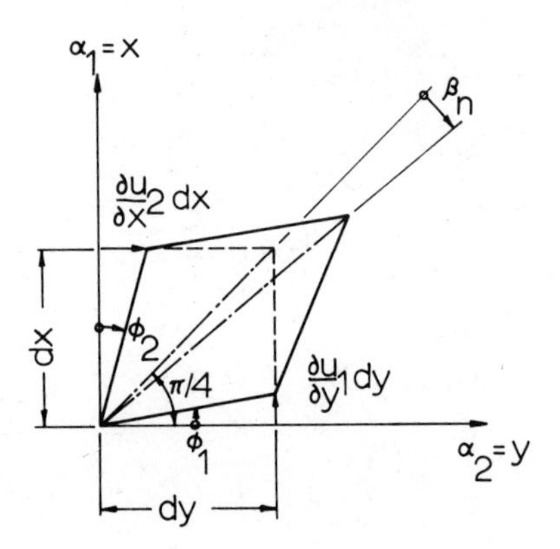

Figure 10.1.2

cartesian coordinates by considering Fig. 10.1.2. First, the rotation of the diagonal of the infinitesimal element is β_n. It is

$$\beta_n = \frac{1}{2}\left(\frac{\pi}{2} - \phi_2 - \phi_1\right) + \phi_2 - \frac{\pi}{4} = \frac{\phi_2 - \phi_1}{2} \tag{10.1.3}$$

But since

$$\phi_2 = \frac{du_2}{dx} \tag{10.1.4}$$

$$\phi_1 = \frac{du_1}{dy} \tag{10.1.5}$$

we obtain

$$\beta_n = \frac{1}{2}\left(\frac{du_2}{dx} - \frac{du_1}{dy}\right) \tag{10.1.6}$$

If we set $A_1 = A_2 = 1$ and $\alpha_1 = x$, $\alpha_2 = y$ in Eq. (10.1.2), we obtain the same result.

Taking the variation of E_m gives

$$\delta E_m = \int_{\alpha_1}\int_{\alpha_2} (T_1 \delta\beta_1 + T_2 \delta\beta_2 + T_n \delta\beta_n) A_1 A_2 \, d\alpha_1 \, d\alpha_2 \tag{10.1.7}$$

where, from Eq. (10.1.2),

$$\delta\beta_n = \frac{1}{2A_1A_2}\left\{\frac{\partial(A_2\delta u_2)}{\partial\alpha_1} - \frac{\partial(A_1\delta u_1)}{\partial\alpha_2}\right\} \tag{10.1.8}$$

Let us integrate the third term

$$\int_{\alpha_1}\int_{\alpha_2} T_n\delta\beta_n A_1A_2\, d\alpha_2\, d\alpha_1 - \int_{\alpha_2}\left(\int_{\alpha_1}\frac{T_n}{2A_1A_2}\,\frac{\partial(A_2\delta u_2)}{\partial\alpha_1}\,A_1\, d\alpha_1\right)A_2\, d\alpha_2$$
$$- \int_{\alpha_1}\left(\int_{\alpha_2}\frac{T_n}{2A_1A_2}\,\frac{\partial(A_1\delta u_1)}{\partial\alpha_2}\,A_2\, d\alpha_2\right)A_1\, d\alpha_1 \tag{10.1.9}$$

Integrating by parts gives

$$\int_{\alpha_1}\int_{\alpha_2} T_n\delta\beta_n A_1A_2\, d\alpha_2\, d\alpha_1 = \frac{1}{2}\int_{\alpha_1}\int_{\alpha_2}\left[\frac{A_1}{A_2}\,\frac{\partial}{\partial\alpha_2}\left(\frac{T_n}{A_1}\right)\delta u_1\right.$$
$$\left. - \frac{A_2}{A_1}\,\frac{\partial}{\partial\alpha_1}\left(\frac{T_n}{A_2}\right)\delta u_2\right]A_1A_2\, d\alpha_1\, d\alpha_2 \tag{10.1.10}$$

since the resulting line integrals also vanish at the boundary. Substituting Eq. (10.1.10) in Eq. 10.1.7 and this equation, in turn, in the development of Sec. 2.7, we obtain as equations of motion

$$-\frac{\partial(N_{11}A_2)}{\partial\alpha_1} - \frac{\partial(N_{12}A_1)}{\partial\alpha_2} - N_{12}\frac{\partial A_1}{\partial\alpha_2} + N_{22}\frac{\partial A_2}{\partial\alpha_1} - A_1A_2\frac{Q_{13}}{R_1}$$
$$+ A_1A_2\rho h\ddot{u}_1 = A_1A_2\left(q_1 + \frac{1}{2A_2}\,\frac{\partial T_n}{\partial\alpha_2}\right) \tag{10.1.11}$$

$$-\frac{\partial(N_{12}A_2)}{\partial\alpha_1} - \frac{\partial(N_{22}A_1)}{\partial\alpha_2} - N_{21}\frac{\partial A_2}{\partial\alpha_1} + N_{11}\frac{\partial A_1}{\partial\alpha_2} - A_1A_2\frac{Q_{23}}{R_2}$$
$$+ A_1A_2\rho h\ddot{u}_2 = A_1A_2\left(q_2 - \frac{1}{2A_1}\,\frac{\partial T_n}{\partial\alpha_1}\right) \tag{10.1.12}$$

$$-\frac{\partial(Q_{13}A_2)}{\partial\alpha_1} - \frac{\partial(Q_{23}A_1)}{\partial\alpha_2} + A_1A_2\left(\frac{N_{11}}{R_1} + \frac{N_{22}}{R_2}\right) + A_1A_2\rho h\ddot{u}_3$$
$$= A_1A_2\left[q_3 + \frac{1}{A_1A_2}\left(\frac{\partial(T_1A_2)}{\partial\alpha_1} + \frac{\partial(T_2A_1)}{\partial\alpha_2}\right)\right] \tag{10.1.13}$$

where Q_{13} and Q_{23} are defined by Eqs. (2.7.23) and (2.7.24).

The admissable boundary conditions are the same.

10.2 THE MODAL EXPANSION SOLUTION

These equations may be written in terms of displacement u_i as

$$L_1\{u_1,u_2,u_3\} - \lambda\dot{u}_1 - \rho h\ddot{u}_1 = -q_1 - \frac{1}{2A_2}\frac{\partial T_n}{\partial \alpha_2} \tag{10.2.1}$$

$$L_2\{u_1,u_2,u_3\} - \lambda\dot{u}_2 - \rho h\ddot{u}_2 = -q_2 + \frac{1}{2A_1}\frac{\partial T_n}{\partial \alpha_1} \tag{10.2.2}$$

$$L_3\{u_1,u_2,u_3\} - \lambda\dot{u}_3 - \rho h\ddot{u}_3 = -q_3 - \frac{1}{A_1A_2}\left(\frac{\partial(T_1A_2)}{\partial \alpha_1} + \frac{\partial(T_2A_1)}{\partial \alpha_1}\right) \tag{10.2.3}$$

The modal expansion series solution is

$$u_i = \sum_{k=1}^{\infty} \eta_k u_{ik} \tag{10.2.4}$$

This gives, after the usual operations,

$$\ddot{\eta}_k + 2\zeta_k\omega_k\dot{\eta}_k + \omega_k^2\eta_k = F_k \tag{10.2.5}$$

where

$$F_k = \frac{1}{\rho h N_k}\int_{\alpha_2}\int_{\alpha_1}\left[q_1U_{1k} + q_2U_{2k} + q_3U_{3k} + \frac{U_{1k}}{2A_2}\frac{\partial T_n}{\partial \alpha_2} - \frac{U_{2k}}{2A_1}\frac{\partial T_n}{\partial \alpha_1}\right.$$
$$\left. + \frac{U_{3k}}{A_1A_2}\left(\frac{\partial(T_1A_2)}{\partial \alpha_1} + \frac{\partial(T_2A_1)}{\partial \alpha_2}\right)\right]A_1A_2\, d\alpha_1\, d\alpha_2 \tag{10.2.6}$$

$$\zeta_k = \frac{\lambda}{2\rho h\omega_k} \tag{10.2.7}$$

$$N_k = \int_{\alpha_2}\int_{\alpha_1}(U_{1k}^2 + U_{2k}^2 + U_{3k}^2)A_1A_2\, d\alpha_1\, d\alpha_2 \tag{10.2.8}$$

Solutions of Eq. (10.2.5) are given in Sec. 8.3.

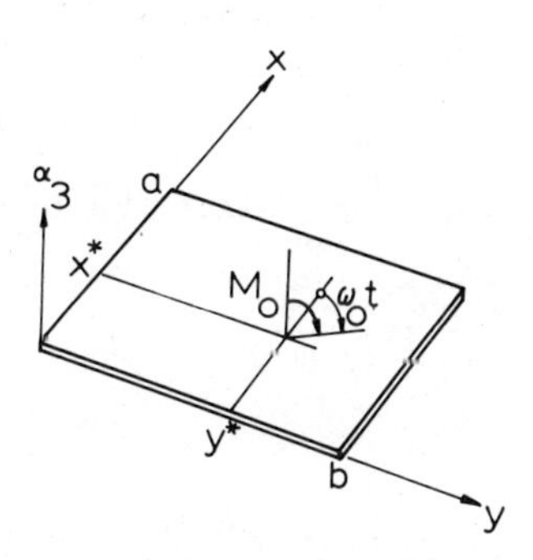

Figure 10.3.1

10.3 ROTATING POINT MOMENT ON A PLATE

For the transverse motion of a plate, the solution to Eq. 10.2.3 reduces to

$$u_3 = \sum_{k=1}^{\infty} \eta_k u_{3k} \tag{10.3.1}$$

where η_k is given by Eq. (10.2.5) and where

$$F_k = \frac{1}{\rho h N_k} \int_{\alpha_1} \int_{\alpha_2} U_{3k} \left[q_3 + \frac{1}{A_1 A_2} \left(\frac{\partial (T_1 A_2)}{\partial \alpha_1} + \frac{\partial (T_2 A_1)}{\partial \alpha_2} \right) \right] A_1 A_2 \, d\alpha_1 \, d\alpha_2 \tag{10.3.2}$$

and $$N_k = \int_{\alpha_2} \int_{\alpha_1} U_{3k}^2 A_1 A_2 \, d\alpha_1 \, d\alpha_2 \tag{10.3.3}$$

Let us take as example a simply supported rectangular plate on which a point moment of magnitude M_o, acting normal to the plate surface, rotates with a constant velocity ω_o [rad/sec]. This is shown in Fig. 10.3.1. In this case $\alpha_1 = x$, $\alpha_2 = y$, $A_1 = A_2 = 1$, and therefore,

$$T_1 = T_x = M_o \cos \omega_o t \; \delta(x - x^*) \; \delta(y - y^*) \tag{10.3.4}$$

$$T_2 = T_y = M_o \sin \omega_o t \; \delta(x - x^*) \; \delta(y - y^*) \tag{10.3.5}$$

$$T_n = 0 \tag{10.3.6}$$

Also, let us assume that $q_3 = 0$. Since the mode shape is given by

$$U_{3k} = U_{3mn} = \sin \frac{m\pi x}{a} \sin \frac{n\pi y}{b} \tag{10.3.7}$$

we obtain

$$F_k = \frac{1}{\rho h N_k} \left[\cos \omega_o t \sin \frac{n\pi y^*}{b} \int_0^a \sin \frac{m\pi x}{a} \frac{\partial}{\partial x} \delta(x - x^*) \, dx \right.$$

$$\left. + \sin \omega_o t \sin \frac{m\pi x^*}{a} \int_0^b \sin \frac{n\pi y}{b} \frac{\partial}{\partial y} \delta(y - y^*) \, dy \right] \tag{10.3.8}$$

Since it can be shown, using integration by parts, that

$$\int_{\alpha_i} F(\alpha_i) \frac{\partial}{\partial \alpha} \left[\frac{1}{A_i} \delta(\alpha_i - \alpha_i^*) \right] d\alpha_i = - \frac{1}{A_i} \left. \frac{\partial F(\alpha_i)}{\partial \alpha_i} \right|_{\alpha_i = \alpha_i^*} \tag{10.3.9}$$

we get

$$F_k = - \frac{1}{\rho h N_k} \left(\cos \omega_o t \frac{m\pi}{a} \sin \frac{n\pi y^*}{b} \cos \frac{m\pi x^*}{a} \right.$$

$$\left. + \sin \omega_o t \frac{n\pi}{b} \sin \frac{m\pi x^*}{a} \cos \frac{n\pi y^*}{b} \right) \tag{10.3.10}$$

where

$$N_k = \frac{ab}{4} \tag{10.3.11}$$

Following Sec. 8.5, we obtain as steady-state solution

$$u_3 = - \frac{4M_o}{ab\rho h} \sum_{m=1}^{\infty} \sum_{n=1}^{\infty} \frac{f(m,n) \sin \frac{m\pi n}{a} \sin \frac{n\pi y}{b} \sin (\omega_o t + \phi_1 - \phi_2)}{\omega_{mn}^2 \sqrt{\left[1 - \left(\frac{\omega}{\omega_{mn}}\right)^2 \right]^2 + 4\zeta_{mn}^2 \left(\frac{\omega}{\omega_{mn}}\right)^2}} \tag{10.3.12}$$

where

$$f(m,n) = \sqrt{\left(\frac{m\pi}{a}\right)^2 \sin^2 \frac{n\pi y^*}{b} \cos^2 \frac{m\pi x^*}{a} + \left(\frac{n\pi}{b}\right)^2 \sin^2 \frac{m\pi x^*}{a} \cos^2 \frac{n\pi y^*}{b}} \tag{10.3.13}$$

$$\phi_1 = \tan^{-1}\left(\frac{m}{n}\,\frac{b}{a}\,\frac{\tan \frac{n\pi y^*}{b}}{\tan \frac{m\pi x^*}{a}}\right) \tag{10.3.14}$$

$$\phi_2 = \tan^{-1} \frac{2\zeta_{mn}\left(\frac{\omega}{\omega_{mn}}\right)}{1 - \left(\frac{\omega}{\omega_{mn}}\right)^2} \tag{10.3.15}$$

We see that in order to minimize the response of a plate to a rotating moment, one ought to avoid having a natural frequency coincide with the rotational speed, $\omega_{mn} \neq \omega_o$. Also, the location of the rotating moment should be on an antinode point of the plate in order to minimize the response of that mode. By antinode points those points are meant where the mode has zero slope in all directions. Location of the rotating moment at a point where two node lines cross maximizes the response of that mode.

At the location of the point moment application, certain assumptions of our theory are again violated, and solutions that consider details of the actual hardware have to be found [10.3]. At a distance removed from the application point by more than twice the thickness, our solution is excellent.

10.4 ROTATING POINT MOMENT ON A SHELL

Let us now consider the case where a point moment M_o acting normal to the shell surface rotates with a rotational velocity ω_o as shown in Fig. 10.4.1. For cylindrical coordinates, $R_1 = \infty$, $R_2 = a$, $A_1 = 1$, $A_2 = a$, $\alpha_1 = x$, $\alpha_2 = \theta$. Thus, the moment distribution is given by

$$T_1 = T_x = \frac{M_o}{a} \cos \omega_o t\ \delta(x - x^*)\ \delta(\theta - \theta^*) \tag{10.4.1}$$

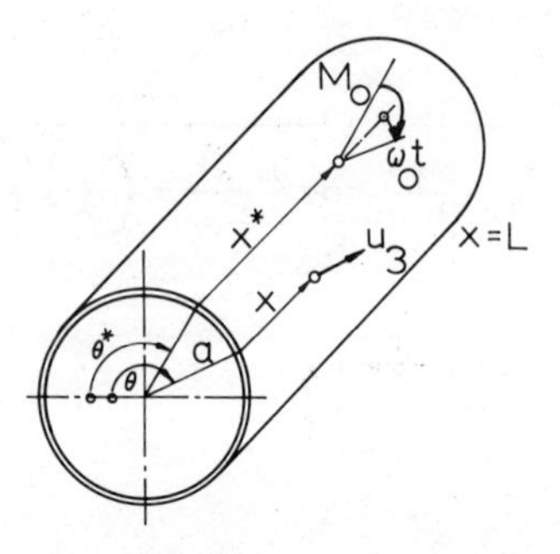

Figure 10.4.1

$$T_2 = T_\theta = \frac{M_o}{a} \sin \omega_o t \ \delta(x - x^*) \ \delta(\theta - \theta^*) \tag{10.4.2}$$

Again, $q_1 = q_2 = q_3 = 0$. The natural modes, for the simply supported circular cylindrical shell treated in Chap. 5 are

$$U_{1k}(x,\theta) = A_{mnp} \cos \frac{m\pi x}{L} \cos n(\theta - \phi) \tag{10.4.3}$$

$$U_{2k}(x,\theta) = B_{mnp} \sin \frac{m\pi x}{L} \sin n(\theta - \phi) \tag{10.4.4}$$

$$U_{3k}(x,\theta) = C_{mnp} \sin \frac{m\pi x}{L} \cos n(\theta - \phi) \tag{10.4.5}$$

where $m = 1, 2, \ldots$, $n = 0, 1, 2, \ldots$, and $p = 1, 2, 3$. The natural frequencies ω_{mnp} are given in Chap. 5. Equation (10.2.6) becomes

$$F_k = - C_{mnp} \frac{M_o \cos \omega_o t}{\rho h N_k} \frac{m\pi}{L} \cos \frac{m\pi x^*}{L} \tag{10.4.6}$$

Since

$$N_k = N_{mnp} = \frac{aL\pi}{2} C_{mnp}^2 k_{mnp} \tag{10.4.7}$$

where

$$k_{mnp} = \begin{cases} (A_{mnp}/C_{mnp})^2 + (B_{mnp}/C_{mnp})^2 + 1 & n \neq 0 \\ 2(A_{mop}/C_{mop})^2 + 2 & n = 0 \end{cases} \tag{10.4.8}$$

we obtain as the solution for the steady-state response

$$\begin{Bmatrix} u_1 \\ u_2 \\ u_3 \end{Bmatrix} = -\frac{2M_o}{aL\pi\rho h} \sum_{p=1}^{3} \sum_{m=1}^{\infty} \sum_{n=1}^{\infty} \frac{f(m)}{k_{mnp}\omega_{mnp}^2[1 - (\omega_o/\omega_{mnp})^2]}$$

$$\begin{Bmatrix} A_{mnp}/C_{mnp} \cos m\pi x/L \cos n(\theta - \theta^*) \\ B_{mnp}/C_{mnp} \sin m\pi x/L \sin n(\theta - \theta^*) \\ \sin m\pi x/L \cos n(\theta - \theta^*) \end{Bmatrix} \cos \omega_o t \tag{10.4.9}$$

where

$$f(m) = \frac{m\pi}{L} \cos \frac{m\pi x^*}{L} \tag{10.4.10}$$

The similarity to the result for the plate is apparent. The function f is now only dependent on m. The reason is that the response will orient itself on the location of the rotating moment, no matter where it is located in the circumferential direction. It is, in general, not possible to have a true antinode at the point of moment application for closed shells of revolution. The modes have no preference as far as circumferential direction is concerned and orient themselves such that they present the least resistance to the action of a rotating moment. However, it is possible to locate the moment such that

$$f(m) = 0 \tag{10.4.11}$$

to remove selectively certain modes from the response. In the example case, we have to make

$$\frac{m\pi x^*}{L} = \frac{\pi}{2}, \frac{3\pi}{2}, \frac{5\pi}{2}, \ldots \tag{10.4.12}$$

10.5 RECTANGULAR PLATE EXCITED BY A LINE MOMENT

Let us consider the case where a line moment varying sinusoidally with time and of uniform magnitude M_o' in newton-meters per meter (N·m/m) is distributed across a simply supported rectangular plate

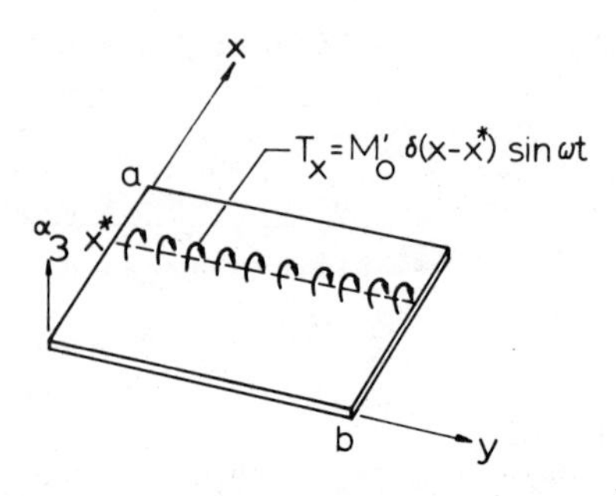

Figure 10.5.1

along the line x = x* as shown in Fig. 10.5.1. Letting $\alpha_1 = x$, $A_1 = 1$, $\alpha_2 = y$, $A_2 = 1$, we express the distributed moments as

$$T_x = M_o' \delta(x - x^*) \sin \omega t \tag{10.5.1}$$

$$T_y = 0 \tag{10.5.2}$$

This gives

$$F_k = \frac{4M_o'}{ab\rho h} \int_0^a \int_0^b \sin \frac{m\pi x}{a} \sin \frac{n\pi y}{b} \frac{\partial}{\partial x} \delta(x - x^*)\, dx\, dy \sin \omega t \tag{10.5.3}$$

or $$F_k = \frac{-4M_o'm}{a^2\rho hn} (1 - \cos n\pi) \cos \frac{m\pi x^*}{a} \sin \omega t \tag{10.5.4}$$

This means that only the modes where n = 1, 3, 5... are excited. For instance, the steady-state response becomes, using Eqs. (8.5.4), (8.5.6), and (8.5.7),

$$u_3(x,y,t) = -\frac{8M_o'}{a^2\rho h} \sum_{n=1,3,\ldots}^{\infty} \sum_{m=1}^{\infty} \frac{\frac{m}{n} \cos \frac{m\pi x^*}{a} \sin \frac{n\pi y}{b} \sin \frac{m\pi x}{a}}{\omega_{mn}^2 \sqrt{\left[1 - \left(\frac{\omega}{\omega_{mn}}\right)^2\right]^2 + 4\zeta_{mn}^2 \left(\frac{\omega}{\omega_{mn}}\right)^2}} \sin(\omega t - \phi_{mn}) \tag{10.5.5}$$

where

$$\phi_{mn} = \tan^{-1} \frac{2\zeta_{mn}(\omega/\omega_{mn})}{1 - (\omega/\omega_{mn})^2} \tag{10.5.6}$$

Note that if we let $x^* = 0$ or $x^* = a$, we obtain the solutions for harmonic edge moments.

Solutions to harmonic edge moments are important when investigating composite structures like two plates joined together at one edge, except that for this type of application we have to evaluate the solution for a line distribution where the moment magnitude varies sinusoidally along the line

$$T_x = M_o' \sin \frac{p\pi y}{b}\, \delta(x - x^*) \sin \omega t \tag{10.5.7}$$

$$T_y = 0 \tag{10.5.8}$$

We obtain

$$F_k = -\frac{4M_o'\pi m}{a^2 b\rho h}\left(\int_0^b \sin \frac{n\pi y}{b} \sin \frac{p\pi y}{b}\, dy\right) \cos \frac{m\pi x^*}{a} \sin \omega t \tag{10.5.9}$$

Since this expression is zero if $p \neq n$, we obtain, for $n = p$,

$$F_k = F_{mp} = -\frac{2M_o'\pi m}{a^2\rho h} \cos \frac{m\pi x^*}{a} \sin \omega t \tag{10.5.10}$$

and the steady-state solution is then

$$u_3(x,y,t) = -\frac{2M_o'\pi}{a^2\rho h}$$

$$\sum_{m=1}^{\infty} \frac{m \cos(m\pi x^*/a) \sin \frac{p\pi y}{b} \sin \frac{m\pi x}{a}}{\omega_{mp}^2 \sqrt{\left[1 - \left(\frac{\omega}{\omega_{mp}}\right)^2\right]^2 + 4\zeta_{mn}^2 \left(\frac{\omega}{\omega_{mp}}\right)^2}} \sin(\omega t - \phi_{mp}) \tag{10.5.11}$$

This is an interesting result since it shows that a sinusoidally distributed line moment will only excite the modes that have shapes that match the distribution shape of the line moment. Using a Fourier sine series description, we can use this solution to generate the solutions for all other line distribution shapes. Such an approach can, of course, also be used for line force distributions.

REFERENCES

10.1 U. Bolleter and W. Soedel, "Dynamic Green's function technique applied to shells loaded by dynamic moments," *J. Acoustical Soc. Amer., vol. 49*, 1971, pp. 753-758.

10.2 W. Soedel, "Shells and plates loaded by dynamic moments with special attention to rotating point moments," *J. Sound Vibration, vol. 48*, no. 2, 1976, pp. 179-188.

10.3 I. Dyer, "Moment impedance of plates," *J. Acoustical Soc. Amer., vol. 32*, 1960, pp. 1220-1297.

11

VIBRATION OF SHELLS AND MEMBRANES UNDER THE INFLUENCE OF INITIAL STRESSES

All cases treated so far were rigid structures which, when in their equilibrium position, have a zero or neglegible stress level. A very different category are skin structures where the forces that restore the displaced skin to its equilibrium position are caused by initially present stresses that can be assumed to be independent of the motion. Such skin structures are referred to as *membranes*. They can be flat like classical membranes or stretched over three-dimensional frames. Note that the restoring effect is entirely different from the membrane approximation for shells that was discussed earlier, where the restoring forces are caused by the changing membrane stresses as the deflection occurs. There is a slight problem in semantics, of course, since the word *membrane* is used in the context of both categories.

A third possibility and a very likely one in engineering is that a combination of the two restoring effects has to be accounted for. Every time a structure is spinning like a turbine blade, a circular saw, etc., the centrifugal forces will introduce an initial stress field that is always present, vibration or no vibration, and which acts as an additional restoring mechanism. Shells loaded by high static pressures have a static or initial stress field that resists or aids deflection. For instance, a spherical shell located internally by static pressure like a boiler will experience a gain in effective stiffness which will increase natural frequencies, while a spherical shell that is loaded externally by static pressure like a diving vessel will experience a decrease of effective stiffness, lowering the natural frequencies. Shells that were fabricated by deep drawing or welding and were not annealed afterward can have large equilibrium stresses that are referred to as *residual stresses*. These stresses can often be so high that a pronounced static buckling or warping occurs.

In the following, Love's equation will be extended to account for the initial stress effect. Next, it will be reduced to the category of pure membrane or skin structures. Examples follow.

Note that the earliest theoretical investigation of the initial stress influence, going beyond that of a string or pure membrane, is probably due to Lamb [11.1] in 1921 for a circular plate under initial tension. One of the first shell solutions of this type was given by Federhofer [11.2] in 1936 for the axially compressed circular cylindrical shell. A good discussion of more recent work on initial stress problems is given by Leissa [11.3].

11.1 STRAIN-DISPLACEMENT RELATIONSHIPS

In order to investigate how shells vibrate under the influence of initial stresses, we have to include some terms in the strain-displacement equation that had been neglected in Sec. 2.3. In Eq. 2.3.9, we retain all terms

$$(d\alpha_i + d\xi_i)^2 = (d\alpha_i)^2 + 2d\alpha_i d\xi_i + (d\xi_i)^2 \tag{11.1.1}$$

Thus, instead of (Eq. 2.3.12), we obtain

$$(d\alpha_i + d\xi_i)^2 = (d\alpha_i)^2 + 2d\alpha_i \sum_{j=1}^{3} \frac{\partial \xi_i}{\partial \alpha_j} d\alpha_j + \left(\sum_{j=1}^{3} \frac{\partial \xi_i}{\partial \alpha_j} d\alpha_j \right)^2 \tag{11.1.2}$$

This means that Eq. (2.3.15) becomes

$$(ds')^2 = \sum_{i=1}^{3}\left[\left(g_{ii} + \sum_{j=1}^{3}\frac{\partial g_{ii}}{\partial \alpha_j}\xi_j\right)(d\alpha_i)^2 + 2d\alpha_i g_{ii}\sum_{j=1}^{3}\frac{\partial \xi_i}{\partial \alpha_j}d\alpha_j\right.$$

$$+ 2d\alpha_i \sum_{j=1}^{3}\frac{\partial g_{ii}}{\partial \alpha_j}\xi_j \sum_{j=1}^{3}\frac{\partial \xi_i}{\partial \alpha_j}d\alpha_j + g_{ii}\left(\sum_{j=1}^{3}\frac{\partial \xi_i}{\partial \alpha_j}\right)^2(d\alpha_j)^2$$

$$\left. + \sum_{j=1}^{3}\frac{\partial g_{ii}}{\partial \alpha_j}\xi_j\left(\sum_{j=1}^{3}\frac{\partial \xi_i}{\partial \alpha_j}d\alpha_j\right)^2\right] \quad (11.1.3)$$

Neglecting, as before, the third term, and of the two new terms the last one gives, after introducing the Kronecker delta notation

$$(ds')^2 = \sum_{i=1}^{3}\sum_{j=1}^{3}\left[\left(g_{ii} + \sum_{k=1}^{3}\frac{\partial g_{ii}}{\partial \alpha_k}\xi_k\right)\delta_{ij}\, d\alpha_i\, d\alpha_j\right.$$

$$+ \left(g_{ii}\frac{\partial \xi_i}{\partial \alpha_j} + g_{jj}\frac{\partial \xi_j}{\partial \alpha_i}\right)d\alpha_i\, d\alpha_j$$

$$\left. + \left(\sum_{k=1}^{3} g_{kk}\frac{\partial \xi_k}{\partial \alpha_i}\frac{\partial \xi_k}{\partial \alpha_j}\right)d\alpha_i\, d\alpha_j\right] \quad (11.1.4)$$

This equation replaces Eq. (2.3.20). We see that our new G_{ij} is

$$G_{ij} = \left(g_{ii} + \sum_{k=1}^{3}\frac{\partial g_{ii}}{\partial \alpha_k}\xi_k\right)\delta_{ij} + g_{ii}\frac{\partial \xi_i}{\partial \alpha_j} + g_{jj}\frac{\partial \xi_j}{\partial \alpha_i}$$

$$+ \sum_{k=1}^{3} g_{kk}\frac{\partial \xi_k}{\partial \alpha_i}\frac{\partial \xi_k}{\partial \alpha_j} \quad (11.1.5)$$

From here on, the derivation proceeds as outlined in Sec. 2.3. We get

$$\varepsilon_{11} = \frac{1}{A_1}\left(\frac{\partial U_1}{\partial \alpha_1} + \frac{U_2}{A_2}\frac{\partial A_1}{\partial \alpha_2} + U_3\frac{A_1}{R_1}\right) + \frac{1}{2A_1^2}\left[\left(\frac{\partial U_1}{\partial \alpha_1} - \frac{U_1}{A_1}\frac{\partial A_1}{\partial \alpha_1}\right)^2\right.$$

$$\left. + \left(\frac{\partial U_2}{\partial \alpha_1} - \frac{U_2}{A_2}\frac{\partial A_2}{\partial \alpha_1}\right)^2 + \left(\frac{\partial U_3}{\partial \alpha_1}\right)^2\right] \quad (11.1.6)$$

$$\varepsilon_{22} = \frac{1}{A_2}\left(\frac{\partial U_2}{\partial \alpha_2} + \frac{U_1}{A_1}\frac{\partial A_2}{\partial \alpha_1} + U_3\frac{A_2}{R_2}\right) + \frac{1}{2A_2^2}\left[\left(\frac{\partial U_2}{\partial \alpha_2} - \frac{U_2}{A_2}\frac{\partial A_2}{\partial \alpha_2}\right)^2 + \left(\frac{\partial U_1}{\partial \alpha_2} - \frac{U_1}{A_1}\frac{\partial A_1}{\partial \alpha_2}\right)^2 + \left(\frac{\partial U_3}{\partial \alpha_2}\right)^2\right] \tag{11.1.7}$$

$$\varepsilon_{33} = \frac{\partial U_3}{\partial \alpha_3} \tag{11.1.8}$$

$$\begin{aligned}\varepsilon_{12} = {} & \frac{A_1(1 + \frac{\alpha_3}{R_1})}{A_2(1 + \frac{\alpha_3}{R_2})}\frac{\partial}{\partial \alpha_2}\left(\frac{U_1}{A_1(1 + \frac{\alpha_3}{R_1})}\right) + \frac{A_2(1 + \frac{\alpha_3}{R_2})}{A_1(1 + \frac{\alpha_3}{R_1})}\frac{\partial}{\partial \alpha_1}\left(\frac{U_2}{A_2(1 + \frac{\alpha_3}{R_2})}\right) \\ & + \frac{A_1(1 + \frac{\alpha_3}{R_1})}{A_2(1 + \frac{\alpha_3}{R_2})}\frac{\partial}{\partial \alpha_1}\left(\frac{U_1}{A_1(1 + \frac{\alpha_3}{R_1})}\right)\frac{\partial}{\partial \alpha_2}\left(\frac{U_1}{A_1(1 + \frac{\alpha_3}{R_1})}\right) \\ & + \frac{A_2(1 + \frac{\alpha_3}{R_2})}{A_1(1 + \frac{\alpha_3}{R_1})}\frac{\partial}{\partial \alpha_1}\left(\frac{U_2}{A_2(1 + \frac{\alpha_3}{R_2})}\right)\frac{\partial}{\partial \alpha_2}\left(\frac{U_2}{A_2(1 + \frac{\alpha_3}{R_2})}\right) \\ & + \frac{1}{A_1A_2(1 + \frac{\alpha_3}{R_1})(1 + \frac{\alpha_3}{R_2})}\frac{\partial U_3}{\partial \alpha_1}\frac{\partial U_3}{\partial \alpha_2}\end{aligned} \tag{11.1.9}$$

$$\begin{aligned}\varepsilon_{13} = {} & A_1\left(1 + \frac{\alpha_3}{R_1}\right)\frac{\partial}{\partial \alpha_3}\left(\frac{U_1}{A_1(1 + \frac{\alpha_3}{R_1})}\right) + \frac{1}{A_1(1 + \frac{\alpha_3}{R_1})}\frac{\partial U_3}{\partial \alpha_1} \\ & + A_1\left(1 + \frac{\alpha_3}{R_1}\right)\frac{\partial}{\partial \alpha_1}\left(\frac{U_1}{A_1(1 + \frac{\alpha_3}{R_1})}\right)\frac{\partial}{\partial \alpha_3}\left(\frac{U_1}{A_1(1 + \frac{\alpha_3}{R_1})}\right) \\ & + \frac{A_2^2(1 + \frac{\alpha_3}{R_2})^2}{A_1(1 + \frac{\alpha_3}{R_1})}\frac{\partial}{\partial \alpha_1}\left(\frac{U_2}{A_2(1 + \frac{\alpha_3}{R_2})}\right)\frac{\partial}{\partial \alpha_3}\left(\frac{U_2}{A_2(1 + \frac{\alpha_3}{R_2})}\right) \\ & + \frac{1}{A_1(1 + \frac{\alpha_3}{R_1})}\frac{\partial U_3}{\partial \alpha_1}\frac{\partial U_3}{\partial \alpha_3}\end{aligned} \tag{11.1.10}$$

$$\varepsilon_{23} = A_2\left(1 + \frac{\alpha_3}{R_2}\right) \frac{\partial}{\partial\alpha_3}\left(\frac{U_2}{A_2(1 + \frac{\alpha_3}{R_2})}\right) + \frac{1}{A_2(1 + \frac{\alpha_3}{R_2})} \frac{\partial U_3}{\partial\alpha_2}$$

$$+ A_2\left(1 + \frac{\alpha_3}{R_2}\right) \frac{\partial}{\partial\alpha_2}\left(\frac{U_2}{A_2(1 + \frac{\alpha_3}{R_2})}\right) \frac{\partial}{\partial\alpha_3}\left(\frac{U_2}{A_2(1 + \frac{\alpha_3}{R_2})}\right)$$

$$+ \frac{A_1^2(1 + \frac{\alpha_3}{R_1})^2}{A_2(1 + \frac{\alpha_3}{R_2})} \frac{\partial}{\partial\alpha_2}\left(\frac{U_1}{A_1(1 + \frac{\alpha_3}{R_1})}\right) \frac{\partial}{\partial\alpha_3}\left(\frac{U_1}{A_1(1 + \frac{\alpha_3}{R_1})}\right)$$

$$+ \frac{1}{A_2(1 + \frac{\alpha_3}{R_2})} \frac{\partial U_3}{\partial\alpha_2} \frac{\partial U_3}{\partial\alpha_3} \tag{11.1.11}$$

Note that U_3 is not a function of α_3 and each last term in Eqs. (11.1.10) and (11.1.11) is therefore zero. We simplify further by neglecting all square terms involving U_1 and U_2 since it is reasonable to be expected that they will always be small compared to the transverse deflections U_3. Equations (11.1.6) to (11.1.11) become, therefore,

$$\varepsilon_{11} = \frac{1}{A_1(1 + \frac{\alpha_3}{R_1})}\left(\frac{\partial U_1}{\partial\alpha_1} + \frac{U_2}{A_2}\frac{\partial A_1}{\partial\alpha_2} + U_3\frac{A_1}{R_1}\right) + \frac{1}{2A_1^2}\left(\frac{\partial U_3}{\partial\alpha_1}\right)^2 \tag{11.1.12}$$

$$\varepsilon_{22} = \frac{1}{A_2(1 + \frac{\alpha_3}{R_2})}\left(\frac{\partial U_2}{\partial\alpha_2} + \frac{U_1}{A_1}\frac{\partial A_2}{\partial\alpha_1} + U_3\frac{A_2}{R_2}\right) + \frac{1}{2A_2^2}\left(\frac{\partial U_3}{\partial\alpha_2}\right)^2 \tag{11.1.13}$$

$$\varepsilon_{33} = 0 \tag{11.1.14}$$

$$\varepsilon_{12} = \frac{A_1(1 + \frac{\alpha_3}{R_1})}{A_2(1 + \frac{\alpha_3}{R_2})} \frac{\partial}{\partial\alpha_2}\left(\frac{U_1}{A_1(1 + \frac{\alpha_3}{R_1})}\right) + \frac{A_2(1 + \frac{\alpha_3}{R_2})}{A_1(1 + \frac{\alpha_3}{R_1})} \frac{\partial}{\partial\alpha_1}\left(\frac{U_2}{A_2(1 + \frac{\alpha_3}{R_2})}\right)$$

$$+ \frac{1}{A_1A_2(1 + \frac{\alpha_3}{R_1})(1 + \frac{\alpha_3}{R_2})} \frac{\partial U_3}{\partial\alpha_1} \frac{\partial U_3}{\partial\alpha_2} \tag{11.1.15}$$

$$\varepsilon_{13} = A_1\left(1 + \frac{\alpha_3}{R_1}\right)\frac{\partial}{\partial\alpha_3}\left(\frac{U_1}{A_1(1 + \frac{\alpha_3}{R_1})}\right) + \frac{1}{A_1(1 + \frac{\alpha_3}{R_1})}\frac{\partial U_3}{\partial\alpha_1} \tag{11.1.16}$$

$$\varepsilon_{23} = A_2\left(1 + \frac{\alpha_3}{R_2}\right)\frac{\partial}{\partial\alpha_3}\left(\frac{U_2}{A_2(1 + \frac{\alpha_3}{R_2})}\right) + \frac{1}{A_2(1 + \frac{\alpha_3}{R_2})}\frac{\partial U_3}{\partial\alpha_2} \tag{11.1.17}$$

Next, if we follow Sec. 2.4 where we have introduced the assumptions that displacements U_1 and U_2 are a linear function of α_3, we obtain

$$\varepsilon_{11} = \varepsilon_{11}^{\circ} + \alpha_3 k_{11} \tag{11.1.18}$$

$$\varepsilon_{22} = \varepsilon_{22}^{\circ} + \alpha_3 k_{22} \tag{11.1.19}$$

$$\varepsilon_{12} = \varepsilon_{12}^{\circ} + \alpha_3 k_{12} \tag{11.1.20}$$

where the membrane strains become

$$\varepsilon_{11}^{\circ} = \frac{1}{A_1}\frac{\partial u_1}{\partial\alpha_1} + \frac{u_2}{A_1A_2}\frac{\partial A_1}{\partial\alpha_2} + \frac{u_3}{R_1} + \frac{1}{2A_1^2}\left(\frac{\partial u_3}{\partial\alpha_1}\right)^2 \tag{11.1.21}$$

$$\varepsilon_{22}^{\circ} = \frac{1}{A_2}\frac{\partial u_2}{\partial\alpha_2} + \frac{u_1}{A_1A_2}\frac{\partial A_2}{\partial\alpha_1} + \frac{u_3}{R_2} + \frac{1}{2A_2^2}\left(\frac{\partial u_3}{\partial\alpha_2}\right)^2 \tag{11.1.22}$$

$$\varepsilon_{12}^{\circ} = \frac{A_2}{A_1}\frac{\partial}{\partial\alpha_1}\left(\frac{u_2}{A_2}\right) + \frac{A_1}{A_2}\frac{\partial}{\partial\alpha_2}\left(\frac{u_1}{A_1}\right) + \frac{1}{A_1A_2}\left(\frac{\partial u_3}{\partial\alpha_1}\right)\left(\frac{\partial u_3}{\partial\alpha_2}\right) \tag{11.1.23}$$

The bending strains are the same as given by Eqs. (2.4.22) to (2.4.24). The definitions of β_1 and β_2 remain the same also.

11.2 THE EQUATIONS OF MOTION

While the general expression for strain energy given in Eq. (2.6.3) is still true, the value of F becomes now (the superscript r means "residual")

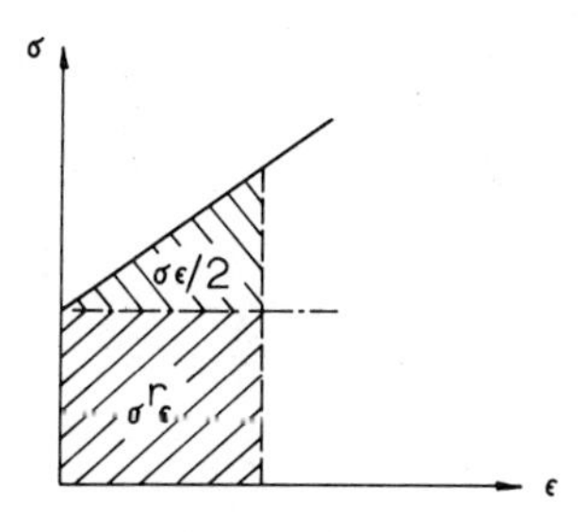

Figure 11.2.1

$$F = \frac{1}{2}(\sigma_{11}\varepsilon_{11} + \sigma_{22}\varepsilon_{22} + \sigma_{12}\varepsilon_{12} + \sigma_{13}\varepsilon_{13} + \sigma_{23}\varepsilon_{23})$$
$$+ \sigma_{11}^r\varepsilon_{11} + \sigma_{22}^r\varepsilon_{22} + \sigma_{12}^r\varepsilon_{12} + \sigma_{13}^r\varepsilon_{13} + \sigma_{23}^r\varepsilon_{23} \tag{11.2.1}$$

This is illustrated in Fig. 11.2.1. The residual stress σ^r is constant, independent of the vibration-induced strain ε, but energy is stored by the deflection and is equal to the rectangular area $\sigma^r\varepsilon$. The variation δF is therefore

$$\delta F = (\sigma_{11} + \sigma_{11}^r)\delta\varepsilon_{11} + (\sigma_{22} + \sigma_{22}^r)\delta\varepsilon_{22} + (\sigma_{12} + \sigma_{12}^r)\delta\varepsilon_{12}$$
$$+ (\sigma_{13} + \sigma_{13}^r)\delta\varepsilon_{13} + (\sigma_{23} + \sigma_{23}^r)\delta\varepsilon_{23} \tag{11.2.2}$$

Recognizing the fact that the integration over the shell thickness of the residual stresses gives rise to force and moment resultants

$$N_{11}^r = \int_{-\frac{h}{2}}^{\frac{h}{2}} \sigma_{11}^r \, d\alpha_3 \tag{11.2.3}$$

$$M_{11}^r = \int_{-\frac{h}{2}}^{\frac{h}{2}} \sigma_{11}^r\alpha_3 \, d\alpha_3 \tag{11.2.4}$$

and so on for N_{12}^r, N_{22}^r, M_{12}^r, M_{22}^r, Q_{13}^r, and Q_{23}^r, we obtain the equations of motion in a similar way as described in Sec. 2.7. Let us consider, for instance, the first term in the strain energy expression. Since

$$\varepsilon_{11} = \left[\cdots + \frac{1}{2A_1^2}\left(\frac{\partial u_3}{\partial \alpha_1}\right)^2\right] \tag{11.2.5}$$

or

$$\delta\varepsilon_{11} = \left(\cdots + \frac{1}{A_1^2}\frac{\partial u_3}{\partial \alpha_1}\frac{\partial(\delta u_3)}{\partial \alpha_1}\right) \tag{11.2.6}$$

this first term becomes

$$\int_{\alpha_1}\int_{\alpha_2}\int_{\alpha_3} (\sigma_{11} + \sigma_{11}^r)\left(\cdots + \frac{A_2}{A_1}\frac{\partial u_3}{\partial \alpha_1}\frac{\partial(\delta u_3)}{\partial \alpha_1}\right) d\alpha_1\, d\alpha_2\, d\alpha_3 \tag{11.2.7}$$

Integrating over the thickness gives

$$\int_{\alpha_1}\int_{\alpha_2} (N_{11} + N_{11}^r)\left(\cdots + \frac{A_2}{A_1}\frac{\partial u_3}{\partial \alpha_1}\frac{\partial(\delta u_3)}{\partial \alpha_1}\right) d\alpha_1\, d\alpha_2 + \int_{\alpha_1}\int_{\alpha_2} (M_{11} + M_{11}^r)(\cdots)\, d\alpha_1\, d\alpha_2 \tag{11.2.8}$$

The notation $(\cdots)$ indicates that all other terms are the same as those in Chap. 2.

If we integrate by parts with respect to α_1, we get

$$\cdots \int_{\alpha_2} (N_{11} + N_{11}^r)\frac{A_2}{A_1}\frac{\partial u_3}{\partial \alpha_1}\delta u_3\, d\alpha_2 + \int_{\alpha_1}\int_{\alpha_2} \frac{\partial}{\partial \alpha_1}\left[(N_{11} + N_{11}^r)\frac{A_2}{A_1}\frac{\partial u_3}{\partial \alpha_1}\right]\delta u_3\, d\alpha_1\, d\alpha_2 \cdots \tag{11.2.9}$$

Proceeding similarly with the other new terms and collecting coefficients of the virtual displacements gives finally the new set of equations of motion.

$$- \frac{\partial}{\partial \alpha_1} [(N_{11} + N_{11}^r)A_2] - \frac{\partial}{\partial \alpha_2} [(N_{21} + N_{21}^r)A_1] - (N_{12} + N_{12}^r) \frac{\partial A_1}{\partial \alpha_2}$$

$$+ (N_{22} + N_{22}^r) \frac{\partial A_2}{\partial \alpha_1} - (Q_{13} + Q_{13}^r) \frac{A_1 A_2}{R_1} + A_1 A_2 \rho h \ddot{u}_1 = A_1 A_2 q_1$$

(11.2.10)

$$- \frac{\partial}{\partial \alpha_1} [(N_{12} + N_{12}^r)A_2] - \frac{\partial}{\partial \alpha_2} [(N_{22} + N_{22}^r)A_1] - (N_{21} + N_{21}^r) \frac{\partial A_2}{\partial \alpha_1}$$

$$+ (N_{11} + N_{11}^r) \frac{\partial A_1}{\partial \alpha_2} - (Q_{23} + Q_{23}^r) \frac{A_1 A_2}{R_2} + A_1 A_2 \rho h \ddot{u}_2 = A_1 A_2 q_2$$

(11.2.11)

$$- \frac{\partial}{\partial \alpha_1} [(Q_{13} + Q_{13}^r)A_2] - \frac{\partial}{\partial \alpha_2} [(Q_{23} + Q_{23}^r)A_1] + (N_{11} + N_{11}^r) \frac{A_1 A_2}{R_1}$$

$$+ (N_{22} + N_{22}^r) \frac{A_1 A_2}{R_2} - \frac{\partial}{\partial \alpha_1} \left[(N_{11} + N_{11}^r) \frac{A_2}{A_1} \frac{\partial u_3}{\partial \alpha_1} \right]$$

$$- \frac{\partial}{\partial \alpha_2} \left[(N_{22} + N_{22}^r) \frac{A_1}{A_2} \frac{\partial u_3}{\partial \alpha_2} \right] - \frac{\partial}{\partial \alpha_1} \left[(N_{21} + N_{21}^r) \frac{\partial u_3}{\partial \alpha_2} \right]$$

$$- \frac{\partial}{\partial \alpha_2} \left[(N_{12} + N_{12}^r) \frac{\partial u_3}{\partial \alpha_1} \right] + A_1 A_2 \rho h \ddot{u}_3 = A_1 A_2 q_3 \qquad (11.2.12)$$

where

$$(Q_{13} + Q_{13}^r)A_1 A_2 = \frac{\partial}{\partial \alpha_1} [(M_{11} + M_{11}^r)A_2] + \frac{\partial}{\partial \alpha_2} [(M_{21} + M_{21}^r)A_1]$$

$$+ (M_{12} + M_{12}^r) \frac{\partial A_1}{\partial \alpha_2} - (M_{22} + M_{22}^r) \frac{\partial A_2}{\partial \alpha_1}$$

(11.2.13)

$$(Q_{23} + Q_{23}^r)A_1 A_2 = \frac{\partial}{\partial \alpha_1} [(M_{12} + M_{12}^r)A_2] + \frac{\partial}{\partial \alpha_2} [(M_{22} + M_{22}^r)A_1]$$

$$+ (M_{21} + M_{21}^r) \frac{\partial A_2}{\partial \alpha_1} - (M_{11} + M_{11}^r) \frac{\partial A_1}{\partial \alpha_2}$$

(11.2.14)

We recognize that the equations divide into a set that defines initial stresses as function of static loads q_1^r, q_2^r, q_3^r and a set that defines the vibration behavior as function of dynamic loads q_1, q_2, q_3. The first set becomes

$$-\frac{\partial}{\partial\alpha_1}(N^r_{11}A_2) - \frac{\partial}{\partial\alpha_2}(N^r_{21}A_1) - N^r_{12}\frac{\partial A_1}{\partial\alpha_2} + N^r_{22}\frac{\partial A_2}{\partial\alpha_1} - Q^r_{13}\frac{A_1A_2}{R_1} = A_1A_2q^r_1 \tag{11.2.15}$$

$$-\frac{\partial}{\partial\alpha_1}(N^r_{12}A_2) - \frac{\partial}{\partial\alpha_2}(N^r_{22}A_1) - N^r_{21}\frac{\partial A_2}{\partial\alpha_1} + N^r_{11}\frac{\partial A_1}{\partial\alpha_2} - Q^r_{23}\frac{A_1A_2}{R_2} = A_1A_2q^r_2 \tag{11.2.16}$$

$$-\frac{\partial}{\partial\alpha_1}(Q^r_{13}A_2) - \frac{\partial}{\partial\alpha_2}(Q^r_{23}A_1) + A_1A_2\left(\frac{N^r_{11}}{R_1} + \frac{N^r_{22}}{R_2}\right) = A_1A_2q^r_3 \tag{11.2.17}$$

where

$$Q^r_{13}A_1A_2 = \frac{\partial}{\partial\alpha_1}(M^r_{11}A_2) + \frac{\partial}{\partial\alpha_2}(M^r_{21}A_1) + M^r_{12}\frac{\partial A_1}{\partial\alpha_2} - M^r_{22}\frac{\partial A_2}{\partial\alpha_1} \tag{11.2.18}$$

$$Q^r_{23}A_1A_2 = \frac{\partial}{\partial\alpha_1}(M^r_{12}A_2) + \frac{\partial}{\partial\alpha_2}(M^r_{22}A_1) + M^r_{21}\frac{\partial A_2}{\partial\alpha_1} - M^r_{11}\frac{\partial A_1}{\partial\alpha_2} \tag{11.2.19}$$

This set is usually solved by way of stress functions for the unknown residual stresses. Once these are known, the following vibration equations are solved:

$$-\frac{\partial}{\partial\alpha_1}(N_{11}A_2) - \frac{\partial}{\partial\alpha_2}(N_{21}A_1) - N_{12}\frac{\partial A_1}{\partial\alpha_2} + N_{22}\frac{\partial A_2}{\partial\alpha_1} - Q_{13}\frac{A_1A_2}{R_1} + A_1A_2\rho h\ddot{u}_1 = A_1A_2q_1 \tag{11.2.20}$$

$$- \frac{\partial}{\partial\alpha_1} (N_{12}A_2) - \frac{\partial}{\partial\alpha_2} (N_{22}A_1) - N_{21} \frac{\partial A_2}{\partial\alpha_1} + N_{11} \frac{\partial A_1}{\partial\alpha_2}$$

$$- Q_{23} \frac{A_1A_2}{R_2} + A_1A_2\rho\ddot{u}_2 = A_1A_2q_2 \qquad (11.2.21)$$

$$- \frac{\partial}{\partial\alpha_1} (Q_{13}A_2) - \frac{\partial}{\partial\alpha_2} (Q_{23}A_1) + A_1A_2\left(\frac{N_{11}}{R_1} + \frac{N_{22}}{R_2}\right)$$

$$- \frac{\partial}{\partial\alpha_1}\left(N_{11}^r \frac{A_2}{A_1} \frac{\partial u_3}{\partial\alpha_1}\right) - \frac{\partial}{\partial\alpha_2}\left(N_{22}^r \frac{A_1}{A_2} \frac{\partial u_3}{\partial\alpha_2}\right)$$

$$- \frac{\partial}{\partial\alpha_1}\left(N_{21}^r \frac{\partial u_3}{\partial\alpha_2}\right) - \frac{\partial}{\partial\alpha_2}\left(N_{12}^r \frac{\partial u_3}{\partial\alpha_1}\right)$$

$$+ A_1A_2\rho\ddot{u}_3 = A_1A_2q_3 \qquad (11.2.22)$$

and where

$$Q_{13}A_1A_2 = \frac{\partial}{\partial\alpha_1} (M_{11}A_2) + \frac{\partial}{\partial\alpha_2} (M_{21}A_1) + M_{12} \frac{\partial A_1}{\partial\alpha_2} - M_{22} \frac{\partial A_2}{\partial\alpha_1} \qquad (11.2.23)$$

$$Q_{23}A_1A_2 = \frac{\partial}{\partial\alpha_1} (M_{12}A_2) + \frac{\partial}{\partial\alpha_2} (M_{22}A_1) + M_{21} \frac{\partial A_2}{\partial\alpha_1} - M_{11} \frac{\partial A_1}{\partial\alpha_2} \qquad (11.2.24)$$

Note that in the third to sixth term of Eq. (11.2.22) we have neglected N_{11}, N_{22}, and N_{12} since they are small in effect as compared to N_{11}^r, N_{22}^r, and N_{12}^r. Not only is this a good assumption for most engineering problems of this type, but also a necessary one if we wish to preserve the linearity of our equations. If this assumption is not possible, for instance, in dynamic buckling problems, a nonlinear set of equations results.

11.3 PURE MEMBRANES

Equations (11.2.10) to (11.2.12) give us the opportunity to obtain the equations of motion of pure membranes. By pure membrane is meant that the structure is a skin stretched over a frame under initial tension N_{11}^r, N_{22}^r, and N_{12}^r. There is a school of thought that holds that pure membranes are not able to support shear forces ($N_{12}^r = 0$), but this is technically not supportable. That shear can exist without buckling the membrane can be shown by stretching a rubber skin over a rectangular frame and stretching it at different nonuniform amounts. In special cases, like the uniformly stretched drum head (circular membrane), shear is zero. The only restriction on a pure membrane is that it cannot support compressive membrane stresses.

A membrane skin has negligible bending resistance ($D = 0$). This implies that all bending moments are zero. Equations (11.2.20) and (11.2.21) become

$$-\frac{\partial}{\partial\alpha_1}(N_{11}A_2) - \frac{\partial}{\partial\alpha_2}(N_{21}A_1) - N_{12}\frac{\partial A_1}{\partial\alpha_2} + N_{22}\frac{\partial A_2}{\partial\alpha_1} + A_1A_2\rho h\ddot{u}_1 = A_1A_2q_1 \tag{11.3.1}$$

$$-\frac{\partial}{\partial\alpha_1}(N_{12}A_2) - \frac{\partial}{\partial\alpha_2}(N_{22}A_1) - N_{21}\frac{\partial A_2}{\partial\alpha_1} + N_{11}\frac{\partial A_1}{\partial\alpha_2} + A_1A_2\rho h\ddot{u}_2 = A_1A_2q_2 \tag{11.3.2}$$

These equations describe the motion of the membrane material in the tangent plane. This motion is independent from the initial stress state. Equation (11.2.22) becomes

$$A_1A_2\left(\frac{N_{11}}{R_1} + \frac{N_{22}}{R_2}\right) - \frac{\partial}{\partial\alpha_1}\left(N_{11}^r\frac{A_2}{A_1}\frac{\partial u_3}{\partial\alpha_1}\right) - \frac{\partial}{\partial\alpha_2}\left(N_{22}^r\frac{A_1}{A_2}\frac{\partial u_3}{\partial\alpha_2}\right) - \frac{\partial}{\partial\alpha_1}\left(N_{21}^r\frac{\partial u_3}{\partial\alpha_2}\right) - \frac{\partial}{\partial\alpha_2}\left(N_{12}^r\frac{\partial u_3}{\partial\alpha_1}\right) + A_1A_2\rho h\ddot{u}_3 = A_1A_2q_3 \tag{11.3.3}$$

This is the membrane equation for transverse vibration. For membranes that are curved surfaces, the three equations of motion are coupled. For flat membranes, Eq. (11.3.3) is independent of Eqs. (11.3.1) and (11.3.2).

The initial stresses in the membrane are calculated from Eqs. (11.2.15) to (11.2.17), which become (q_1^r, q_2^r, and q_3^r are in this case static loads)

$$-\frac{\partial}{\partial\alpha_1}(N_{11}^r A_2) - \frac{\partial}{\partial\alpha_2}(N_{21}^r A_1) - N_{12}^r \frac{\partial A_1}{\partial\alpha_2} + N_{22}^r \frac{\partial A_2}{\partial\alpha_1} = A_1 A_2 q_1^r \tag{11.3.4}$$

$$-\frac{\partial}{\partial\alpha_1}(N_{12}^r A_2) - \frac{\partial}{\partial\alpha_2}(N_{22}^r A_1) - N_{21}^r \frac{\partial A_2}{\partial\alpha_1} + N_{11}^r \frac{\partial A_1}{\partial\alpha_2} = A_1 A_2 q_2^r \tag{11.3.5}$$

$$A_1 A_2 \left(\frac{N_{11}^r}{R_1} + \frac{N_{22}^r}{R_2}\right) = A_1 A_2 q_3^r \tag{11.3.6}$$

11.4 EXAMPLE: THE CIRCULAR MEMBRANE

Let us look at the classical membrane problem. It is the case of a drum skin shown in Fig. 11.4.1, stretched uniformly around the periphery such that it creates a uniform boundary tension.

$$N_{rr}^r = N_{rr}^* \tag{11.4.1}$$

At this point, we could state that the resulting tension is uniform through the entire membrane and proceed to the vibration problem, but let us prove this using the static equations.

The fundamental form is

$$(ds)^2 = (dr)^2 + r^2(d\theta)^2 \tag{11.4.2}$$

Thus, $A_1 = 1$, $A_2 = r$, $d\alpha_1 = dr$, and $d\alpha_2 = d\theta$. Let us now calculate the initial stress distribution in the interior of the membrane. Equations (11.3.4) and (11.3.5) become [Eq. (11.3.6) is inapplicable since a flat membrane has no curvature]

$$-\frac{\partial}{\partial r}(N_{rr}^r r) - \frac{\partial}{\partial\theta}(N_{\theta r}^r) + N_{\theta\theta}^r = 0 \tag{11.4.3}$$

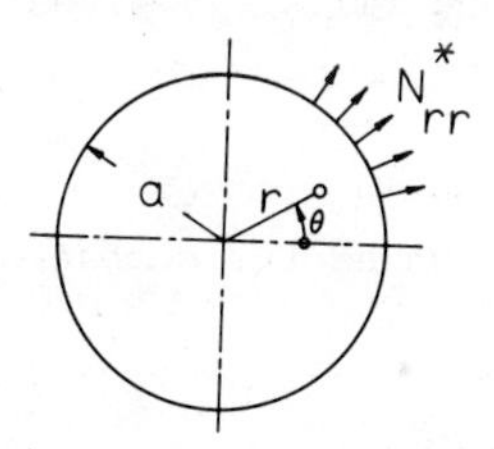

Figure 11.4.1

$$- \frac{\partial}{\partial r} (N^r_{r\theta} r) - \frac{\partial}{\partial \theta} (N^r_{\theta\theta}) - N^r_{\theta r} = 0 \tag{11.4.4}$$

with the boundary condition, at r = a,

$$N^r_{rr} = N^*_{rr} \tag{11.4.5}$$

Since the loading is axisymmetric, the stress state has to be axisymmetric. Thus,

$$\frac{\partial(\cdot)}{\partial \theta} = 0 \tag{11.4.6}$$

and $$N^r_{r\theta} = 0 \tag{11.4.7}$$

This gives

$$- \frac{\partial}{\partial r} (N^r_{rr} r) + N^r_{\theta\theta} = 0 \tag{11.4.8}$$

This equation is satisfied if

$$F = N^r_{rr} r \tag{11.4.9}$$

$$\frac{dF}{dr} = N^r_{\theta\theta} \tag{11.4.10}$$

Since

$$\varepsilon^r_{rr} = \frac{du_r}{dr} \tag{11.4.11}$$

and $$\varepsilon^r_{\theta\theta} = \frac{u_r}{r} \tag{11.4.12}$$

or $$\frac{d}{dr} (r\varepsilon^r_{\theta\theta}) = \frac{du_r}{dr} \tag{11.4.13}$$

equating Eqs. (11.4.13) and (11.4.11) gives

$$\frac{d}{dr} (r\varepsilon^r_{\theta\theta}) - \varepsilon^r_{rr} = 0 \tag{11.4.14}$$

This equation is known as the *compatibility equation*. Furthermore, from

$$N^r_{rr} = K(\varepsilon^r_{rr} + \mu\varepsilon^r_{\theta\theta}) \tag{11.4.15}$$

$$N_{\theta\theta}^r = K(\varepsilon_{\theta\theta}^r + \mu\varepsilon_{rr}^r) \tag{11.4.16}$$

we get

$$\varepsilon_{rr}^r = \frac{1}{K}(N_{rr}^r + \mu N_{\theta\theta}^r) = \frac{1}{K}\left(\frac{F}{r} + \mu\frac{dF}{dr}\right) \tag{11.4.17}$$

and

$$\varepsilon_{\theta\theta}^r = \frac{1}{K}(N_{\theta\theta}^r + \mu N_{rr}^r) = \frac{1}{K}\left(\frac{dF}{dr} + \mu\frac{F}{r}\right) \tag{11.4.18}$$

Equation (11.4.14) becomes, therefore,

$$\frac{d^2F}{dr^2} + \frac{1}{r}\frac{dF}{dr} - \frac{F}{r^2} = 0 \tag{11.4.19}$$

or

$$\frac{d}{dr}\left[\frac{1}{r}\frac{d(Fr)}{dr}\right] = 0 \tag{11.4.20}$$

Integrating gives

$$F = C_1 r + C_2\frac{1}{r} \tag{11.4.21}$$

We have, at $r = a$,

$$F = N_{rr}^* a \tag{11.4.22}$$

Also, at $r = 0$, a singularity cannot exist, which implies that $C_2 = 0$. Thus

$$C_1 = N_{rr}^* \tag{11.4.23}$$

Therefore

$$F = N_{rr}^* r \tag{11.4.24}$$

or

$$N_{rr}^r = \frac{F}{r} = N_{rr}^* \tag{11.4.25}$$

$$N_{\theta\theta}^r = \frac{dF}{dr} = N_{rr}^* \tag{11.4.26}$$

This result implies that a circular membrane under uniform boundary tension has equal and constant stress resultants in the radial and circumferential directions at any point in its interior, i.e.,

$$N_{rr}^{r} = N_{\theta\theta}^{r} = N_{rr}^{*} \tag{11.4.27}$$

Note that this is the case only for uniform boundary tension. If the boundary tension is nonuniform, nonuniform interior membrane forces result that include shear forces.

We may now proceed to Eq. (11.3.3), which describes the transverse vibration of the membrane. It becomes, for our example,

$$- N_{rr}^{*}\left(\frac{\partial^2 u_3}{\partial r^2} + \frac{1}{r}\frac{\partial u_3}{\partial r} + \frac{1}{r^2}\frac{\partial^2 u_3}{\partial \theta^2}\right) + \rho h \ddot{u}_3 = q_3 \tag{11.4.28}$$

To solve the eigenvalue problem, we set $q_3 = 0$ and write

$$u_3(r,\theta,t) = U_{3k}e^{j\omega_k t} \tag{11.4.29}$$

Substituting this in the equation of motion gives

$$- N_{rr}^{*}\left(\frac{\partial^2 U_{3k}}{\partial r^2} + \frac{1}{r}\frac{\partial U_{3k}}{\partial r} + \frac{1}{r^2}\frac{\partial^2 U_{3k}}{\partial \theta^2}\right) - \rho h \omega_k^2 U_{3k} = 0 \tag{11.4.30}$$

Suspecting that we may be able to separate variables, we write

$$U_{3k}(r,\theta) = R(r)\Theta(\theta) \tag{11.4.31}$$

Substituting this gives

$$\frac{r^2}{R}\frac{d^2R}{dr^2} + \frac{r}{R}\frac{dR}{dr} + r^2\frac{\omega_k^2\rho h}{N_{rr}^{*}} = -\frac{1}{\Theta}\frac{d^2\Theta}{d\theta^2} \tag{11.4.32}$$

The left and right side of Eq. (11.4.32) must be equal to the same constant, which we call p^2. This gives

$$\frac{d^2\Theta}{d\theta^2} + p^2\Theta = 0 \tag{11.4.33}$$

and

$$r^2\frac{d^2R}{dr^2} + r\frac{dR}{dr} + \left(\frac{\omega_k^2\rho h}{N_{rr}^{*}}r^2 - p^2\right)R = 0 \tag{11.4.34}$$

The solution of Eq. (11.4.33) is

$$\Theta = A \cos p(\theta - \phi) \tag{11.4.35}$$

where ϕ is an arbitrary angle and A is an arbitrary constant. However, for the closed drumhead, Θ must be a periodic function of period 2π to preserve continuity of deflection. Thus

$$p = n \quad (n = 0, 1, 2, \ldots) \tag{11.4.36}$$

Defining

$$\lambda^2 = \frac{\omega_k^2 \rho h}{N_{rr}^*} \tag{11.4.37}$$

and a new variable

$$\psi^2 = \lambda^2 r^2 \tag{11.4.38}$$

we may write Eq. (11.4.34) as

$$\frac{d^2R}{d\psi^2} + \frac{1}{\psi}\frac{dR}{d\psi} + \left(1 - \frac{n^2}{\psi^2}\right) R = 0 \tag{11.4.39}$$

This type of equation is called *Bessel's differential equation of integer order n,* as we have already seen in Chap. 5 for the circular plate. It can be solved using a power series, which can be formulated into what are known as *Bessel functions*. For each value of the integer n there are two linearly independent solutions of Bessel's equation. One of them is the Bessel function of the first kind of order n, denoted as $J_n(\psi)$. The other is the Bessel function of the second kind of order n, denoted by $Y_n(\psi)$. Thus, the solution of Eq. (11.4.39) is, for each n,

$$R_n = B_n J_n(\psi) + C_n Y_n(\psi) \tag{11.4.40}$$

Note that

$$Y_n(0) = \infty \tag{11.4.41}$$

Thus, since it is physically impossible for the drumhead to have an infinite deflection at its center, it follows that

$$C_n = 0 \tag{11.4.42}$$

The other condition that has to be satisfied is that at $r = a$

$$u_3 = 0 \tag{11.4.43}$$

which implies that

$$R(\psi = \lambda a) = 0 \tag{11.4.44}$$

or that

$$J_n(\lambda a) = 0 \tag{11.4.45}$$

For a given n, this equation has an infinite number of roots $(\lambda a)_{mn}$, identified by m = 0, 1, 2, ... in ascending order. A few of these are listed in Table 11.4.1.

Table 11.4.1 Values for $(\lambda a)_{mn}$

m \ n	0	1	2	3
0	2.404	5.520	8.654	11.792
1	3.832	7.016	10.173	13.323
2	5.135	8.417	11.620	14.796
3	6.379	9.760	13.017	16.224

The natural frequencies of the membrane are therefore given by

$$\omega_k = \omega_{mn} = \frac{(\lambda a)_{mn}}{a}\sqrt{\frac{N^*_{rr}}{\rho h}} \tag{11.4.46}$$

To obtain the natural mode, we substitute in turn each natural frequency back into Eq. (11.4.31) and get, utilizing Eqs. (11.4.35) and (11.4.40),

$$U_k = U_{mn} = J_n(\lambda_{mn} r) \cos n(\theta - \phi) \tag{11.4.47}$$

This problem was first solved by Pagani [11.4].

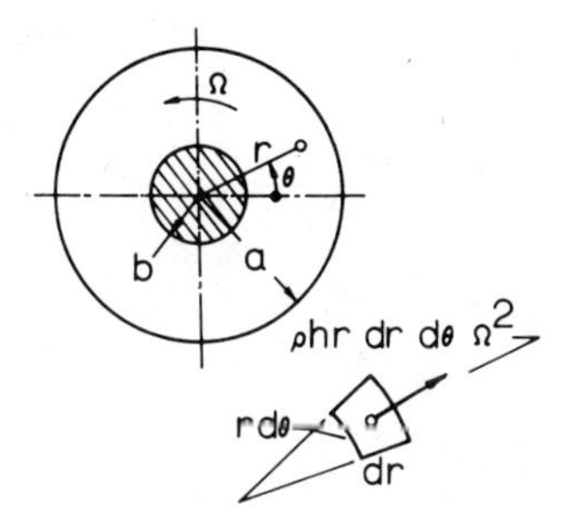

Figure 11.5.1

11.5 SPINNING SAW BLADE

When a circular saw blade as shown in Fig. 11.5.1 is spinning with a constant rotational speed Ω in radians per second, centrifugal forces create a stress field that acts like an initial stress in raising natural frequencies. The first treatment of a saw-blade-like case was given by Southwell [11.5]. Other variations include the effect of temperature distribution and purposely induced residual stresses, as typically studied by Mote [11.6].

Let us first obtain the centrifugal stresses. Since we have in this case an axisymmetric situation, all derivations with respect to θ vanish. For $\alpha_1 = r$, $A_1 = 1$, $\alpha_2 = \theta$, $A_2 = r$, Eq. (11.2.15) becomes

$$-\frac{d}{dr}(rN_{rr}^{r}) + N_{\theta\theta}^{r} = rq_r \tag{11.5.1}$$

where the load q_r is the centrifugal force created by a mass element as it is spinning at radius r divided by the area $rd\theta\, dr$:

$$q_r = \rho h \Omega^2 r \tag{11.5.2}$$

Since the membrane strains are

$$\varepsilon_{rr}^{\circ} = \frac{du_r^r}{dr} \tag{11.5.3}$$

$$\varepsilon_{\theta\theta}^{\circ} = \frac{u_r^r}{r} \tag{11.5.4}$$

where the superscript r indicates again residual, we obtain

$$r \frac{d^2 u_r^r}{dr^2} + \frac{du_r^r}{dr} - \frac{u_r^r}{r} = -\frac{\rho h \Omega^2 r^2}{K} \tag{11.5.5}$$

This can be written as

$$r \frac{d}{dr}\left[\frac{1}{r}\frac{d}{dr}(ru_r^r)\right] = -\frac{\rho h \Omega^2 r^2}{K} \tag{11.5.6}$$

Integrating this gives the radial displacement due to the centrifugal force as

$$u_r^r = -\frac{\rho h \Omega^2 r^3}{8K} + C_1 r + \frac{C_2}{r} \tag{11.5.7}$$

The integration constants have to be evaluated from the boundary conditions. In case of the freely spinning saw blade, these boundary conditions are

$$u_r^r(r = b) = 0 \tag{11.5.8}$$

$$N_{rr}^r(r = a) = K\left(\frac{du_r^r}{dr} + \mu \frac{u_r^r}{r}\right)_{r=a} = 0 \tag{11.5.9}$$

Since

$$N_{rr}^r = (1 + \mu)KC_1 - \frac{(1 - \mu)K}{r^2} C_2 - \frac{\rho h \Omega^2 (3 + \mu)}{8} r^2 \tag{11.5.10}$$

we obtain

$$\begin{bmatrix} (1 + \mu)a^2 & (\mu - 1) \\ b^2 & 1 \end{bmatrix} \begin{Bmatrix} C_1 \\ C_2 \end{Bmatrix} = \frac{\rho h \Omega^2}{8K} \begin{Bmatrix} a^4(3 + \mu) \\ b^4 \end{Bmatrix} \tag{11.5.11}$$

From this we find

$$C_1 = \frac{\rho h \Omega^2}{8K} \frac{a^4(3 + \mu) + (1 - \mu)b^4}{(1 + \mu)a^2 + (1 - \mu)b^2} \tag{11.5.12}$$

$$C_2 = \frac{\rho h\Omega^2}{8K} \frac{(1 + \mu)a^2b^4 - (3 + \mu)a^4b^2}{(1 + \mu)a^2 + (1 - \mu)b^2} \tag{11.5.13}$$

The membrane stress resultant N^r_{rr} is therefore defined. The membrane stress resultant $N^r_{\theta\theta}$ is given by

$$N^r_{\theta\theta} = (1 + \mu)KC_1 + \frac{(1 - \mu)K}{r^2} C_2 - \frac{\rho h\Omega^2(3\mu + 1)}{8} r^2 \tag{11.5.14}$$

The equation of motion is defined by Eq. (11.2.22). From it we obtain as the equation of motion of the spinning saw blade

$$D\nabla^4 u_3 - \frac{1}{r} \frac{\partial}{\partial r}\left(N^r_{rr} r \frac{\partial u_3}{\partial r}\right) - \frac{1}{r} \frac{\partial}{\partial \theta}\left(N^r_{\theta\theta} \frac{1}{r} \frac{\partial u_3}{\partial \theta}\right) + \rho h \ddot{u}_3 = q_3 \tag{11.5.15}$$

To obtain the natural frequencies, we set $q_3 = 0$ and let

$$u_3 = U_3 e^{j\omega t} \tag{11.5.16}$$

This gives

$$D\nabla^4 U_3 - \frac{1}{r} \frac{\partial}{\partial r}\left(N^r_{rr} r \frac{\partial U_3}{\partial r}\right) - \frac{1}{r} \frac{\partial}{\partial \theta}\left(N^r_{\theta\theta} \frac{1}{r} \frac{\partial U_3}{\partial \theta}\right) - \rho h\omega^2 U_3 = 0 \tag{11.5.17}$$

A general exact solution in closed form has not been found yet for the spinning saw problem. Both Southwell [11.5] and Mote [11.6] approached it with variational methods. Mote used the Raleigh-Ritz technique with the function

$$U_3(r,\theta) = (r - b) \sum_{i=0}^{i=k} a_i (r - b)^i \cos n\theta \tag{11.5.18}$$

Numerical values for the case where a = 80 mm, b = 20 mm, h = 1 mm, the material is carbon steel, and the rotational speed is in rotations per minute, are shown in Fig. 11.5.2. All results are for modes that have no nodal circles, only nodal diameters. Nodal circles occur only at still higher frequencies for this case. It

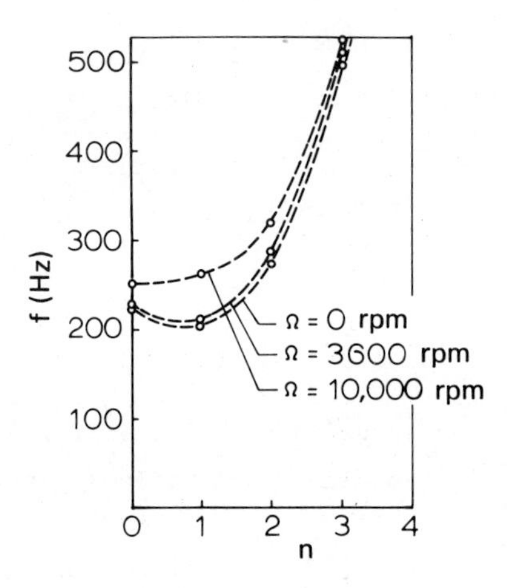

Figure 11.5.2

is interesting to note that the influence of the centrifugal force can be ignored at two-pole-induction motor speed (3600 rpm). This seems also to be the general finding for saw blades of other normally used dimensions. However, as the speed increases beyond that, the centrifugal effect gains in importance because its influence grows with the square of the rotational speed.

Treatment of other membrane stress-producing effects like heating (thermal stresses) or "tensioning," a process where beneficial residual stresses are hammered into the blade, can be found in Refs. 11.7 to 11.9.

11.6 DONNELL-MUSHTARI-VLASOV EQUATIONS EXTENDED TO INCLUDE INITIAL STRESSES

Since we have already seen that the Donnell-Mushtari-Vlasov equations are a useful simplification of the Love's equations, let us proceed through an identical simplification process with Eqs. (11.2.20) to (11.2.24). We obtain

$$D\nabla^4 u_3 + \nabla_k^2 \phi - \nabla_r^2 u_3 + \rho h \ddot{u}_3 = q_3 \tag{11.6.1}$$

and $$Eh\nabla_k^2 u_3 - \nabla^4 \phi = 0 \tag{11.6.2}$$

where

$$\nabla_r^2(\cdot) = \frac{1}{A_1A_2}\left[\frac{\partial}{\partial\alpha_1}\left(N_{11}^r \frac{A_2}{A_1} \frac{\partial(\cdot)}{\partial\alpha_1}\right) + \frac{\partial}{\partial\alpha_2}\left(N_{22}^r \frac{A_1}{A_2} \frac{\partial(\cdot)}{\partial\alpha_2}\right)\right.$$
$$\left. + \frac{\partial}{\partial\alpha_1}\left(N_{21}^r \frac{\partial(\cdot)}{\partial\alpha_2}\right) + \frac{\partial}{\partial\alpha_2}\left(N_{12}^r \frac{\partial(\cdot)}{\partial\alpha_1}\right)\right] \tag{11.6.3}$$

and where $\nabla_k^2(\cdot)$ is defined by Eq. (6.7.11) and $\nabla^2(\cdot)$ by Eq. (4.4.21).

To find the natural frequencies and modes, we set $q_3 = 0$ and

$$u_3(\alpha_1,\alpha_2,t) = U_3(\alpha_1,\alpha_2)e^{j\omega t} \tag{11.6.4}$$

$$\phi(\alpha_1,\alpha_2,t) = \Phi(\alpha_1,\alpha_2)e^{j\omega t} \tag{11.6.5}$$

This gives

$$D\nabla^4U_3 + \nabla_k^2\Phi - \nabla_r^2U_3 - \rho h\omega^2U_3 = 0 \tag{11.6.6}$$

and $$Eh\nabla_k^2U_3 - \nabla^4\Phi = 0 \tag{11.6.7}$$

Operating with ∇^4 and ∇_k^2 on these equations allows again a combination

$$D\nabla^8U_3 + Eh\nabla_k^4U_3 - \nabla^4\nabla_r^2U_3 - \rho h\omega^2\nabla^4U_3 = 0 \tag{11.6.8}$$

Let us now investigate, as an example, a closed circular cylindrical shell that is under a uniform boundary tension T in newtons per meter (N/m) as shown in Fig. 11.6.1. It can be shown, by solving the static set of equations, that in this case we have throughout the shell $N_{xx}^r = T$ and $N_{\theta\theta}^r = N_{x\theta}^r = 0$.

Let us first evaluate the operators. They become

$$\nabla_r^2(\cdot) = T\frac{\partial^2(\cdot)}{\partial x^2} \tag{11.6.9}$$

$$\nabla^2(\cdot) = \frac{1}{a^2}\frac{\partial^2(\cdot)}{\partial\theta^2} + \frac{\partial^2(\cdot)}{\partial x^2} \tag{11.6.10}$$

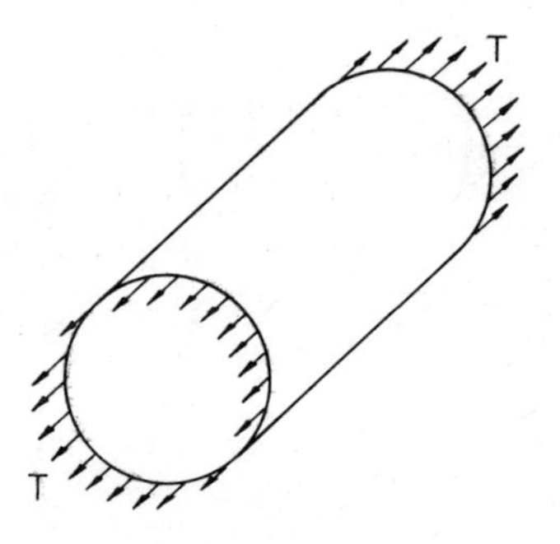

Figure 11.6.1

$$\nabla_k^2(\cdot) = \frac{1}{a^2}\ \frac{\partial^2(\cdot)}{\partial x^2} \tag{11.6.11}$$

If the shell is simply supported at both ends, we find that the mode shape

$$U_3(x,\theta) = \sin\frac{m\pi x}{L}\cos n(\theta - \phi) \tag{11.6.12}$$

satisfies both the boundary conditions and Eq. (11.6.8). This equation becomes

$$D\left[\left(\frac{m\pi}{L}\right)^2 + \left(\frac{n}{a}\right)^2\right]^4 + \frac{Eh}{a^2}\left(\frac{m\pi}{L}\right)^4 + \left[T\left(\frac{m\pi}{L}\right)^2 - \rho h\omega^2\right]\left[\left(\frac{m\pi}{L}\right)^2 + \left(\frac{n}{a}\right)^2\right]^2 = 0 \tag{11.6.13}$$

Solving for ω, we may write the result as

$$\omega_{mnT}^2 = \omega_{mno}^2 + \frac{T}{\rho h}\left(\frac{m\pi}{L}\right)^2 \tag{11.6.14}$$

where ω_{mno} is the natural frequency of the shell when there is no boundary tension, as given by Eq. (6.12.3), and ω_{mnT} is the natural frequency when tension or compression (since T could be negative) are present. Note that the natural frequency increases as the tension T is increased. If there is compression, $T < 0$, the natural frequency is decreased from that of the compression-free shell. As a matter of fact, when

$$T = -\left(\frac{L}{m\pi}\right)^2 \omega_{mno}^2 \rho h \tag{11.6.15}$$

We reach a point where the natural frequency for a particular m,n mode becomes zero. This is because at that value of T, the critical buckling load for a buckling mode identical in shape to this particular m,n mode has been reached and the shell has zero stiffness for this particular mode.

REFERENCES

11.1 H. Lamb, "The vibrations of a spinning disc," *Proc. Roy. Soc. London,* (Ser. A), *vol. 99*, 1921, pp. 272-280.

11.2 K. Federhofer, "Über die Eigenschwingungen der axial gedrückten Kreiszylinderschale," *Sitzungsberichte der Akad. Wiss. Wien, vol. 145*, 1936, pp. 681-688.

11.3 A. W. Leissa, *Vibrations of Shells,* NASA SP-288, U.S. Government Printing Office, Washington, D.C., 1973.

11.4 M. Pagani, "Note sur le mouvement vibratoire d'une membrane elastique de forme circulaire," Royal Academy of Science at Brussels, 1829.

11.5 R. V. Southwell, "On the free transverse vibrations of a uniform circular disc clamped at its center; and on the effects of rotation," *Proc. Roy. Soc. London* (Ser. A), *vol. 101,* 1922, pp. 133-153.

11.6 C. D. Mote, Jr., "Free vibrations of initially stressed circular disks," *J. Eng. Ind.*, Trans. ASME (Ser. B), *vol. 87,* no. 2, 1965, pp. 258-264.

11.7 C. D. Mote, Jr., "Effects of tensioning on buckling and vibration of circular saw blades," ASME Paper 73-DE-M, 1973.

11.8 C. D. Mote, Jr., "Theory of thermal natural frequency variations in disks," *Int. J. Mech. Sci., vol. 8,* 1966, pp. 547-557.

11.9 D. S. Dugdale, "Theory of circular saw tensioning," *Int. J. Products Res., vol. 4,* no. 3, 1966, pp. 237-248.

12

SHELL EQUATIONS WITH SHEAR DEFORMATION AND ROTATORY INERTIA

In all previous development we have taken the shear deformation to be zero. This assumption allowed us to obtain β_1 and β_2 as functions of the displacements u_1, u_2, and u_3. However, for shells where the thickness is large as compared to either the overall dimension or to the wavelength of the highest frequency of interest, shear deformation cannot be neglected any longer. We must allow for the fact that $\varepsilon_{13} \neq 0$ and $\varepsilon_{23} \neq 0$. This means that we will have two additional unknowns, namely β_1 and β_2.

12.1 EQUATIONS OF MOTION

From Eqs. (2.3.52) and (2.3.53) we have

$$\varepsilon_{13} = A_1 \frac{\partial}{\partial \alpha_3}\left(\frac{U_1}{A_1}\right) + \frac{1}{A_1} \frac{\partial U_3}{\partial \alpha_1} \tag{12.1.1}$$

$$\varepsilon_{23} = A_2 \frac{\partial}{\partial \alpha_3}\left(\frac{U_2}{A_2}\right) + \frac{1}{A_2} \frac{\partial U_3}{\partial \alpha_2} \tag{12.1.2}$$

Substituting Eqs. (2.4.1) to (2.4.3) gives

$$\varepsilon_{13} = \beta_1 - \frac{u_1}{R_1} + \frac{1}{A_1}\frac{\partial u_3}{\partial \alpha_1} \tag{12.1.3}$$

$$\varepsilon_{23} = \beta_2 - \frac{u_2}{R_2} + \frac{1}{A_2}\frac{\partial u_3}{\partial \alpha_2} \tag{12.1.4}$$

In Sec. 2.4 we were able to solve these two equations for β_1 and β_2. This is now not possible. β_1 and β_2 have to be treated as unknowns.

All this implies, of course, that we should now use the definitions of shear stress in a more direct way:

$$\sigma_{13} = G\varepsilon_{13} \tag{12.1.5}$$

$$\sigma_{23} = G\varepsilon_{23} \tag{12.1.6}$$

We have to consider ε_{13} and ε_{23}, and therefore σ_{13} and σ_{23}, to be the values at the neutral surface. Since the free surfaces of the shell can clearly not support a shear stress, the average values of σ_{13} and σ_{23} are less. If σ_{13}^a and σ_{23}^a are the average shear stress, then

$$\sigma_{13}^a = k'\sigma_{13} \tag{12.1.7}$$

$$\sigma_{23}^a = k'\sigma_{23} \tag{12.1.8}$$

The factor k' depends on the actual distribution of shear stress in the α_3 direction. If the distribution is parabolic as sketched in Fig. 12.1.1, $k' = 2/3$.

Summing up over the shell thickness, we obtain the shear force resultants Q_{13} and Q_{23}.

$$Q_{13} = \int_{-\frac{h}{2}}^{\frac{h}{2}} \sigma_{13}^a \left(1 + \frac{\alpha_3}{R_2}\right) d\alpha_3 \tag{12.1.9}$$

or $$Q_{13} = k'\sigma_{13}h = k'\varepsilon_{13}Gh \tag{12.1.10}$$

Similarly,

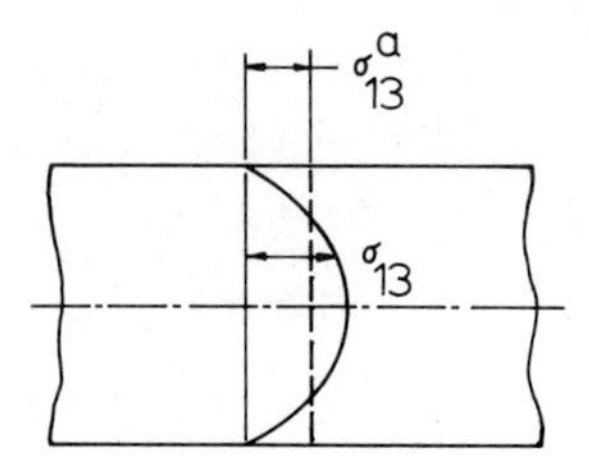

Figure 12.1.1

$$Q_{23} = k'\varepsilon_{23}Gh \tag{12.1.11}$$

Since rotatory inertia effects become noticeable at frequencies where shear deflections have to be considered, it is advisable to include rotatory inertia at this point. For this purpose we use as kinetic energy expression Eq. (2.6.7), with the $\dot{\beta}_1$ and $\dot{\beta}_2$ terms not neglected.

The equations of motion become, therefore,

$$-\frac{\partial(N_{11}A_2)}{\partial\alpha_2} - \frac{\partial(N_{21}A_1)}{\partial\alpha_2} - N_{12}\frac{\partial A_1}{\partial\alpha_2} + N_{22}\frac{\partial A_2}{\partial\alpha_1}$$

$$- A_1A_2\frac{k'\varepsilon_{13}Gh}{R_1} + A_1A_2\rho h\ddot{u}_1 = A_1A_2q_1 \tag{12.1.12}$$

$$-\frac{\partial(N_{12}A_2)}{\partial\alpha_1} - \frac{\partial(N_{22}A_1)}{\partial\alpha_2} - N_{21}\frac{\partial A_2}{\partial\alpha_1} + N_{11}\frac{\partial A_1}{\partial\alpha_2}$$

$$- A_1A_2\frac{k'\varepsilon_{23}Gh}{R_2} + A_1A_2\rho h\ddot{u}_2 = A_1A_2q_2 \tag{12.1.13}$$

$$- k'Gh\frac{\partial(\varepsilon_{13}A_2)}{\partial\alpha_1} - k'Gh\frac{\partial(\varepsilon_{23}A_1)}{\partial\alpha_2} + A_1A_2\left(\frac{N_{11}}{R_1} + \frac{N_{22}}{R_2}\right)$$

$$+ A_1A_2\rho h\ddot{u}_3 = A_1A_2q_3 \tag{12.1.14}$$

$$\frac{\partial(M_{11}A_2)}{\partial\alpha_1} + \frac{\partial(M_{21}A_1)}{\partial\alpha_2} + M_{12}\frac{\partial A_1}{\partial\alpha_2} - M_{22}\frac{\partial A_2}{\partial\alpha_1}$$

$$- Ghk'\varepsilon_{13}A_1A_2 - A_1A_2\frac{\rho h^3}{12}\ddot{\beta}_1 = 0 \tag{12.1.15}$$

$$\frac{\partial(M_{12}A_2)}{\partial\alpha_1} + \frac{\partial(M_{22}A_1)}{\partial\alpha_2} + M_{21}\frac{\partial A_2}{\partial\alpha_1} - M_{11}\frac{\partial A_1}{\partial\alpha_2}$$

$$- Ghk'\varepsilon_{23}A_1A_2 - A_1A_2\frac{\rho h^3}{12}\ddot{\beta}_2 = 0 \qquad (12.1.16)$$

We have five equations and five unknowns: u_1, u_2, u_3, β_1, and β_2.

Summarizing the strain-displacement relationships, we have now

$$\varepsilon^\circ_{11} = \frac{1}{A_1}\frac{\partial u_1}{\partial\alpha_1} + \frac{u_2}{A_1A_2}\frac{\partial A_1}{\partial\alpha_2} + \frac{u_3}{R_1} \qquad (12.1.17)$$

$$\varepsilon^\circ_{22} = \frac{1}{A_2}\frac{\partial u_2}{\partial\alpha_2} + \frac{u_1}{A_1A_2}\frac{\partial A_2}{\partial\alpha_1} + \frac{u_3}{R_2} \qquad (12.1.18)$$

$$\varepsilon^\circ_{12} = \frac{A_2}{A_1}\frac{\partial}{\partial\alpha_1}\left(\frac{u_2}{A_2}\right) + \frac{A_1}{A_2}\frac{\partial}{\partial\alpha_2}\left(\frac{u_1}{A_1}\right) \qquad (12.1.19)$$

$$k_{11} = \frac{1}{A_1}\frac{\partial\beta_1}{\partial\alpha_1} + \frac{\beta_2}{A_1A_2}\frac{\partial A_1}{\partial\alpha_2} \qquad (12.1.20)$$

$$k_{22} = \frac{1}{A_2}\frac{\partial\beta_2}{\partial\alpha_2} + \frac{\beta_1}{A_1A_2}\frac{\partial A_2}{\partial\alpha_1} \qquad (12.1.21)$$

$$k_{12} = \frac{A_2}{A_1}\frac{\partial}{\partial\alpha_1}\left(\frac{\beta_2}{A_2}\right) + \frac{A_1}{A_2}\frac{\partial}{\partial\alpha_2}\left(\frac{\beta_1}{A_1}\right) \qquad (12.1.22)$$

$$\varepsilon_{13} = \frac{1}{A_1}\frac{\partial u_3}{\partial\alpha_1} - \frac{u_1}{R_1} + \beta_1 \qquad (12.1.23)$$

$$\varepsilon_{23} = \frac{1}{A_2}\frac{\partial u_3}{\partial\alpha_2} - \frac{u_2}{R_2} + \beta_2 \qquad (12.1.24)$$

The necessary boundary conditions become (the subscripts n and t denote normal to boundary and tangential to the boundary, respectively)

$$\begin{aligned} N_{nn} &= N^*_{nn} \quad \text{or} \quad u_n = u^*_n \\ N_{nt} &= N^*_{nt} \quad \text{or} \quad u_t = u^*_t \\ Q_{n3} &= Q^*_{n3} \quad \text{or} \quad u_3 = u^*_3 \\ M_{nn} &= M^*_{nn} \quad \text{or} \quad \beta_n = \beta^*_n \\ M_{nt} &= M^*_{nt} \quad \text{or} \quad \beta_t = \beta^*_t \end{aligned} \qquad (12.1.25)$$

12.2 BEAMS WITH SHEAR DEFLECTION AND ROTATORY INERTIA

As our first reduction, let us derive the equation of motion of a transversely vibrating beam, commonly known as the *Timoshenko beam equation* [12.1]. We set $\alpha_1 = x$, $A_1 = 1$, $\alpha_2 = y$, $A_2 = 1$, $\partial(\cdot)/\partial\alpha_2 = 0$. The equations of motion reduce then to

$$- Ghk' \frac{\partial \varepsilon_{x3}}{\partial x} + \rho h\ddot{u}_3 = q_3 \tag{12.2.1}$$

$$\frac{\partial M_{xx}}{\partial x} - Ghk'\varepsilon_{x3} - \frac{\rho h^3}{12} \ddot{\beta}_x = 0 \tag{12.2.2}$$

Stress-displacement relationships reduce to

$$k_{xx} = \frac{\partial \beta_x}{\partial x} \tag{12.2.3}$$

$$\varepsilon_{x3} = \frac{\partial u_3}{\partial x} + \beta_x \tag{12.2.4}$$

Since the equation for M_{xx} becomes

$$M_{xx} = D \frac{\partial \beta_x}{\partial x} \tag{12.2.5}$$

Eqs. (12.2.1) and (12.2.2) become

$$- Ghk' \frac{\partial^2 u_3}{\partial x^2} - Ghk' \frac{\partial \beta_x}{\partial x} + \rho h\ddot{u}_3 = q_3 \tag{12.2.6}$$

$$D \frac{\partial^2 \beta_x}{\partial x^2} - Ghk'\beta_x - Ghk' \frac{\partial u_3}{\partial x} - \frac{\rho h^3}{12} \ddot{\beta}_x = 0 \tag{12.2.7}$$

Differentiating Eq. (12.2.7) with respect to x gives

$$D \frac{\partial^3 \beta_x}{\partial x^3} - Ghk' \frac{\partial \beta_x}{\partial x} - Ghk' \frac{\partial^2 u_3}{\partial x^2} - \frac{\rho h^3}{12} \frac{\partial(\ddot{\beta}_x)}{\partial x} = 0 \tag{12.2.8}$$

From Eq. (12.2.6) we obtain

$$\frac{\partial \beta_x}{\partial x} = \frac{\rho}{Gk'} \ddot{u}_3 - \frac{\partial^2 u_3}{\partial x^2} - \frac{q_3}{Ghk'} \tag{12.2.9}$$

Substituting this into Eq. (12.2.8) gives

$$D\frac{\partial^4 u_3}{\partial x^4} + \rho h\frac{\partial^2 u_3}{\partial t^2} - \left(\frac{D\rho}{Gk'} + \frac{\rho h^3}{12}\right)\frac{\partial^4 u_3}{\partial x^2 \partial t^2} + \frac{\rho^2 h^3}{12Gk'}\frac{\partial^4 u_3}{\partial t^4}$$

$$= q_3 + \frac{\rho h^2}{12Gk'}\frac{\partial^2 q_3}{\partial t^2} - \frac{D}{Ghk'}\frac{\partial^2 q_3}{\partial x^2} \tag{12.2.10}$$

Multiplying the equation by the width of the beam, we recognize that

$$Db = EI \tag{12.2.11}$$

$$\rho hb = \rho A \tag{12.2.12}$$

$$q_3 b = p \tag{12.2.13}$$

where I is the area moment of inertia, A is the cross-sectional area, and p is the force per unit length. Therefore, Eq. (12.2.10) becomes

$$EI\frac{\partial^4 u_3}{\partial x^4} + \rho A\frac{\partial^2 u_3}{\partial t^2} - \left(\frac{EI\rho}{Gk'} + \rho I\right)\frac{\partial^4 u_3}{\partial x^2 \partial t^2} + \frac{\rho^2 I}{Gk'}\frac{\partial^4 u_3}{\partial t^4}$$

$$= p + \frac{\rho I}{Ghk'}\frac{\partial^2 p}{\partial t^2} - \frac{EI}{Ghk'}\frac{\partial^2 p}{\partial x^2} \tag{12.2.14}$$

This is Timoshenko's beam equation [12.1].

However, it is probably better to work directly with Eqs. (12.2.6) and (12.2.7), which become

$$-GAk'\left(\frac{\partial^2 u_3}{\partial x^2} + \frac{\partial \beta_x}{\partial x}\right) + \rho A\ddot{u}_3 = p \tag{12.2.15}$$

$$EI\frac{\partial^2 \beta_x}{\partial x^2} - GAk'\left(\frac{\partial u_3}{\partial x} + \beta_x\right) - \frac{\rho A h^2}{12}\ddot{\beta}_x = 0 \tag{12.2.16}$$

For example, let us solve this equation for its natural frequencies and modes for the simply supported beam. Setting $p = 0$ and

$$u_3(x,t) = U_3(x)e^{j\omega t} \tag{12.2.17}$$

$$\beta_x(x,t) = B_x(x)e^{j\omega t} \tag{12.2.18}$$

where

$$U_3 = A \sin \frac{m\pi x}{L} \tag{12.2.19}$$

$$B_x = B \cos \frac{m\pi x}{L} \tag{12.2.20}$$

satisfy the boundary conditions which are, at x = 0 and L,

$$U_3 = 0 \tag{12.2.21}$$

$$M_{xx} = 0 \tag{12.2.22}$$

Substituting this in Eqs. (12.2.15) and (12.2.16) gives

$$\begin{bmatrix} a_{11} - \rho A\omega^2 & a_{12} \\ a_{21} & a_{22} - \frac{\rho A h^2}{12}\omega^2 \end{bmatrix} \begin{Bmatrix} A \\ B \end{Bmatrix} = 0 \tag{12.2.23}$$

where

$$a_{11} = GAk'\left(\frac{m\pi}{L}\right)^2 \tag{12.2.24}$$

$$a_{12} = a_{21} = GAk'\left(\frac{m\pi}{L}\right) \tag{12.2.25}$$

$$a_{22} = GAk' + EI\left(\frac{m\pi}{L}\right)^2 \tag{12.2.26}$$

Since A and B cannot be equal to zero, the determinant has to be zero to satisfy the equation. This gives a second-order algebraic equation in ω^2

$$A_1\omega^4 + A_2\omega^2 + A_3 = 0 \tag{12.2.27}$$

where

$$A_1 = (\rho A)^2 \frac{h^2}{12} \tag{12.2.28}$$

$$A_2 = -\rho A\left(a_{11}\frac{h^2}{12} + a_{22}\right) \tag{12.2.29}$$

$$A_3 = a_{11}a_{22} - a_{12}a_{21} \tag{12.2.30}$$

and thus we obtain two natural frequencies for every value of m. The lower one is associated with a mode that is dominated by transverse deflection and can be compared to the natural frequency obtained from the classical beam equation without shear deflection. The higher one is associated with a shear-type vibration and is of much lesser technical interest.

Let us define

$$\omega_{ms} = \xi\omega_m \tag{12.2.31}$$

where ω_m is the m-th natural frequency when shear and rotatory inertia is not considered and ω_{ms} is the m-th natural frequency when they are considered. ξ is a correction factor. It turns out that $\xi < 1.0$ since the addition of a shear deflection makes the system behave as if it is less stiff and the addition of a rotatory inertia increases the mass effect. Both tend to decrease the calculated natural frequency.

Utilizing our solution, we may plot the correction factor ξ as shown in Fig. 12.2.1. It shows that the error that is introduced when we neglect shear and rotatory inertia is sizable for low values of L/h and for high values of m. What is required in summary is that the thickness h is small compared to the length between nodes of the highest mode of interest. The length between nodes is, exactly for the simply supported case and approximately for other

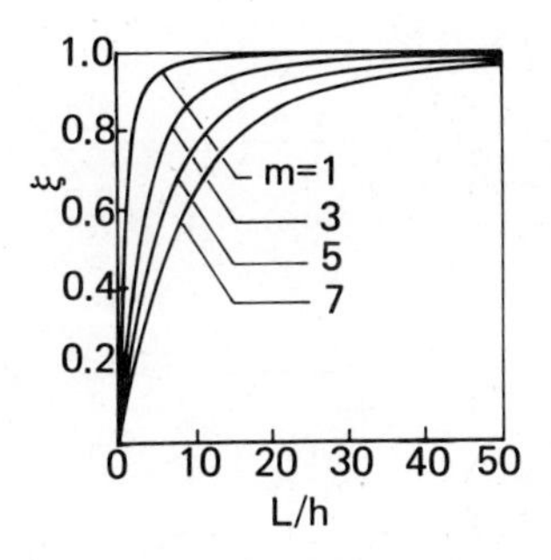

Figure 12.2.1

cases, $\lambda = L/m$. Thus, only if

$$h \ll L/m \tag{12.2.32}$$

can we ignore shear and rotatory inertia safely. For instance, if h is 10% of L/m, the frequency error is approximately 2%.

In Ref. 12.2, this question is discussed at length for both beams and plates.

12.3 PLATES WITH TRANSVERSE SHEAR DEFLECTION AND ROTATORY INERTIA

For plates, we set $1/R_1 = 1/R_2 = 0$. Equations (12.1.12) and (12.1.13) uncouple from the other equations. They are of little interest in this chapter.

Equation (12.1.14) becomes

$$-k'Gh\frac{\partial(\varepsilon_{13}A_2)}{\partial\alpha_1} - k'Gh\frac{\partial(\varepsilon_{23}A_1)}{\partial\alpha_2} + A_1A_2\rho h\ddot{u}_3 = A_1A_2q_3 \tag{12.3.1}$$

and Eqs. (12.1.15) and (12.1.16) remain the same. The strain-displacement relations needed are Eqs. (12.1.20) to (12.1.22), and Eqs. (12.1.23) and (12.1.24), which become

$$\varepsilon_{13} = \frac{1}{A_1}\frac{\partial u_3}{\partial\alpha_1} + \beta_1 \tag{12.3.2}$$

$$\varepsilon_{23} = \frac{1}{A_2}\frac{\partial u_3}{\partial\alpha_2} + \beta_2 \tag{12.3.3}$$

Substituting these in Eqs. (12.3.1), (12.1.15), and (12.1.16) gives

$$-k'Gh\left[\frac{\partial}{\partial\alpha_1}\left(\frac{A_2}{A_1}\frac{\partial u_3}{\partial\alpha_1}\right) + \frac{\partial}{\partial\alpha_2}\left(\frac{A_1}{A_2}\frac{\partial u_3}{\partial\alpha_2}\right) + \frac{\partial}{\partial\alpha_1}(A_2\beta_1 + A_1\beta_2)\right] + A_1A_2\rho h\ddot{u}_3 = A_1A_2q_3 \tag{12.3.4}$$

$$D(1-\mu)\left[(k_{11} - k_{22})\frac{\partial A_2}{\partial\alpha_1} + k_{12}\frac{\partial A_1}{\partial\alpha_2} + \frac{A_1}{2}\frac{\partial k_{21}}{\partial\alpha_2} + \frac{A_2}{1-\mu}\left(\frac{\partial k_{11}}{\partial\alpha_1} + \mu\frac{\partial k_{12}}{\partial\alpha_1}\right)\right] - k'GhA_1A_2\left(\frac{1}{A_1}\frac{\partial u_3}{\partial\alpha_1} + \beta_1\right) - A_1A_2\frac{\rho h^3}{12}\ddot{\beta}_1 = 0 \tag{12.3.5}$$

$$D(1-\mu)\left[(k_{22}-k_{11})\frac{\partial A_1}{\partial \alpha_2}+k_{21}\frac{\partial A_2}{\partial \alpha_1}+\frac{A_2}{2}\frac{\partial k_{12}}{\partial \alpha_1}\right.$$
$$\left.+\frac{A_1}{1-\mu}\left(\frac{\partial k_{22}}{\partial \alpha_2}+\mu\frac{\partial k_{11}}{\partial \alpha_2}\right)\right]-k'GhA_1A_2\left(\frac{1}{A_2}\frac{\partial u_3}{\partial \alpha_2}+\beta_2\right)$$
$$-A_1A_2\frac{\rho h^3}{12}\ddot{\beta}_2-0 \qquad (12.3.6)$$

where

$$k_{11}=\frac{1}{A_1}\frac{\partial \beta_1}{\partial \alpha_1}+\frac{\beta_2}{A_1A_2}\frac{\partial A_1}{\partial \alpha_2} \qquad (12.3.7)$$

$$k_{22}=\frac{1}{A_2}\frac{\partial \beta_2}{\partial \alpha_2}+\frac{\beta_1}{A_1A_2}\frac{\partial A_2}{\partial \alpha_1} \qquad (12.3.8)$$

$$k_{12}=\frac{A_2}{A_1}\frac{\partial}{\partial \alpha_1}\left(\frac{\beta_2}{A_2}\right)+\frac{A_1}{A_2}\frac{\partial}{\partial \alpha_2}\left(\frac{\beta_1}{A_1}\right) \qquad (12.3.9)$$

These are three equations and three unknowns: β_1, β_2, and u_3. The necessary boundary conditions are

$$Q_{n3}=Q_{n3}^* \quad \text{or} \quad u_3=u_3^* \qquad (12.3.10)$$

$$M_{nn}=M_{nn}^* \quad \text{or} \quad \beta_n=\beta_n^* \qquad (12.3.11)$$

$$M_{nt}=M_{nt}^* \quad \text{or} \quad \beta_t=\beta_t^* \qquad (12.3.12)$$

Let us now look at the special case of a rectangular plate. Since $A_1 = 1$, $A_2 = 1$, $d\alpha_1 = dx$, and $d\alpha_2 = dy$, we obtain

$$-k'Gh\left(\frac{\partial^2 u_3}{\partial x^2}+\frac{\partial^2 u_3}{\partial y^2}+\frac{\partial \beta_x}{\partial x}+\frac{\partial \beta_y}{\partial y}\right)+\rho h\ddot{u}_3=q_3 \qquad (12.3.13)$$

$$D(1-\mu)\left[\frac{1}{2}\frac{\partial k_{21}}{\partial y}+\frac{1}{1-\mu}\left(\frac{\partial k_{11}}{\partial x}+\mu\frac{\partial k_{22}}{\partial x}\right)\right]$$
$$-k'Gh\left(\frac{\partial u_3}{\partial x}+\beta_x\right)-\frac{\rho h^3}{12}\ddot{\beta}_x=0 \qquad (12.3.14)$$

$$D(1-\mu)\left[\frac{1}{2}\frac{\partial k_{12}}{\partial x}+\frac{1}{1-\mu}\left(\frac{\partial k_{22}}{\partial y}+\mu\frac{\partial k_{11}}{\partial y}\right)\right]$$
$$-k'Gh\left(\frac{\partial u_3}{\partial y}+\beta_y\right)-\frac{\rho h^3}{12}\ddot{\beta}_y=0 \qquad (12.3.15)$$

$$k_{11} = \frac{\partial \beta_x}{\partial x} \tag{12.3.16}$$

$$k_{22} = \frac{\partial \beta_y}{\partial y} \tag{12.3.17}$$

$$k_{12} = \frac{\partial \beta_y}{\partial x} + \frac{\partial \beta_x}{\partial y} \tag{12.3.18}$$

This gives

$$- k'Gh\left(\frac{\partial^2 u_3}{\partial x^2} + \frac{\partial^2 u_3}{\partial y^2} + \frac{\partial \beta_x}{\partial x} + \frac{\partial \beta_y}{\partial y}\right) + \rho h\ddot{u}_3 = q_3 \tag{12.3.19}$$

$$D\left(\frac{1 + \mu}{2} \frac{\partial^2 \beta_y}{\partial x \partial y} + \frac{1 - \mu}{2} \frac{\partial^2 \beta_x}{\partial y^2} + \frac{\partial^2 \beta_x}{\partial x^2}\right) - k'Gh\left(\frac{\partial u_3}{\partial x} + \beta_x\right) - \frac{\rho h^3}{12} \ddot{\beta}_x = 0 \tag{12.3.20}$$

$$D\left(\frac{1 + \mu}{2} \frac{\partial^2 \beta_x}{\partial x \partial y} + \frac{1 - \mu}{2} \frac{\partial^2 \beta_y}{\partial x^2} + \frac{\partial^2 \beta_y}{\partial y^2}\right) - k'Gh\left(\frac{\partial u_3}{\partial y} + \beta_y\right) - \frac{\rho h^3}{12} \ddot{\beta}_y = 0 \tag{12.3.21}$$

These equations are consistent with the Timoshenko beam equation since they reduce to it if we set $\frac{\partial(\cdot)}{y} = 0$ and $\beta_y = 0$.

By introducing the Laplacian operator

$$\nabla^2(\cdot) = \frac{\partial(\cdot)}{\partial x^2} + \frac{\partial(\cdot)}{\partial y^2} \tag{12.3.22}$$

we may write these equations also as

$$-k'Gh\left(\nabla^2 u_3 + \frac{\partial \beta_x}{\partial x} + \frac{\partial \beta_y}{\partial y}\right) + \rho h\ddot{u}_3 = q_3 \tag{12.3.23}$$

$$\frac{D}{2}\left[(1 - \mu)\nabla^2 \beta_x + (1 + \mu) \frac{\partial}{\partial x}\left(\frac{\partial \beta_x}{\partial x} + \frac{\partial \beta_y}{\partial y}\right)\right] - k'Gh\left(\frac{\partial u_3}{\partial x} + \beta_x\right) - \frac{\rho h^3}{12} \ddot{\beta}_x = 0 \tag{12.3.24}$$

$$\frac{D}{2}\left[(1 - \mu)\nabla^2\beta_y + (1 + \mu)\frac{\partial}{\partial y}\left(\frac{\partial\beta_x}{\partial x} + \frac{\partial\beta_y}{\partial y}\right)\right] - k'Gh\left(\frac{\partial u_3}{\partial y} + \beta_y\right)$$

$$- \frac{\rho h^3}{12}\ddot{\beta}_y = 0 \qquad (12.3.25)$$

Equations (12.3.23) to (12.3.25) are the equations of motion for the plate in cartesian coordinates. They were first derived by Mindlin [12.3].

Let us, for example, investigate the natural frequencies of a simply supported plate. We let

$$u_3 = U_3 e^{j\omega t} \qquad (12.3.26)$$

$$\beta_x = B_x e^{j\omega t} \qquad (12.3.27)$$

$$\beta_y = B_y e^{j\omega t} \qquad (12.3.28)$$

where, by inspection, we find that

$$U_3 = A \sin\frac{m\pi x}{a}\sin\frac{n\pi y}{b} \qquad (12.3.29)$$

$$B_x = B \cos\frac{m\pi x}{a}\sin\frac{n\pi y}{b} \qquad (12.3.30)$$

$$B_y = C \sin\frac{m\pi x}{a}\cos\frac{n\pi y}{b} \qquad (12.3.31)$$

satisfy the boundary conditions that, at $x = 0$ and a,

$$U_3 = 0 \qquad (12.3.32)$$

$$B_y = 0 \qquad (12.3.33)$$

$$M_{xx} = 0 \qquad (12.3.34)$$

and at $y = 0$ and b,

$$U_3 = 0 \qquad (12.3.35)$$

$$B_x = 0 \qquad (12.3.36)$$

$$M_{yy} = 0 \qquad (12.3.37)$$

Substituting this in Eqs. (12.3.23) to (12.3.25) gives

$$\begin{bmatrix} a_{11} - \rho h\omega^2 & a_{12} & a_{13} \\ a_{21} & a_{22} - \frac{\rho h^3}{12}\omega^2 & a_{23} \\ a_{31} & a_{32} & a_{33} - \frac{\rho h^3}{12}\omega^2 \end{bmatrix} \begin{Bmatrix} A \\ B \\ C \end{Bmatrix} = 0 \tag{12.3.38}$$

where

$$a_{11} = k'Gh\left[\left(\frac{m\pi}{a}\right)^2 + \left(\frac{n\pi}{b}\right)^2\right] \tag{12.3.39}$$

$$a_{22} = D\left[\left(\frac{m\pi}{a}\right)^2 + \frac{1-\mu}{2}\left(\frac{n\pi}{b}\right)^2\right] + k'Gh \tag{12.3.40}$$

$$a_{33} = D\left[\frac{1-\mu}{2}\left(\frac{m\pi}{a}\right)^2 + \left(\frac{n\pi}{b}\right)^2\right] + k'Gh \tag{12.3.41}$$

$$a_{12} = a_{21} = k'Gh\,\frac{m\pi}{a} \tag{12.3.42}$$

$$a_{13} = a_{31} = k'Gh\,\frac{n\pi}{b} \tag{12.3.43}$$

$$a_{23} = a_{32} = D\,\frac{1+\mu}{2}\,\frac{m\pi}{a}\,\frac{n\pi}{b} \tag{12.3.44}$$

Setting the determinant to zero will give us a cubic equation in ω^2. For each m,n combination we obtain three natural frequencies. The lowest of these is the one of most interest since it corresponds to the mode where the transverse deflection dominates. The other two frequencies are much higher and correspond to shear modes.

The error that is introduced by neglecting shear and rotatory inertia is again a function of the thickness and the distance between nodal lines. It is similar in magnitude to the beam analysis error [12.2].

12.4 CIRCULAR CYLINDRICAL SHELLS WITH TRANSVERSE SHEAR DEFLECTION AND ROTATORY INERTIA

In this case, $A_1 = 1$, $A_2 = a$, $d\alpha_1 = dx$, $d\alpha_2 = d\theta$, $1/R_1 = 0$, and $R_2 = a$. The equations of motion become, therefore,

$$-a\frac{\partial N_{xx}}{\partial x} - \frac{\partial N_{\theta x}}{\partial \theta} + a\rho h\ddot{u}_x = aq_x \tag{12.4.1}$$

$$-a\frac{\partial N_{x\theta}}{\partial x} - \frac{\partial N_{\theta\theta}}{\partial \theta} - k'Gh\varepsilon_{\theta 3} + a\rho h\ddot{u}_\theta = aq_\theta \tag{12.4.2}$$

$$-ak'Gh\frac{\partial \varepsilon_{x3}}{\partial x} - k'Gh\frac{\partial \varepsilon_{\theta 3}}{\partial \theta} + N_{\theta\theta} + a\rho h\ddot{u}_3 = aq_3 \tag{12.4.3}$$

$$a\frac{\partial M_{xx}}{\partial x} + \frac{\partial M_{\theta x}}{\partial \theta} - k'Gha\varepsilon_{x3} - \frac{a\rho h^3}{12}\ddot{\beta}_x = 0 \tag{12.4.4}$$

$$a\frac{\partial M_{x\theta}}{\partial x} + \frac{\partial M_{\theta\theta}}{\partial \theta} - k'Gha\varepsilon_{\theta 3} - \frac{a\rho h^3}{12}\ddot{\beta}_\theta = 0 \tag{12.4.5}$$

where

$$\varepsilon_{x3} = \frac{\partial u_3}{\partial x} + \beta_x \tag{12.4.6}$$

$$\varepsilon_{\theta 3} = \frac{1}{a}\frac{\partial u_3}{\partial \theta} - \frac{u_\theta}{a} + \beta_\theta \tag{12.4.7}$$

$$\varepsilon^\circ_{xx} = \frac{\partial u_x}{\partial x} \tag{12.4.8}$$

$$\varepsilon^\circ_{\theta\theta} = \frac{1}{a}\frac{\partial u_\theta}{\partial \theta} + \frac{u_3}{a} \tag{12.4.9}$$

$$\varepsilon^\circ_{x\theta} = \frac{\partial u_\theta}{\partial x} + \frac{1}{a}\frac{\partial u_x}{\partial \theta} \tag{11.4.10}$$

$$k_{xx} = \frac{\partial \beta_x}{\partial x} \tag{12.4.11}$$

$$k_{\theta\theta} = \frac{1}{a}\frac{\partial \beta_\theta}{\partial \theta} \tag{12.4.12}$$

$$k_{x\theta} = \frac{\partial \beta_\theta}{\partial x} + \frac{1}{a}\frac{\partial \beta_x}{\partial \theta} \tag{12.4.13}$$

Let us examine the simply supported shell. We assume the following solution for the vibration at a natural frequency:

$$u_x(x,\theta,t) = U_x(x,\theta)e^{j\omega t} \tag{12.4.14}$$

$$u_\theta(x,\theta,t) = U_\theta(x,\theta)e^{j\omega t} \tag{12.4.15}$$

$$u_3(x,\theta,t) = U_3(x,\theta)e^{j\omega t} \tag{12.4.16}$$

$$\beta_x(x,\theta,t) = B_x(x,\theta)e^{j\omega t} \tag{12.4.17}$$

$$\beta_\theta(x,\theta,t) = B_\theta(x,\theta)e^{j\omega t} \tag{12.4.18}$$

where

$$U_x(x,\theta) = A \cos\frac{m\pi x}{L} \cos n(\theta - \phi) \tag{12.4.19}$$

$$U_\theta(x,\theta) = B \sin\frac{m\pi x}{L} \sin n(\theta - \phi) \tag{12.4.20}$$

$$U_3(x,\theta) = C \sin\frac{m\pi x}{L} \cos n(\theta - \phi) \tag{12.4.21}$$

$$B_x(x,\theta) = F \cos\frac{m\pi x}{L} \cos n(\theta - \phi) \tag{12.4.22}$$

$$B_\theta(x,\theta) = G \sin\frac{m\pi x}{L} \sin n(\theta - \phi) \tag{12.4.23}$$

These equations satisfy the 10 boundary conditions

$$u_3(0,\theta,t) = 0 \tag{12.4.24}$$

$$u_\theta(0,\theta,t) = 0 \tag{12.4.25}$$

$$M_{xx}(0,\theta,t) = 0 \tag{12.4.26}$$

$$N_{xx}(0,\theta,t) = 0 \tag{12.4.27}$$

$$\beta_\theta(0,\theta,t) = 0 \tag{12.4.28}$$

and

$$u_3(L,\theta,t) = 0 \tag{12.4.29}$$

$$u_\theta(L,\theta,t) = 0 \tag{12.4.30}$$

$$M_{xx}(L,\theta,t) = 0 \tag{12.4.31}$$

$$N_{xx}(L,\theta,t) = 0 \tag{12.4.32}$$

$$\beta_\theta(L,\theta,t) = 0 \tag{12.4.33}$$

Substituting these equations in Eqs. (12.4.1) to (12.4.5) gives

$$\begin{bmatrix} a_{11} - \rho h\omega^2 & a_{12} & a_{13} & 0 & 0 \\ a_{21} & a_{22} - \rho h\omega^2 & a_{23} & 0 & a_{25} \\ a_{31} & a_{32} & a_{33} - \rho h\omega^2 & a_{34} & a_{35} \\ 0 & 0 & a_{43} & a_{44} - \frac{\rho h^3}{12}\omega^2 & a_{45} \\ 0 & a_{52} & a_{53} & a_{54} & a_{55} - \frac{\rho h^3}{12}\omega^2 \end{bmatrix} \begin{Bmatrix} A \\ B \\ C \\ F \\ G \end{Bmatrix} = 0 \tag{12.4.34}$$

where

$$a_{11} = K\left[\left(\frac{m\pi}{L}\right)^2 + \frac{1-\mu}{2}\left(\frac{n}{a}\right)^2\right] \tag{12.4.35}$$

$$a_{12} = a_{21} = -K\,\frac{1+\mu}{2}\,\frac{n}{a}\,\frac{m\pi}{L} \tag{12.4.36}$$

$$a_{13} = a_{31} = -K\,\frac{\mu}{a}\,\frac{m\pi}{L} \tag{12.4.37}$$

$$a_{22} = K\left[\frac{1-\mu}{2}\left(\frac{m\pi}{L}\right)^2 + \left(\frac{n}{a}\right)^2\right] + k'\,\frac{Gh}{a^2} \tag{12.4.38}$$

$$a_{23} = a_{32} = \frac{K}{a}\,\frac{n}{a} + \frac{n}{a}\,\frac{k'Gh}{a} \tag{12.4.39}$$

$$a_{25} = a_{52} = -\frac{k'Gh}{a} \tag{12.4.40}$$

$$a_{33} = k'Gh\left[\left(\frac{m\pi}{L}\right)^2 + \left(\frac{n}{a}\right)^2\right] + \frac{K}{a^2} \tag{12.4.41}$$

$$a_{34} = a_{43} = k'Gh\,\frac{m\pi}{L} \tag{12.4.42}$$

$$a_{35} = a_{53} = -k'Gh\,\frac{n}{a} \tag{12.4.43}$$

$$a_{44} = k'Gh + D\left[\frac{1-\mu}{2}\left(\frac{n}{a}\right)^2 + \left(\frac{m\pi}{L}\right)^2\right] \tag{12.4.44}$$

$$a_{45} = a_{54} = -D\,\frac{1+\mu}{2}\,\frac{n}{a}\,\frac{m\pi}{L} \tag{12.4.45}$$

$$a_{55} = k'Gh + D\left[\frac{1-\mu}{2}\left(\frac{m\pi}{L}\right)^2 + \left(\frac{n}{a}\right)^2\right] \tag{12.4.46}$$

Since the matrix equation is only satisfied, in general, if the determinant of the matrix is zero, we obtain a fifth-order algebraic equation in ω^2. The roots of this equation are the natural frequencies. There will be five distinct frequencies for every m and n combination. The mode shapes are obtained by substituting each root back into the matrix equations and by solving for four of the constants in terms of the fifth. For instance, we may solve for A/C, B/C, F/C, and G/C. We find, as before for the cylindrical shell in Chap. 5, that we have modes where transverse deflections dominate and modes where in-plane deflections dominate. In addition, there will now be two modes where shear-type deflections dominate. The influence of the inclusion of shear deformation and rotatory inertia on natural frequencies for modes where transverse motion dominates is similar to that discussed for the rectangular plate and the beam.

Other shell geometries have been analyzed for the influence of rotatory inertia and shear and conclusions are similar. For conical shells see, for instance, Garnet and Kempner [12.4] and Naghdi [12.5]; for spherical shells see, for instance, Kalnins and Kraus [12.6]. Summarizing discussions are given by Leissa [12.7, 12.8].

REFERENCES

12.1 S. Timoshenko, "On the correction for shear of the differential equations for transverse vibrations of prismatic bars," *Phil. Mag., ser. 6, vol. 41*, 1921, pp. 742-748.

12.2 U. R. Kristiansen, W. Soedel, and J. F. Hamilton, "An investigation of scaling laws for vibrating beams and plates with special attention to the effects of shear and rotatory inertia," *J. Sound Vibration, vol. 20*, no. 1, 1972, pp. 113-122.

12.3 R. D. Mindlin, "Influence of rotatory inertia and shear on flexural motions of isotropic elastic plates," *J. Appl. Mech., vol. 18*, no. 1, 1951, pp. 31-38.

12.4 H. Garnet and J. Kempner, "Axisymmetric free vibrations of conical shells," *J. Appl. Mech., vol. 31*, no. 3, 1964, pp. 458-466.

12.5 P. M. Naghdi, "On the theory of thin elastic shells," *Quart. Appl. Math., vol. 14*, no. 4, 1957, pp. 369-380.

12.6 A. Kalnins and H. Kraus, "Effect of transverse shear and rotatory inertia on vibration of spherical shells," *Proceedings of the 5th U.S. National Congress of Applied Mechanics*, Minneapolis, 1966, p. 134.

12.7 A. W. Leissa, *Vibration of Plates*, NASA SP-160, U.S. Government Printing Office, Washington, D.C., 1969.

12.8 A. W. Leissa, *Vibration of Shells*, NASA SP-288, U.S. Government Printing Office, Washington, D.C., 1973.

13

COMBINATIONS OF STRUCTURES

Most shell structures are combinations of basic shell elements such as cylindrical and conical shells, spherical caps, and carry stiffeners, lumped masses, and so on. This has prompted the development of several methods of calculating the eigenvalues of such combinations. One way is, of course, the finite element method.

A useful approach, discussed for cases other than plates and shells by Bishop and Johnson [13.1], is the receptance method. With the receptance method, vibrational characteristics of a combined system, for instance, a shell stiffened by a rib, are calculated from characteristics of the component systems, in this case the shell and the stiffening rib. The number of variables is minimized, since only the displacements at the points of interaction of the subsystems are a part of the solution.

A feature of the receptance method of practical value to the design process is that the receptances of the component systems may be determined by any method that is sufficiently accurate. In the following, the receptances are written in terms of the natural frequencies and modes, when appropriate. These can be obtained experimentally, from finite element programs, etc.

Even in finite element programming, the receptance method opens a way to reduce program size since each component can be evaluated by itself for its eigenvalues. The eigenvalues of the combined structure are given by the receptance method. For example, a rectangular box can be thought of as being assembled from six rectangular plates. A missile structure can be thought of as being composed of a circular cylindrical shell, a circular conical shell and a spherical cap.

13.1 THE RECEPTANCE METHOD

Basically, a *receptance* is defined as the ratio of a deflection response at a certain point to a harmonic force or moment input at the same or at a different point:

$$\alpha_{ij} = \frac{\text{deflection response of system A at location i}}{\text{harmonic force or moment input to system A at location j}}$$

The response may be either a line deflection or a slope. Usually, the subsystems are labeled A, B, C, etc., and the receptances are labeled α, β, γ, etc.

Note that it follows from Maxwell's reciprocity theorem that $\alpha_{ij} = \alpha_{ji}$. Note also that a receptance can be interpreted as the inverse of a mobility. The method could be set up in terms of such mobilities, which is especially interesting to acousticians since they are used to working with mobilities, but there is advantage in using the receptance notation since it makes some of the operations simpler to describe.

Let us now take the simplest conceivable case of two systems being joined through a single displacement. This is shown schematically in Fig. 13.1.1. For instance, a spring attached to a shell panel falls into this category. The receptance of system A at the attachment point is α_{11} and is evaluated first. The input is a harmonic force of amplitude F_{A1}.

$$f_{A1} = F_{A1} e^{j\omega t} \tag{13.1.1}$$

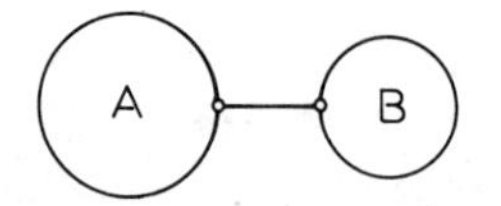

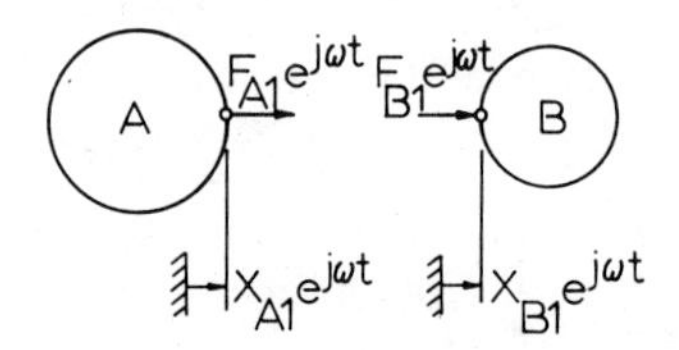

Figure 13.1.1

The output of the undamped system is

$$x_{A1} = X_{A1}e^{j\omega t} \tag{13.1.2}$$

Note that it is essential that we treat the undamped system in order to obtain the undamped eigenvalues of the combined system.

The receptance is

$$\alpha_{11} = \frac{x_{A1}}{f_{A1}} = \frac{X_{A1}}{F_{A1}} \tag{13.1.3}$$

The receptance of system B at the point of attachment, by similar reasoning, is

$$\beta_{11} = \frac{x_{B1}}{f_{B1}} = \frac{X_{B1}}{F_{B1}} \tag{13.1.4}$$

Now, if we join system A and system B, we obtain the condition that

$$X_{A1} = X_{B1} \tag{13.1.5}$$

and $$F_{A1} + F_{B1} = 0 \tag{13.1.6}$$

Combining these equations and applying the definitions of the receptances gives

$$\alpha_{11} + \beta_{11} = 0 \tag{13.1.7}$$

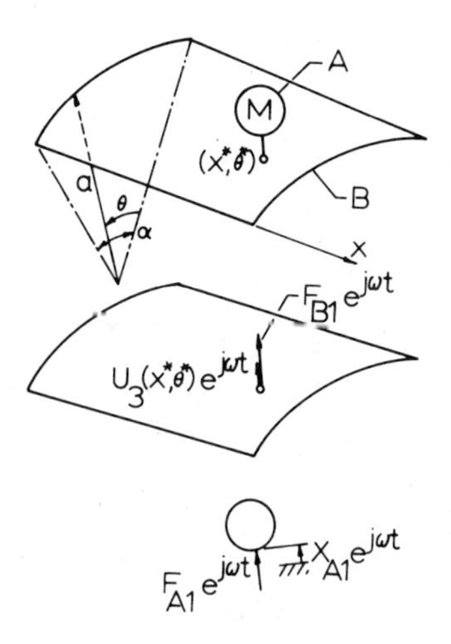

Figure 13.2.1

Frequencies at which this equation is satisfied are natural frequencies of the combined system.

13.2 MASS ATTACHED TO CYLINDRICAL PANEL

The receptance of a mass is found by finding the steady-state response of a mass to a harmonic force input. From Figure 13.2.1 we obtain

$$M\ddot{x}_{A1} = F_{A1}e^{j\omega t} \tag{13.2.1}$$

or

$$-M\omega^2 X_{A1} = F_{A1} \tag{13.2.2}$$

or

$$\alpha_{11} = -\frac{1}{M\omega^2} \tag{13.2.3}$$

The harmonic response at point (x^*,θ^*) due to a point force on the panel is given by Eq. (8.8.8) and is $(\zeta_k = 0)$

$$u_3(x^*,\theta^*,t) = \frac{4F_{B1}e^{j\omega t}}{\rho hL a\alpha} \sum_{m=1}^{\infty}\sum_{n=1}^{\infty} \frac{1}{(\omega_{mn}^2 - \omega^2)} \sin^2\frac{m\pi x^*}{L}\sin^2\frac{n\pi\theta^*}{\alpha} \tag{13.2.4}$$

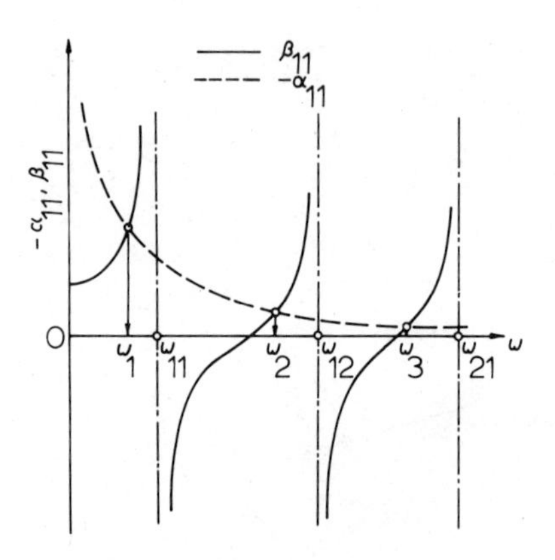

Figure 13.2.2

where ω_{mn} is given by Eq. 6.12.5.

Thus, the receptance is

$$\beta_{11} = \frac{4}{\rho hLa\alpha} \sum_{m=1}^{\infty} \sum_{n=1}^{\infty} \frac{1}{\omega_{mn}^2 - \omega^2} \sin^2 \frac{m\pi x^*}{L} \sin^2 \frac{n\pi\theta^*}{\alpha} \tag{13.2.5}$$

The characteristic equation for the total system is, according to Eq. (13.1.7),

$$\frac{4}{\rho hLa\alpha} \sum_{m=1}^{\infty} \sum_{n=1}^{\infty} \frac{1}{\omega_{mn}^2 - \omega^2} \sin^2 \frac{m\pi x^*}{L} \sin^2 \frac{n\pi\theta^*}{\alpha} - \frac{1}{M\omega^2} = 0 \tag{13.2.6}$$

This equation has to be solved for its roots, $\omega = \omega_k$, by a numerical procedure.

We can also do it graphically. We plot α_{11} as function of ω and then β_{11} as function of ω. According to Eq. (13.2.6), it is required that $\alpha_{11} = -\beta_{11}$. A typical plot is shown in Fig. 13.2.2. As expected, the natural frequencies are in general lowered by the attached mass.

In order to discuss the characteristic behavior of such an equation, let us assume the special case where the influence of the mass is such that the new natural frequencies ω_k are not too different from the original natural frequencies ω_{mn}. In this case, we will find that for a particular root ω_k, one term in the series dominates all the others so that the equation can be written approximately

$$\frac{4}{\rho hL\alpha(\omega_{mn}^2 - \omega_k^2)} \sin^2 \frac{m\pi x^*}{L} \sin^2 \frac{n\pi\theta^*}{\alpha} - \frac{1}{M\omega_k^2} = 0 \qquad (13.2.7)$$

or, solving for ω_k,

$$\omega_k^2 = \frac{\omega_{mn}^2}{\frac{4M}{M_s} \sin^2 \frac{m\pi x^*}{L} \sin^2 \frac{n\pi\theta^*}{\alpha} + 1} \qquad (13.2.8)$$

where M_s is the mass of the entire panel

$$M_s = \rho hL\alpha \qquad (13.2.9)$$

First we notice the obvious, namely that if we set $M = 0$, nothing changes. Secondly, if the mass happens to be attached to what is a node line of the massless plate, the mass has no influence on that particular frequency. Finally, if the mass is located on an anti-node, we have

$$\omega_k^2 = \frac{\omega_{mn}^2}{1 + 4\frac{M}{M_s}} \qquad (13.2.10)$$

This would be the largest influence the mass can have on a particular mode, subject to the restrictions imposed by the simplifying assumption.

What does the new mode shape look like? Obviously, the mode shape of the plate with a mass attached must be different from that of the plate without mass. The solution is found by arguing that the response of the plate to a harmonic point input of frequency ω_k is given by

$$u_3(x,\theta,t) = \frac{4F_{B1}e^{j\omega_k t}}{\rho hL\alpha} \sum_{m=1}^{\infty} \sum_{n=1}^{\infty} \frac{\sin \frac{m\pi x^*}{L} \sin \frac{n\pi\theta^*}{\alpha}}{\omega_{mn}^2 - \omega_k^2} \sin \frac{m\pi x}{L} \sin \frac{n\pi\theta}{\alpha} \qquad (13.2.11)$$

Since ω_k is the natural frequency of the combined system, this equation must describe the new mode shape. Since the amplitude is arbitrary, we obtain

$$U_{3k}(x,\theta) = \sum_{m=1}^{\infty}\sum_{n=1}^{\infty} \frac{\sin\frac{m\pi x^*}{L}\sin\frac{n\pi\theta^*}{\alpha}}{\omega_{mn}^2 - \omega_k^2}\sin\frac{m\pi x}{L}\sin\frac{n\pi\theta}{\alpha} \tag{13.2.12}$$

We see that as more different ω_k is from the original ω_{mn}, as more the new mode shape is a distortion of the original mode shape. We also recognize that if the mass is attached to a node line of an original mode shape (m,n) this particular mode shape is preserved intact in its original form, since in this case the double series becomes dominated by the one particular (m,n) combination as $\omega_k \to \omega_{mn}$.

13.3 SPRING ATTACHED TO SHALLOW CYLINDRICAL PANEL

The receptance of a spring attached to ground is obtained from (Fig. 13.3.1)

$$Kx_{A1} = F_{A1}e^{j\omega t} \tag{13.3.1}$$

or

$$KX_{A1} = F_{A1} \tag{13.3.2}$$

This gives

$$\alpha_{11} = \frac{1}{K} \tag{13.3.3}$$

Using the receptance β_{11} of Eq. (13.2.5) for a simply supported circular cylindrical panel gives

$$\frac{4}{M_s}\sum_{m=1}^{\infty}\sum_{n=1}^{\infty}\frac{1}{\omega_{mn}^2 - \omega^2}\sin^2\frac{m\pi x^*}{L}\sin^2\frac{n\pi\theta^*}{\alpha} + \frac{1}{K} = 0 \tag{13.3.4}$$

where M_s is the mass of the panel and is given by Eq. (13.2.9). The expected solution of this equation is illustrated graphically in Fig. 13.3.2 where the solution $\alpha_{11} = -\beta_{11}$ is shown. As expected, we observe that the presence of a grounded spring will in general increase all natural frequencies, provided that the spring can be

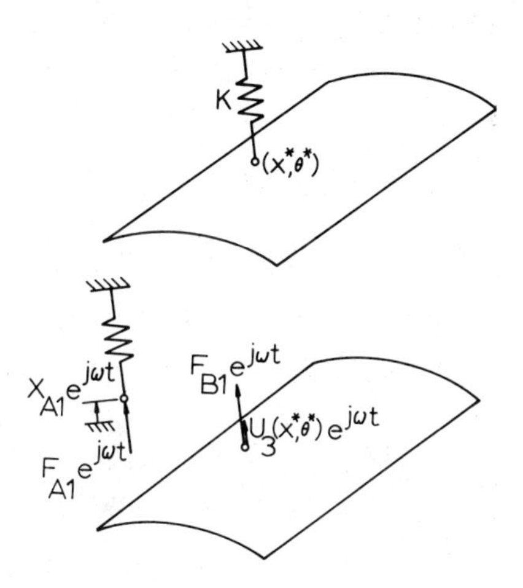

Figure 13.3.1

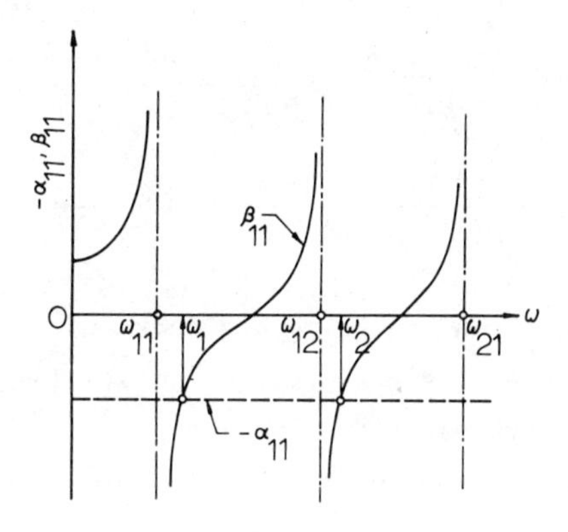

Figure 13.3.2

approximated as massless.

If we make the same assumption as in the previous chapter, namely that for small deviations the series is dominated by the term for which $\omega_{mn}^2 - \omega^2$ is a minimum, we obtain approximately

$$\frac{4}{M_s(\omega_{mn}^2 - \omega^2)} \sin^2 \frac{m\pi x^*}{L} \sin^2 \frac{n\pi\theta^*}{\alpha} + \frac{1}{K} = 0 \tag{13.3.5}$$

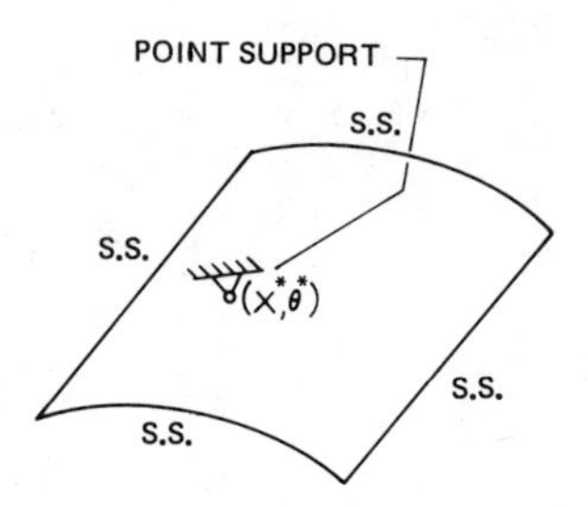

Figure 13.3.3

or $$\omega_k^2 = \omega_{mn}^2 + \frac{4K}{M_s} \sin^2 \frac{m\pi x^*}{L} \sin^2 \frac{n\pi\theta^*}{\alpha} \tag{13.3.6}$$

Note that this case allows us to solve the problem of finding the natural frequencies and modes of a simply supported panel that has also a point support at (x^*, θ^*), as shown in Fig. 13.3.3. All what we have to do is let K approach infinity. The single-term solution cannot any longer be used, rather, we have to solve Eq. (13.3.4) for its roots with $1/K = 0$.

$$\frac{4}{M_s} \sum_{m=1}^{\infty} \sum_{n=1}^{\infty} \frac{1}{\omega_{mn}^2 - \omega^2} \sin^2 \frac{m\pi x^*}{L} \sin^2 \frac{n\pi\theta^*}{\alpha} = 0 \tag{13.3.7}$$

The solutions are given by the points where curve A in Fig. 13.3.2 crosses the abscissa. As expected, we notice in general a substantial increase in natural frequency, except for cases where the point support location coincides with a nodal line of one of the original modes.

13.4 THE DYNAMIC ABSORBER

Let us now examine the case where a spring mass system is attached to the panel as shown in Fig. 13.4.1. To formulate the receptance β_{11}, we write the equations of motion of system B

$$M\ddot{y} + Ky = Kx_{A1} \tag{13.4.1}$$

and $$(y - x_{A1})K + F_{A1}e^{j\omega t} = 0 \tag{13.4.2}$$

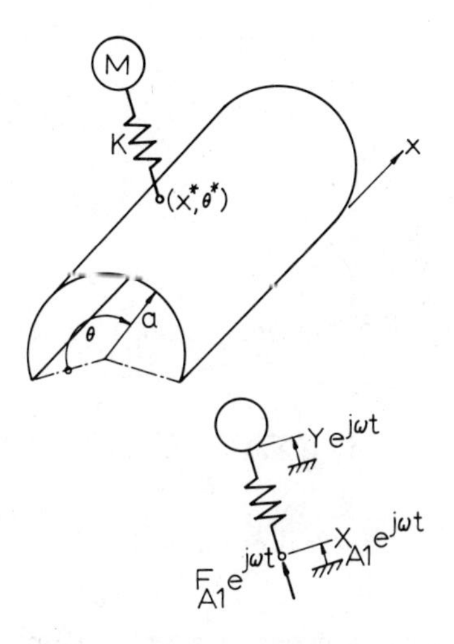

Figure 13.4.1

since

$$y = Ye^{j\omega t} \tag{13.4.3}$$

$$x_{A1} = X_{A1}e^{j\omega t} \tag{13.4.4}$$

we obtain

$$\begin{bmatrix} K - M\omega^2 & -K \\ -K & K \end{bmatrix} \begin{Bmatrix} Y \\ X_{A1} \end{Bmatrix} = \begin{Bmatrix} 0 \\ F_{A1} \end{Bmatrix} \tag{13.4.5}$$

or

$$X_{A1} = \frac{F_{A1}(K - M\omega^2)}{-KM\omega^2} \tag{13.4.6}$$

Thus, the receptance β_{11} is

$$\beta_{11} = -\frac{1}{M\omega^2}\left[1 - \left(\frac{\omega}{\omega_1}\right)^2\right] \tag{13.4.7}$$

where $\omega_1^2 = \frac{K}{M}$.

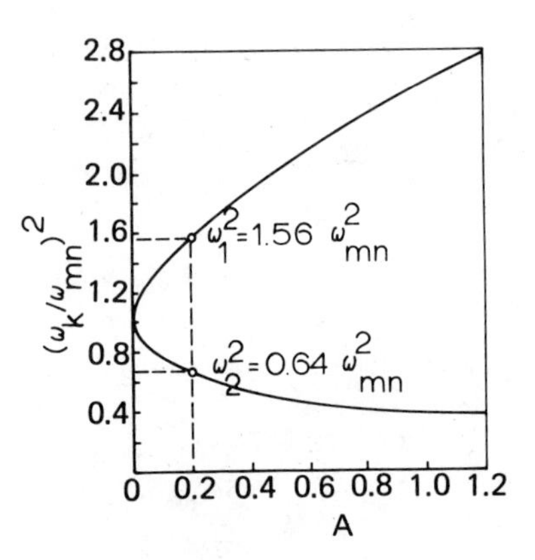

Figure 13.4.2

The frequency equation becomes, therefore,

$$\frac{4}{M_s}\sum_{m=1}^{\infty}\sum_{n=1}^{\infty}\frac{1}{\omega_{mn}^2-\omega^2}\sin^2\frac{m\pi x^*}{L}\sin^2\frac{n\pi\theta^*}{\alpha}-\frac{\omega_1^2-\omega^2}{M\omega^2\omega_1^2}=0 \tag{13.4.8}$$

Let us again examine the approximate case where the roots of this equation are perturbations of the original ω_{mn}^2 values. We obtain

$$\frac{4}{M_s}\frac{1}{\omega_{mn}^2-\omega^2}\sin^2\frac{m\pi x^*}{L}\sin^2\frac{n\pi\theta^*}{\alpha}-\frac{\omega_1^2-\omega^2}{M\omega^2\omega_1^2}=0 \tag{13.4.9}$$

The new natural frequencies $\omega = \omega_k$ are therefore

$$\omega_k^2=\frac{\omega_{mn}^2}{2}\left[\left(\frac{\omega_1}{\omega_{mn}}\right)^2(1+A)+1\right][1\pm\sqrt{1-\varepsilon_{mn}}] \tag{13.4.10}$$

where $$\varepsilon_{mn}=\frac{4(\omega_1/\omega_{mn})^2}{[(\omega_1/\omega_{mn})^2(1+A)+1]^2} \tag{13.4.11}$$

$$A=\frac{4M}{M_s}\sin^2\frac{m\pi x^*}{L}\sin^2\frac{n\pi\theta^*}{\alpha} \tag{13.4.12}$$

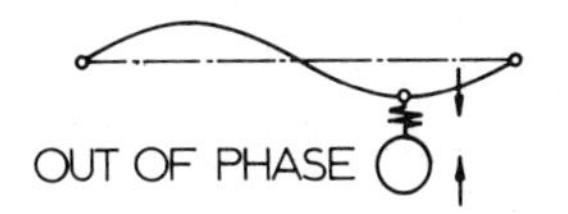

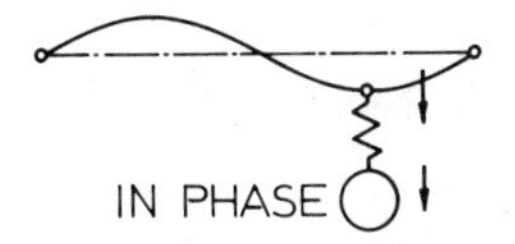

Figure 13.4.3

As a check, when $\omega_1 = 0$, which means K = 0 and thus no attachment, the equations gives $\omega_k^2 = \omega_{mn}^2$, 0, as expected. The zero has in this case no physical meaning. When $K \to \infty$, which implies that $\omega_1 \to \infty$, we obtain from Eq. (13.4.9) the result of Eq. (13.2.8). When A = 0, which means that either M = 0 or the spring-mass system is attached to a node line for the particular m,n mode, we obtain $\omega_k^2 = \omega_{mn}^2$, ω_1^2. This is correct because the two subsystem are in this case uncoupled.

In the general case, we get for every m,n combination two natural frequencies. Let us for instance take the case where $\omega_1 = \omega_{mn}$. We get

$$\omega_k^2 = \omega_{mn}^2 \left[\left(1 + \frac{A}{2}\right) \pm \sqrt{\left(\frac{A}{2}\right)^2 + A}\right]$$

This relationship is plotted in Fig. 13.4.2. It can be shown that in general the higher frequency branch will correspond to a mode where the motion of the mass will be out of phase with the motion of the pannel and that the lower frequency branch will correspond to a system mode where the motion of the mass will be in phase with the motion of the panel. This is sketched in Fig. 13.4.3.

13.5 STIFFENING OF SHELLS

Let us now investigate requirements for stiffening a shell on the example of a simply supported circular cylindrical panel as shown in Fig. 13.5.1. We are faced here with the requirement that we have

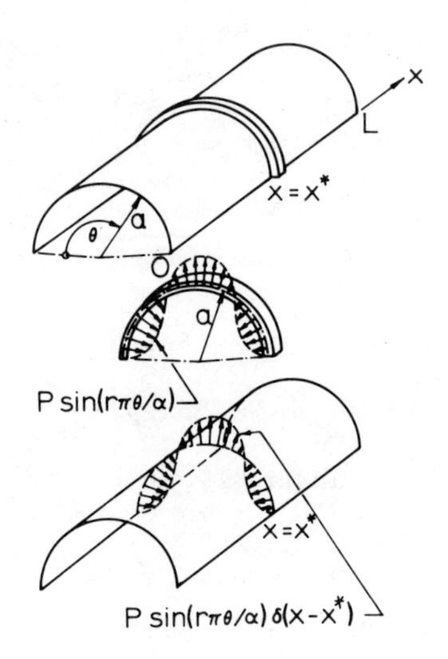

Figure 13.5.1

to join system A, the panel, and system B, the ring, along a continuous line. A way to do this and still utilize Eq. (13.1.7) was worked out by Sakharov [13.2] for the case of a closed circular cylindrical shell with stiffening rings at both ends. It requires that the two systems have the same mode shapes along the line at which they are joined. In our case, the mode shape of the panel is

$$U_3(x,0) = \sin \frac{m\pi x}{L} \sin \frac{n\pi\theta}{\alpha} \tag{13.5.1}$$

while the mode shape of the stiffening ring is

$$U_3(\theta) = \sin \frac{n\pi\theta}{\alpha} \tag{13.5.2}$$

This allows us to formulate a line receptance that is defined as the response along the line to a harmonic line load that is distributed sinusoidally along the line.

$$q_3(x,\theta,t) = P \sin \frac{r\pi\theta}{\alpha} \, \delta(x - x^*) \, e^{j\omega t} \tag{13.5.3}$$

where P is the line load amplitude in [N/m].

Taking advantage of the work described in Sec. 8.9, we note

that $Q_3^* = P \sin \frac{r\pi\theta}{\alpha}$, $Q_1^* = 0$, $Q_2^* = 0$. Thus

$$F_k^* = F_{mn}^* = \frac{P}{\rho h N_k} \sin \frac{m\pi x^*}{L} \int_0^{\alpha} \sin \frac{r\pi\theta}{\alpha} \sin \frac{n\pi\theta}{\alpha} a \, d\theta \tag{13.5.4}$$

where

$$N_k = \frac{a\alpha L}{4} \tag{13.5.5}$$

Evaluating the integral gives zero unless n = r. In that case we obtain

$$F_{mn}^* = \begin{cases} \frac{2P}{\rho h L} \sin \frac{m\pi x^*}{L} & n = r \\ 0 & n \neq r \end{cases} \tag{13.5.6}$$

Since in this case

$$S(t) = e^{j\omega t} \tag{13.5.7}$$

we obtain the steady-state solution of the modal participation factor from Eq. (8.5.4) as

$$\eta_k = \frac{F_{mn}^*}{\omega_{mn}^2 - \omega^2} e^{j\omega t} \tag{13.5.8}$$

and, therefore,

$$u_3(x,\theta,t) = \frac{2Pe^{j\omega t}}{\rho h L} \sum_{m=1}^{\infty} \frac{\text{Sin} \frac{m\pi x^*}{L} \sin \frac{m\pi x}{L} \sin \frac{r\pi\theta}{\alpha}}{\omega_m^2 - \omega^2} \tag{13.5.9}$$

and, at x = x*,

$$u_3(x^*,\theta,t) = \frac{2Pe^{j\omega t}}{\rho h L} \sum_{m=1}^{\infty} \frac{1}{\omega_{mn}^2 - \omega^2} \sin^2 \frac{m\pi x^*}{L} \sin \frac{r\pi\theta}{\alpha} \tag{13.5.10}$$

Formulating as receptance

$$\alpha_{11} = \frac{u_3(x^*, \theta, t)}{P \sin \frac{r\pi\theta}{\alpha} e^{j\omega t}} \tag{13.5.11}$$

gives

$$\alpha_{11} = \frac{2}{\rho h L} \sum_{m=1}^{\infty} \frac{1}{\omega_{mn}^2 - \omega^2} \sin^2 \frac{m\pi x^*}{L} \tag{13.5.12}$$

Next, to obtain the receptance β_{11} for the ring, we have to solve for the response of a simply supported ring segment to a load

$$q_3'(\theta, t) = P \sin \frac{r\pi\theta}{\alpha} e^{j\omega t} \tag{13.5.13}$$

where q_3' and P have both the unit [N/m]. Note that we assume that the systems are joined along their mid surfaces. In reality, the ring may be joined to the panel either above or below. This influence is generally small, but was investigated in Ref. 13.3.

The transverse mode shape expression is

$$U_{3n} = \sin \frac{n\pi\theta}{\alpha} \tag{13.5.14}$$

The associated natural frequencies ω_n were obtained in Sec. 5.4.

Multiplying and dividing Eq. (8.4.5) by the width b of the ring segment and performing the indicated integration with respect to the width, we obtain

$$F_k = \frac{2Pe^{j\omega t}}{\rho_s A\alpha} \int_0^{\alpha} \sin \frac{r\pi\theta}{\alpha} \sin \frac{n\pi\theta}{\alpha} \, d\theta \tag{13.5.15}$$

where A is the cross-sectional area of the ring and ρ_s is the mass density of the ring material. Evaluating the integral, we obtain

$$F_k = F_n = \begin{cases} \dfrac{Pe^{j\omega t}}{\rho_s A} & n = r \\ 0 & n \neq r \end{cases} \tag{13.5.16}$$

The response is, therefore,

$$u_3(\theta,t) = \frac{Pe^{j\omega t}}{\rho_s A(\omega_n^2 - \omega^2)} \sin\frac{r\pi\theta}{\alpha} \tag{13.5.17}$$

and the receptance, defined as

$$\beta_{11} = \frac{u_3(\theta,t)}{P \sin\frac{r\pi\theta}{\alpha} e^{j\omega t}} \tag{13.5.18}$$

becomes

$$\beta_{11} = \frac{1}{\rho_s A(\omega_n^2 - \omega^2)} \tag{13.5.19}$$

Note that the receptances α_{11} and β_{11} are compatible. Both describe the same displacement divided by the same input. Thus, the characteristic equation whose roots furnish the combined system natural frequencies, becomes

$$\frac{2}{\rho hL} \sum_{m=1}^{\infty} \frac{1}{\omega_{mn}^2 - \omega^2} \sin^2\frac{m\pi x^*}{L} + \frac{1}{\rho_s A(\omega_n^2 - \omega^2)} = 0 \tag{13.5.20}$$

The graphical solution is shown in Fig. 13.5.2 for a stiffener

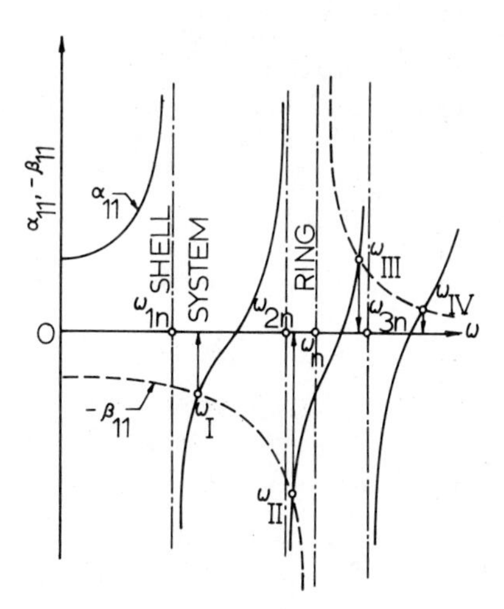

Figure 13.5.2

whose natural frequency corresponding to the n-th mode, ω_n, is lower than the panel natural frequency ω_{3n}, ω_{4n}, ... corresponding to the same circumferential mode shape, but higher than ω_{1n} and ω_{2n}. We see that the natural frequencies of the combined system ω_{III}, ω_{IV}, ..., are lower that the corresponding unstiffened panel frequencies ω_{3n}, ω_{4n}, ..., while the frequencies ω_I and ω_{II} are higher than the corresponding frequencies ω_{1n} and ω_{2n}. This allows us to formulate the rule that, for a circumferential mode shape corresponding to n, all panel frequencies that are lower than the stiffener frequency are raised and all panel frequencies that are higher than the stiffener frequency are lowered.

We can show this by again assuming that the system frequency is only a small perturbation of the original frequency. In this case, we obtain approximately

$$\frac{2}{\rho h L(\omega_{mn}^2 - \omega^2)} \sin^2 \frac{m\pi x^*}{L} + \frac{1}{\rho A(\omega_n^2 - \omega^2)} = 0 \tag{13.5.21}$$

The solutions of this equation are $\omega = \omega_k$ and are given by

$$\omega_k^2 = \omega_{mn}^2 \frac{1 + \frac{2M_s}{M}\left(\frac{\omega_n}{\omega_{mn}}\right)^2 \sin^2 \frac{m\pi x^*}{L}}{1 + \frac{2M_s}{M} \sin^2 \frac{m\pi x^*}{L}} \tag{13.5.22}$$

where M_s is the mass of the stiffener

$$M_s = \rho_s A a \alpha \tag{13.5.23}$$

and where M is the mass of the panel

$$M = \rho h L a \alpha \tag{13.5.24}$$

The approximate solution shows immediately that

$$\omega_k > \omega_{mn} \quad \text{if } \omega_n > \omega_{mn} \tag{13.5.25}$$

and

$$\omega_k < \omega_{mn} \quad \text{if } \omega_n < \omega_{mn} \tag{13.5.26}$$

Based on experimental evidence, these results are generally valid for any kind of shell or plate and any kind of stiffener.

We also notice the expected result that the stiffener will have no effect on modes whose node lines coincide with the stiffener location.

Similar results where obtained in Refs. 13.3 and 13.4 where the eigenvalues of circular cylindrical shells with multiring stiffeners are explored. Stiffening with stringers follows the same rules.

13.6 TWO SYSTEMS JOINED BY TWO OR MORE DISPLACEMENTS

If we want to join a beam to a beam, we need as a minimum to enforce continuity of transverse deflection and slope (two displacements). In cases where axial vibration in each beam member is also of concern, we need to enforce continuity for three displacements. Two shells may be attached to each other at n points, in this case we will have to enforce n displacements.

In the following, let us take the two displacement case as specific example, shown in Fig. 13.6.1, but generalize immediately after each step for the case where two components are joined by n coordinates.

From the basic definition of a receptance, which is like an influence function, we obtain the displacement amplitudes of system

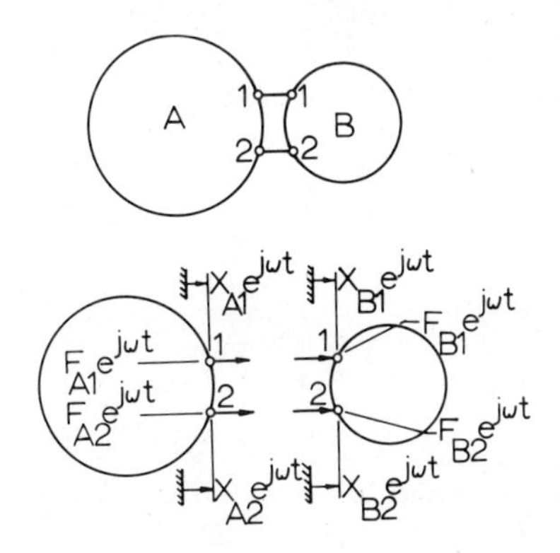

Figure 13.6.1

A as function of harmonic force inputs at these locations as

$$X_{A1} = \alpha_{11}F_{A1} + \alpha_{12}F_{A2} \tag{13.6.1}$$

$$X_{A2} = \alpha_{21}F_{A1} + \alpha_{22}F_{A2} \tag{13.6.2}$$

In general, for n displacements,

$$\{X_A\} = [\alpha]\{F_A\} \tag{13.6.3}$$

Note that we require now that the cross receptances α_{ij} where $i \neq j$ have to be known also. Similarly, for system B,

$$X_{B1} = \beta_{11}F_{B1} + \beta_{12}F_{B2} \tag{13.6.4}$$

$$X_{B2} = \beta_{21}F_{B1} + \beta_{22}F_{B2} \tag{13.6.5}$$

In general,

$$\{X_B\} = [\beta]\{F_B\} \tag{13.6.6}$$

When the two systems are joined, the forces at each displacement junction have to add up to zero, or

$$F_{A1} = -F_{B1} \tag{13.6.7}$$

$$F_{A2} = -F_{B2} \tag{13.6.8}$$

In general,

$$\{F_A\} = -\{F_B\} \tag{13.6.9}$$

Also, the displacements have to be equal because of continuity,

$$X_{A1} = X_{B1} \tag{13.6.10}$$

$$X_{A2} = X_{B2} \tag{13.6.11}$$

or, in general,

$$\{X_A\} = \{X_B\} \tag{13.6.12}$$

We may now combine the equation and obtain

$$(\alpha_{11} + \beta_{11})F_{A1} + (\alpha_{12} + \beta_{12})F_{A2} = 0 \tag{13.6.13}$$

$$(\alpha_{21} + \beta_{21})F_{A1} + (\alpha_{22} + \beta_{22})F_{A2} = 0 \tag{13.6.14}$$

In general, this can be written

$$[[\alpha] + [\beta]]\{F_A\} = 0 \qquad (13.6.15)$$

Since $F_{A1} = F_{A2} = 0$ would be the trivial solution, it must be that

$$\begin{vmatrix} \alpha_{11} + \beta_{11} & \alpha_{12} + \beta_{12} \\ \alpha_{21} + \beta_{21} & \alpha_{22} + \beta_{22} \end{vmatrix} = 0 \qquad (13.6.16)$$

In general, this can be written as

$$|[\alpha] + [\beta]| = 0 \qquad (13.6.17)$$

In expanded form, the two displacement case becomes

$$(\alpha_{11} + \beta_{11})(\alpha_{22} + \beta_{22}) - (\alpha_{12} + \beta_{12})^2 = 0 \qquad (13.6.18)$$

13.7 SUSPENSION OF AN INSTRUMENT PACKAGE IN A SHELL

To illustrate the case of two systems joined by two displacements, let us treat a circular cylindrical shell inside of which an instrument package is supported by way of two equal springs as shown in Fig. 13.7.1. Let us take the case where the springs are attached at opposite points at locations (x^*, θ^*) and $(x^*, \theta^* + \pi)$.

The receptances of system B are obtained by first considering displacement X_{B1}, with $X_{B2} = 0$, in Fig. 13.7.1 and evaluating F_{B1} and F_{B2}.

This gives

$$\beta_{11} = \frac{2[1 - (\frac{\omega}{\omega_1})^2]}{K[1 - 2(\frac{\omega}{\omega_1})^2]} \qquad (13.7.1)$$

where

$$\omega_1^2 = \frac{2K}{M} \qquad (13.7.2)$$

and $$\beta_{12} = \frac{2}{K}\left[1 - (\frac{\omega}{\omega_1})^2\right] \qquad (13.7.3)$$

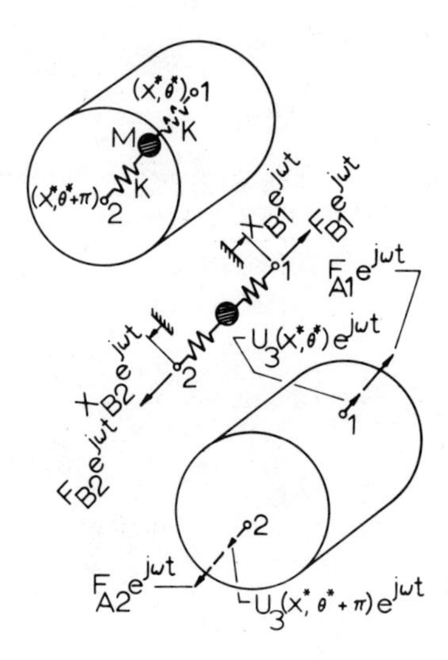

Figure 13.7.1

Next, we set $X_{B2} = 0$ and evaluate the forces F_{B1} and F_{B2} to the harmonic displacement X_{B1}. This gives, because of the symmetry of the problem,

$$\beta_{22} = \frac{2[1 - (\frac{\omega}{\omega_1})^2]}{K[1 - 2(\frac{\omega}{\omega_1})^2]} \tag{13.7.4}$$

$$\beta_{21} = \frac{2}{K}\left[1 - (\frac{\omega}{\omega_1})^2\right] \tag{13.7.5}$$

The receptances for the shell are obtained from the solution for a harmonic point force acting on a circular cylindrical shell as obtained in Sec. 8.8. In Eq. (8.8.29), we evaluate u_3 at (x^*,θ^*) and let $P_3 = F_{A1}$. This gives

$$\alpha_{11} = \frac{4}{M_s} \sum_{m=1}^{\infty} \sum_{n=0}^{\infty} \frac{\sin^2 \frac{m\pi x^*}{L}}{\varepsilon_n(\omega_{mn}^2 - \omega^2)} \tag{13.7.6}$$

and, evaluating u_3 at $(x^*, \theta^* + \pi)$, we obtain

$$\alpha_{21} = \frac{4}{M_s} \sum_{m=1}^{\infty} \sum_{n=0}^{\infty} \frac{\sin^2 \frac{m\pi x^*}{L} \cos n\pi}{\varepsilon_n(\omega_{mn}^2 - \omega^2)} \tag{13.7.7}$$

where

$$M_s = 2\pi\rho haL \tag{13.7.8}$$

Next, applying the load $P_3 = F_{A2}$ at x^*, $\theta^* + \pi$ and evaluating u_3 at $(x^*, \theta^* + \pi)$ gives

$$\alpha_{22} = \frac{4}{M_s} \sum_{m=1}^{\infty} \sum_{n=0}^{\infty} \frac{\sin^2 \frac{m\pi x^*}{L}}{\varepsilon_n(\omega_{mn}^2 - \omega^2)} \tag{13.7.9}$$

and, evaluating u_3 at (x^*, θ^*), we obtain as expected

$$\alpha_{12} = \frac{4}{M_s} \sum_{m=1}^{\infty} \sum_{n=0}^{\infty} \frac{\sin^2 \frac{m\pi x^*}{L} \cos n\pi}{\varepsilon_n(\omega_{mn}^2 - \omega^2)} \tag{13.7.10}$$

The characteristic equation is, since in this case $\alpha_{11} = \alpha_{22}$ and $\beta_{11} = \beta_{22}$,

$$(\alpha_{11} + \beta_{11})^2 - (\alpha_{12} + \beta_{12})^2 = 0 \tag{13.7.11}$$

or

$$\alpha_{11} + \beta_{11} = \pm(\alpha_{12} + \beta_{12}) \tag{13.7.12}$$

This gives

$$\frac{4}{M_s} \sum_{m=1}^{\infty} \sum_{n=0}^{\infty} \frac{\sin^2 \frac{m\pi x^*}{L}}{\varepsilon_n(\omega_{mn}^2 - \omega^2)} (1 \mp \cos n\pi)$$

$$+ \frac{2}{K} \frac{1 - (\frac{\omega}{\omega_1})^2}{1 - 2(\frac{\omega}{\omega_1})^2} \left[1 \mp 1 \pm 2\left(\frac{\omega}{\omega_1}\right)^2\right] = 0 \tag{13.7.13}$$

and may be written in terms of two equations. The first is

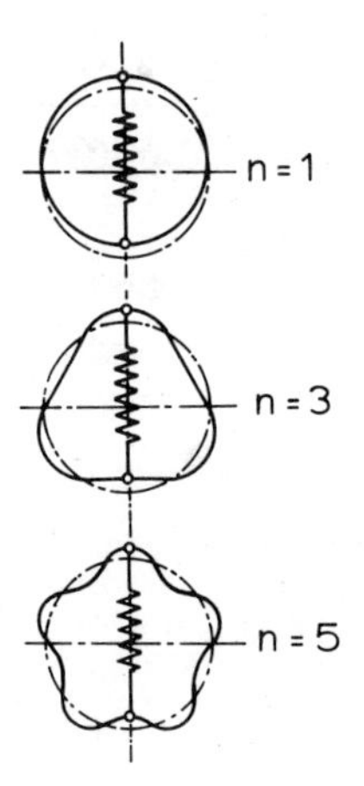

Figure 13.7.2

$$\frac{4}{M_s} \sum_{m=1}^{\infty} \sum_{n=0}^{\infty} \frac{\sin^2 \frac{m\pi x^*}{L}}{\varepsilon_n(\omega_{mn}^2 - \omega^2)} (1 - \cos n\pi)$$

$$+ \frac{4}{K}\left(\frac{\omega}{\omega_1}\right)^2 \frac{1 - (\frac{\omega}{\omega_1})^2}{1 - 2(\frac{\omega}{\omega_1})^2} = 0 \qquad (13.7.14)$$

and the second is

$$\frac{4}{M_s} \sum_{m=1}^{\infty} \sum_{n=0}^{\infty} \frac{\sin^2 \frac{m\pi x^*}{L}}{\varepsilon_n(\omega_{mn}^2 - \omega^2)} (1 + \cos n\pi) + \frac{4}{K} \frac{[1 - (\frac{\omega}{\omega_1})^2]^2}{1 - 2(\frac{\omega}{\omega_1})^2} = 0$$

(13.7.15)

The roots $\omega = \omega_k$ of these equations will be the natural frequencies of the system. Substitution of these into the displacement solution of the shell will give the new mode shapes.

The meaning of these equations can be best explained by examining a reduced system. Let us suppose that the mass approaches zero so that we are left simply with a connecting spring. In this case $1/\omega_1 = 0$. Let us also assume that the influence of the spring

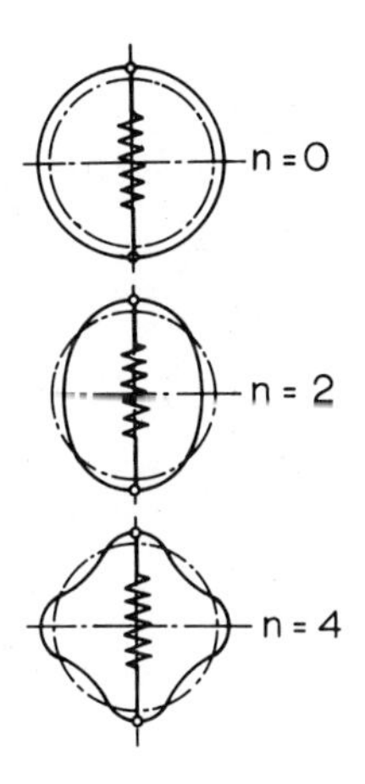

Figure 13.7.3

is small so that we can utilize a one-term solution. This gives, for Eq. (13.7.14),

$$\frac{4}{M_s \varepsilon_n} \sin^2 \frac{m\pi x^*}{L} (1 - \cos n\pi)(\omega_{mn}^2 - \omega^2) = 0 \qquad (13.7.16)$$

This equation is satisfied either if $n = 0, 2, 4$ or if the roots $\omega = \omega_k$ are

$$\omega_k^2 = \omega_{mn}^2 \qquad (13.7.17)$$

It means that the spring is not active for $n = 1, 3, 5, \ldots$ since the displacements at each end of the spring do not produce a net deflection of the spring, as illustrated in Fig. 13.7.2.

From Eq. (13.7.15) we obtain the one-term solution, for $n = 0, 2, 4, \ldots$,

$$\omega_k^2 = \omega_{mn}^2 + \frac{2K}{M_s \varepsilon_n} \sin^2 \frac{m\pi x^*}{L} \qquad (13.7.18)$$

In this case, the spring is compressed equally from both ends, as shown in Fig. 13.7.3.

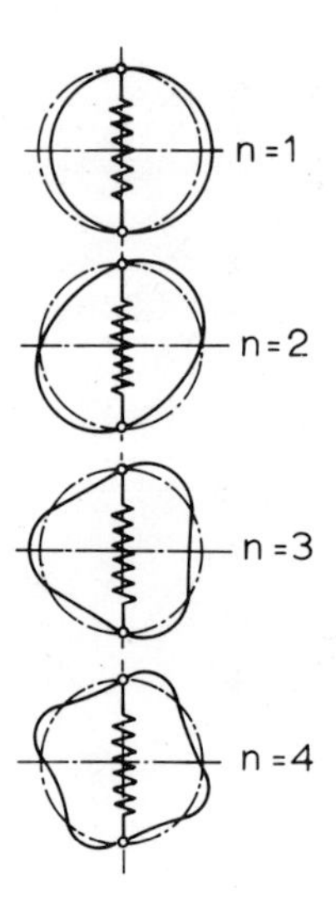

Figure 13.7.4

In addition, it is possible that the spring attachment points lie on a node line for all n, so that $\omega_k^2 = \omega_{mn}^2$. This is sketched in Fig. 13.7.4. Note that this case does not fall out of the receptance mathematics since the assumption was made that the point of attachment is a point of motion. Rather, modes where all points of attachment are node points are found by examining the individual component modes. The presence of the mass will result in further combinations because of in-phase or out-of-phase motion of the modes following similar principles.

REFERENCES

13.1 R. E. D. Bishop and D. C. Johnson, *The Mechanics of Vibration*, Cambridge University Press, London, 1960.

13.2 I. E. Sakharov, "Use of the method of dynamic rigidities for calculating the frequencies of natural vibrations of built-up shells," NASA Technical Translation F-341, pp. 797-805 (from Russian, *Theory of Plates and Shells*, S. M. Durgaryan (ed.), 1962).

13.3 I. D. Wilken and W. Soedel, The receptance method applied to ring-stiffened cylindrical shells: analysis of modal characteristics, *J. Sound Vibration, vol. 44,* no. 4, 1976, pp. 563-576.

13.4 I. D. Wilken and W. Soedel, Simplified prediction of the modal characteristics of ring-stiffened cylindrical shells, *J. Sound Vibration, vol. 44,* no. 4, 1976, pp. 577-589.

13.5 J. T. Weissenburger, Effect of local modifications on the vibration characteristics of linear systems, *J. Appl. Mechs., vol. 35,* no. 2, 1968, pp. 327-332.

14

HYSTERESIS DAMPING

The equivalent viscous damping coefficient that was used in the chapters on the forced response of shell structures is a function of several effects. While there may be truly a motion-resisting force proportional to velocity, we may also have turbulent damping proportional to velocity squared caused by the surrounding media, boundary damping because of either friction in the boundary joints themselves (rivets, clamps, etc.), or because of the elasticity of the boundary, we have to allow for a certain amount of energy to be converted to wave action of the boundary material which is lost to the system that is being investigated, and finally internal damping of the material. Internal damping is characterized by a hysteresis loop. There is also the possibility that damping is introduced by friction between two shell surfaces. For instance, to dampen the hermetically sealed shells of refrigeration machinery, a ring of the same sheet material is loosely pressed inside the mainshell so that the two surfaces can work against each other when vibrating.

Historically, internal damping was first investigated in 1784 by Coulomb [14.1]. He showed experimentally, using his torsional pendulum, that damping was also caused by a micro-structural mechanism and not only by air friction. He recognized that this internal damping, or *hysteresis damping*, as it is often termed, was a function of vibration amplitude. Many investigations on this topic have followed since.

14.1 THE EQUIVALENT VISCOUS DAMPING COEFFICIENT

The forces per unit surface area in the three different directions, when equivalent viscous damping is assumed, are given by (i = 1,2,3)

$$q_i = \lambda \dot{u}_i \tag{14.1.1}$$

and if the motion is harmonic

$$u_i = U_i \sin \omega t \tag{14.1.2}$$

we get

$$q_i = \lambda U_i \omega \cos \omega t \tag{14.1.3}$$

The average dissipated energy per unit surface area and per cycle of harmonic motion is

$$E_d = \frac{1}{A} \iint_A \sum_{i=1}^{3} \int_0^{2\pi/\omega} \lambda U_i^2 \omega^2 \cos^2 \omega t \; dt \; dA \tag{14.1.4}$$

where A is the surface area of the shell. We get

$$E_d = \lambda \pi \omega \frac{1}{A} \iint_A (U_1^2 + U_2^2 + U_3^2) \; dA \tag{14.1.5}$$

Thus, if we can identify by theoretical models or by experiment what the energy dissipated per cycle and unit area is, we may solve for λ and obtain

$$\lambda = \frac{E_d A}{\pi\omega \iint_A (U_1^2 + U_2^2 + U_3^2) \; dA} \tag{14.1.6}$$

When transverse motion is dominant, the most common case, $U_1^2 + U_2^2 << U_3^2$, and we may set

$$\lambda = \frac{E_d A}{\pi\omega \iint_A U_3^2 \, dA} \tag{14.1.7}$$

14.2 HYSTERESIS DAMPING

Structural damping is characterized by the fact that if we cycle a tensil test specimen, we obtain a hysteresis loop as shown in Fig. 14.2.1. The shaded area of the loop is the total energy dissipated per cycle. If we divide the force by the cross section of the specimen and the displacement by the length of the specimen, we get a stress-strain plot of the same phenomenon. The area is now equal to the energy dissipated per cycle and volume.

Unfortunately, no generally acceptable way has been found to utilize this information directly. An approximation is the so-called complex modulus.

We have Hook's law

$$\sigma = E\varepsilon \tag{14.2.1}$$

Substituting

$$\sigma = \sigma_{max} \sin \omega t \tag{14.2.2}$$

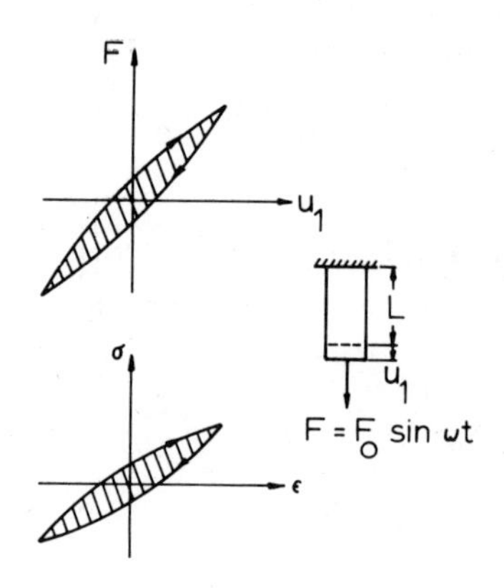

Figure 14.2.1

where

$$\sigma_{max} = \frac{F_o}{A} \tag{14.2.3}$$

gives

$$\varepsilon = \frac{\sigma_{max}}{E} \sin \omega t \tag{14.2.4}$$

Plotting σ as a function of ε gives, as expected, a straight line. For the line to acquire width so that we obtain a resemblance to a hysteresis loop, we have to replace E by $E(1 + j\eta)$, where η is called the hysteresis loss factor. In this case we obtain

$$\varepsilon = \frac{\sigma_{max}}{E(1 + j\eta)} \sin \omega t \tag{14.2.5}$$

This can be written as

$$\varepsilon = \frac{\sigma_{max}}{E\sqrt{1 + \eta^2}} \sin(\omega t - \phi) \tag{14.2.6}$$

where

$$\phi = \tan^{-1}\eta \tag{14.2.7}$$

For typically small values of η we obtain

$$\phi \cong \eta \tag{14.2.8}$$

and

$$\sqrt{1 + \eta^2} \cong 1 \tag{14.2.9}$$

Thus

$$\varepsilon = \frac{\sigma_{max}}{E} \sin(\omega t - \eta) \tag{14.2.10}$$

Plotting σ as function of ε in Figure 14.2.2 gives an ellipse with the approximate half axes

$$a = \frac{\sigma_{max}}{\cos \alpha} \tag{14.2.11}$$

$$b = \frac{\sigma_{max}}{E} \eta \cos \alpha \tag{14.2.12}$$

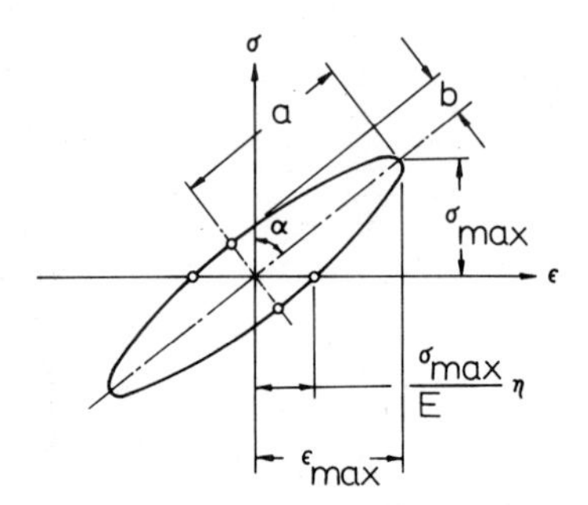

Figure 14.2.2

Since the energy dissipated per cycle and volume is equal to the area of the ellipse, we obtain

$$E_1 = \frac{\pi}{E}\sigma_{max}^2\eta = \pi E\varepsilon_{max}^2\eta \qquad (14.2.13)$$

The total dissipated energy in the specimen is

$$E_T = Lbh\pi E\varepsilon_{max}^2\eta \qquad (14.2.14)$$

Since the maximum strain energy in the test specimen is

$$U_{max} = \frac{Lbh}{2}\sigma_{max}\varepsilon_{max} = \frac{Lbh}{2}E\varepsilon_{max}^2 \qquad (14.2.15)$$

we find that [14.2]

$$\eta = \frac{1}{2\pi}\frac{E_T}{U_{max}} \qquad (14.2.16)$$

This means that $2\pi\eta$ defines the ratio of dissipated energy per cycle to the strain energy at peak amplitude. Thus, in the case of a shell, we can argue that

$$E_T = 2\pi\eta U_{max} \qquad (14.2.17)$$

where U_{max} is the strain energy of the shell at peak amplitude. The energy dissipated per cycle and unit surface is then

$$E_d = \frac{2\pi\eta}{A}U_{max} \qquad (14.2.18)$$

where A is the reference surface area of the shell, plate, or beam. Thus, the equivalent viscous damping coefficient λ is

$$\lambda = \frac{2U_{max}}{\omega \iint_A (U_1^2 + U_2^2 + U_3^2)\, dA} \eta \tag{14.2.19}$$

Remembering the discussion of Rayleigh's method in Sec. 7.4 where it was shown that

$$\omega_k^2 = \frac{2U_{max}}{\rho h \iint_A (U_1^2 + U_2^2 + U_3^2)\, dA} \tag{14.2.20}$$

we may write the equivalent viscous damping coefficient as

$$\lambda = \rho h\, \omega_k \frac{\omega_k}{\omega} \eta \tag{14.2.21}$$

It may be used directly when the forcing is harmonic. For nonharmonic forcing, some choice about a mean value of ω_k^2/ω will have to be made.

14.3 DIRECT UTILIZATION OF HYSTERESIS MODEL IN ANALYSIS

For the technically significant class of cases where the steady-state response to harmonic excitation is to be obtained, one can work directly with the hysteresis model. Introducing the complex modulus into Love's equation gives

$$(1 + j\eta)L_i(u_1,u_2,u_3) - \rho h\ddot{u}_i = -q_i^* e^{j\omega t} \tag{14.3.1}$$

where $L_i(u_1,u_2,u_3)$ represents the same operators as given in Eqs. (8.1.3) to (8.1.5). The general forcing terms in Love's equation are now restricted to harmonic excitation, with q_i^* representing the pressure load distribution.

The modal expansion solution is

$$u_i(\alpha_1,\alpha_2,t) = \sum_{k=1}^{\infty} \eta_k(t)U_{ik}(\alpha_1,\alpha_2) \tag{14.3.2}$$

One should take note that η_k are the modal participation factors while η without subscript is the traditional notation for the hysteresis loss factor.

Substitution into Eq. (14.3.1) gives

$$\sum_{k=1}^{\infty} [(1 + j\eta)\eta_k L_i(U_{1k},U_{2k},U_{3k}) - \rho h\ddot{\eta}_k U_{ik}] = -q_i^* e^{j\omega t} \quad (14.3.3)$$

From the eigenvalue analysis, where $\eta = 0$ and $q_i = 0$, we obtain the identity

$$L_i(U_{1k},U_{2k},U_{3k}) = -\rho h\omega_k^2 U_{ik} \quad (14.3.4)$$

This gives

$$\sum_{k=1}^{\infty} [\rho h\ddot{\eta}_k + \rho h(1 + j\eta)\omega_k^2\eta_k]U_{ik} = q_i^* e^{j\omega t} \quad (14.3.5)$$

Multiplying both sides by a mode U_{ip}, where p may be either equal to k or unequal and writing the relationship for every value of i = 1, 2,3 gives

$$\sum_{k=1}^{\infty} [\rho h\ddot{\eta}_k + \rho h(1 + j\eta)\omega_k^2\eta_k]U_{1k}U_{1p} = q_1^* U_{1p} e^{j\omega t} \quad (14.3.6)$$

$$\sum_{k=1}^{\infty} [\rho h\ddot{\eta}_k + \rho h(1 + j\eta)\omega_k^2\eta_k]U_{2k}U_{2p} = q_2^* U_{2p} e^{j\omega t} \quad (14.3.7)$$

$$\sum_{k=1}^{\infty} [\rho h\ddot{\eta}_k + \rho h(1 + j\eta)\omega_k^2\eta_k]U_{3k}U_{3p} = q_3^* U_{3p} e^{j\omega t} \quad (14.3.8)$$

Adding Eqs. (14.3.6) to (14.3.8) and integrating over the reference surface of the shell gives

$$\sum_{k=1}^{\infty} (\rho h\ddot{\eta}_k + \rho h(1 + j\eta)\omega_k^2\eta_k) \int_{\alpha_2}\int_{\alpha_1} (U_{1k}U_{1p} + U_{2k}U_{2p} + U_{3k}U_{3p})A_1A_2 \, d\alpha_1 \, d\alpha_2$$

$$= \int_{\alpha_2}\int_{\alpha_1} (q_1^* U_{1p} + q_2^* U_{2p} + q_3^* U_{3p})A_1A_2 \, d\alpha_1 \, d\alpha_2 \quad (14.3.9)$$

Utilizing the orthogonality property of natural modes gives

$$\ddot{\eta}_k + (1 + j\eta)\omega_k^2\eta_k = F_k^* e^{j\omega t} \tag{14.3.10}$$

where

$$F_k^* = \frac{1}{\rho h N_k} \int_{\alpha_2}\int_{\alpha_1} (q_1^* U_{1k} + q_2^* U_{2k} + q_3^* U_{3k}) A_1 A_2 \, d\alpha_1 \, d\alpha_2 \tag{14.3.11}$$

$$N_k = \int_{\alpha_2}\int_{\alpha_1} (U_{1k}^2 + U_{2k}^2 + U_{3k}^2) A_1 A_2 \, d\alpha_1 \, d\alpha_2 \tag{14.3.12}$$

This result is comparable to the one given by Eqs. (8.5.2) and (8.5.3).

The steady-state solution will be

$$\eta_k = \Lambda_k e^{j(\omega t - \phi_k)} \tag{14.3.13}$$

Substitution in Eq. (14.3.10) gives

$$\Lambda_k e^{-j\phi_k} = \frac{F_k^*}{(\omega_k^2 - \omega^2) + j\eta\omega_k^2} \tag{14.3.14}$$

The magnitude of the response is, therefore,

$$\Lambda_k = \frac{F_k^*}{\omega_k^2 \sqrt{[1 - (\frac{\omega}{\omega_k})^2]^2 + \eta^2}} \tag{14.3.15}$$

The phase angle is

$$\phi_k = \tan^{-1} \frac{\eta}{1 - (\frac{\omega}{\omega_k})^2} \tag{14.3.16}$$

An interesting by-product of this analysis is the relationship between the modal damping coefficient and the hysteresis loss factor. It must be that

$$2\zeta_k \omega_k \omega = \eta\omega_k^2 \tag{14.3.17}$$

Thus, the equivalent modal damping coefficient becomes

$$\zeta_k = \frac{1}{2} \frac{\omega_k}{\omega} \eta \tag{14.3.18}$$

The equivalent viscous damping coefficient is, therefore.

$$\lambda = \rho h \omega_k \frac{\omega_k}{\omega} \eta \tag{14.3.19}$$

This agrees, as expected, with Eq. (14.2.21).

14.4 HYSTERETICALLY DAMPED PLATE EXCITED BY SHAKER

The following illustrates how the hysteresis loss factor can be obtained from a measurement, using as an example the simply supported plate (Fig. 14.4.1).

For a simply supported rectangular plate, $q_1^* = q_2^* = 0$. The harmonically varying point load of amplitude F in [N], representing the harmonic input from a shaker, is described by

$$q_3^* = F\, \delta(x - x^*)\, \delta(y - y^*) \tag{14.4.1}$$

The eigenvalues are

$$U_{3k} = U_{3mn} = \sin \frac{m\pi x}{a} \sin \frac{n\pi y}{b} \tag{14.4.2}$$

$$\omega_k = \omega_{mn} = \pi^2 \left[\left(\frac{m}{a}\right)^2 + \left(\frac{n}{b}\right)^2\right] \sqrt{\frac{D}{\rho h}} \tag{14.4.3}$$

Equation (14.3.11) becomes

$$F_k^* = \frac{4F}{\rho hab} \sin \frac{m\pi x^*}{a} \sin \frac{n\pi y^*}{b} \tag{14.4.4}$$

The solution is, therefore,

$$u_3(x,y,t) = \frac{4F}{\rho hab} \sum_{m=1}^{\infty} \sum_{n=1}^{\infty} \frac{\sin \frac{m\pi x^*}{a} \sin \frac{m\pi x}{a} \sin \frac{n\pi y^*}{b} \sin \frac{n\pi y}{b}}{\omega_{mn}^2 \sqrt{[1 - (\frac{\omega}{\omega_{mn}})^2]^2 + \eta^2}} e^{j(\omega t - \phi_{mn})} \tag{14.4.5}$$

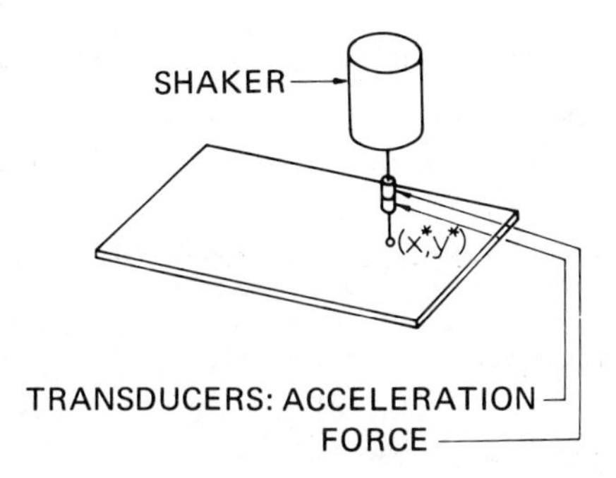

Figure 14.4.1

where

$$\phi_{mn} = \tan^{-1} \frac{\eta}{1 - \left(\frac{\omega}{\omega_{mn}}\right)^2} \tag{14.4.6}$$

Let us now assume that the acceleration response is measured at the point of attachment of the shaker. Also, that the force amplitude is monitored. Furthermore, the measurement is made at each of the natural frequencies ω_{mn}. We have in this case

$$u_3(x^*,y^*,t) = \frac{4F}{\omega^2 \rho hab\eta} \sin^2 \frac{m\pi x^*}{a} \sin^2 \frac{n\pi y^*}{b} e^{j(\omega t - \frac{\pi}{2})} \tag{14.4.7}$$

or, the acceleration is

$$\ddot{u}_3(x^*,y^*,t) = -\frac{4F}{\rho hab\eta} \sin^2 \frac{m\pi x^*}{a} \sin^2 \frac{n\pi y^*}{b} e^{j(\omega t - \frac{\pi}{2})} \tag{14.4.8}$$

Solving for η in terms of the measured acceleration amplitude $|\ddot{u}_3|$ and force amplitude F is

$$\eta = \frac{4}{\rho hab} \sin^2 \frac{m\pi x^*}{a} \sin^2 \frac{n\pi y^*}{b} \frac{F}{|\ddot{u}_3|} \tag{14.4.9}$$

In Ref. 14.3, several other methods of defining η are discussed. Typically, η is not constant with frequency.

REFERENCES

14.1 C. A. Coulomb, "Recherches theoriques et experimentales sur la force torsion et sur l'elasticite des fils de metal," Memoirs of the Paris Academy, 1784.

14.2 D. Ross, E. E. Ungar, and E. M. Kerwin, Jr., Damping of plate flexural vibrations by means of viscoelastic laminae, *Structural Damping*, J. E. Ruzicka (ed.), American Society of Mechanical Engineers, New York, 1959, pp. 49-88.

14.3 R. Plunkett, Measurement of damping, *Structural Damping*, J. E. Ruzicka (ed.), American Society of Mechanical Engineers, New York, 1959, pp. 117-131.

15

SHELLS MADE OF COMPOSITE MATERIAL

In all of the previous chapters, the shell material was assumed to be homogeneous and isotropic. Because of the need for lightweight designs, for instance, in space applications, composite shell materials have become more and more common. One of the advantages of composite materials is that one can design directional properties into them almost on demand. The disadvantage is that structures built with composite materials are more difficult to analyze and even to understand in their idiosyncracies of behavior and failure.

15.1 THE NATURE OF COMPOSITES

In the following, we will concentrate on the most common composite arrangement that one finds in thin-wall structures, namely laminated composite. The composite is in this case built up of sheets (laminae) of uniform thickness. Each lamina may be isotropic, orthotropic, or anisotropic. From a materials composition viewpoint, it may be homogeneous or heterogeneous. Once the lamina are joined to each other, the most general case is what is called *coupled anisotropic*. Some of this is illustrated in Figure 15.1.1

Usually, a lamina or ply is composed of reinforcing material, most commonly fibers, in a supporting matrix. The fibers usually carry the load. The matrix material usually holds the fibers in place so that they are properly spaced, protects them against corrosion, and seals the structure against the escape of gases and liquids. Each lamina usually consists of a set of parallel fibers embedded into the matrix material. The laminae can then be assembled with the fibers of each lamina pointing into different directions in such a way that the desired stiffness properties are obtained.

Most engineering materials are isotropic. This means that the properties are not a function of direction. All planes which pass through a point in the material are planes of material property symmetry. To define the material, we need only two elastic constants, namely Young's modulus and Poisson's ratio. An axially loaded rectangular strip will remain rectangular as it is distorted, as shown in Fig. 15.1.1.

An orthotropic material has three planes of material symmetry. We will see that we need four material constants to describe the plane stress state.

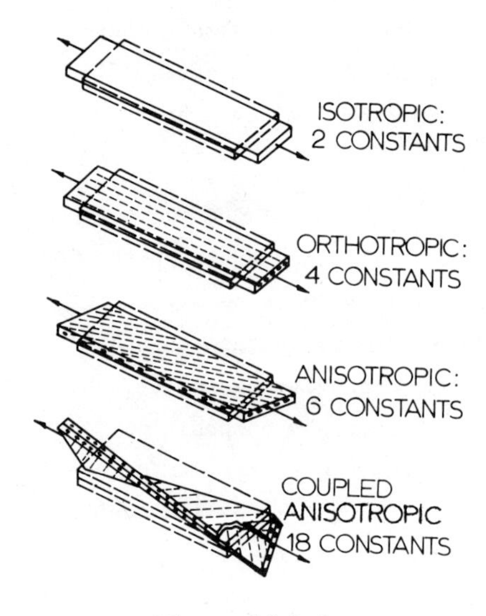

Figure 15.1.1

15.2 LAMINA-CONSTITUTIVE RELATIONSHIP

It is assume that each lamina is in a state of plane stress. For material that is homogeneous and isotropic, we have relations (2.2.10) to (2.2.12). They may be written as

$$\begin{Bmatrix} \sigma_{xx} \\ \sigma_{yy} \\ \sigma_{xy} \end{Bmatrix} = [Q] \begin{Bmatrix} \varepsilon_{xx} \\ \varepsilon_{yy} \\ \varepsilon_{xy} \end{Bmatrix} \tag{15.2.1}$$

where

$$[Q] = \begin{bmatrix} Q_{11} & Q_{12} & 0 \\ Q_{21} & Q_{22} & 0 \\ 0 & 0 & Q_{33} \end{bmatrix} \tag{15.2.2}$$

and where

$$Q_{11} = Q_{22} = \frac{E}{1 - \mu^2} \tag{15.2.3}$$

$$Q_{12} = Q_{21} = \frac{\mu E}{1 - \mu^2} \tag{15.2.4}$$

$$Q_{33} = G = \frac{E}{2(1 + \mu)} \tag{15.2.5}$$

The constitutive relationship for a homogeneous orthotropic lamina in a state of plane stress as shown in Fig. 15.2.1 is also given by Eq. (15.2.1), except that now (the filament direction is the x direction)

$$Q_{11} = \frac{E_{xx}}{1 - \mu_{xy}\mu_{yx}} \tag{15.2.6}$$

$$Q_{22} = \frac{E_{yy}}{1 - \mu_{xy}\mu_{yx}} \tag{15.2.7}$$

$$Q_{12} = \frac{\mu_{yx}E_{xx}}{1 - \mu_{xy}\mu_{yx}} \tag{15.2.8}$$

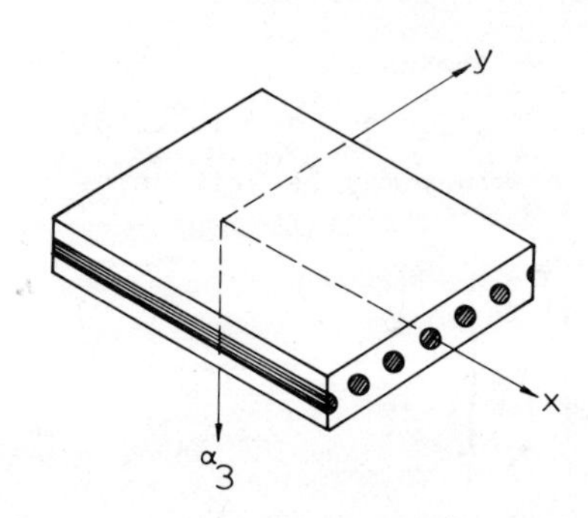

Figure 15.2.1

$$Q_{21} = \frac{\mu_{yx}E_{yy}}{1 - \mu_{xy}\mu_{yx}} \tag{15.2.9}$$

$$Q_{33} = G_{xy} \tag{15.2.10}$$

Because of the requirement that

$$Q_{12} = Q_{21} \tag{15.2.11}$$

we obtain

$$\mu_{yx}E_{xx} = \mu_{xy}E_{yy} \tag{15.2.12}$$

We note that we have four material constants: E_{xx}, E_{yy}, μ_{xy}, and G_{xy}. For filamentary lamina as shown in Fig. 15.2.1, Halpin and Tsai [15.1] suggested the following interpolation, based on the volume ratio of filament to matrix material:

$$E_{xx} = E_f V_f + E_m V_m \tag{15.2.13}$$

$$E_{yy} = E_m \left(\frac{1 + \zeta\alpha V_f}{1 - \zeta V_f} \right) \tag{15.2.14}$$

$$\mu_{xy} = \mu_f V_f + \mu_m V_m \tag{15.2.15}$$

$$G_{xy} = G_m \left(\frac{1 + \zeta\beta V_f}{1 - \zeta V_f} \right) \tag{15.2.16}$$

where

$$\alpha = \frac{E_f/E_m - 1}{E_f/E_m + \zeta} \tag{15.2.17}$$

$$\beta = \frac{G_f/G_m - 1}{G_f/G_m + \zeta} \tag{15.2.18}$$

and where

E_f = modulus of elasticity of fiber [N/m^2]

E_m = modulus of elasticity of matrix [N/m^2]

μ_f = Poisson's ratio of fiber

μ_m = Poisson's ratio of matrix

V_f = volume fraction of fiber

V_m = volume fraction of material (note: $V_f + V_m = 1$)

G_f = shear modulus of fiber [N/m^2]

G_m = shear modulus of matrix [N/m^2]

The factor ζ is an adjustment factor that depends on some extent on the boundary conditions. It can be taken as $\zeta = 1$ for a first approximation.

A special case occurs when very stiff fibers are embedded in a relatively soft matrix. For instance, if we take pneumatic tires as an example, typically $E_f >> E_m$ and $G_f >> G_m$,

$$E_{xx} \cong E_f V_f \tag{15.2.19}$$

$$E_{yy} \cong E_m \tag{15.2.20}$$

$$\mu_{xy} \cong \mu_m V_m \tag{15.2.21}$$

$$G_{xy} \cong G_m \tag{15.2.22}$$

since $\alpha = \beta \cong 1$.

Since the lamina may not always be oriented so that its principal stiffness directions coincide with the coordinates, relation (15.2.1) has to be transformed to account for a possible θ rotation as shown in Fig. 15.2.2. By analyzing the equilibrium of an infinitesimal element as shown in Fig. 15.2.3, we obtain

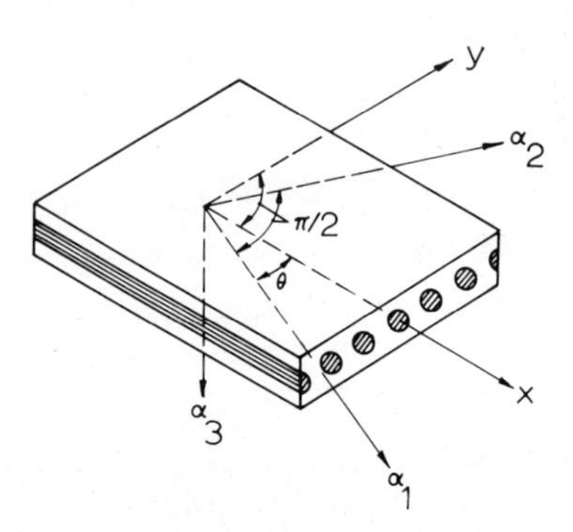

Figure 15.2.2

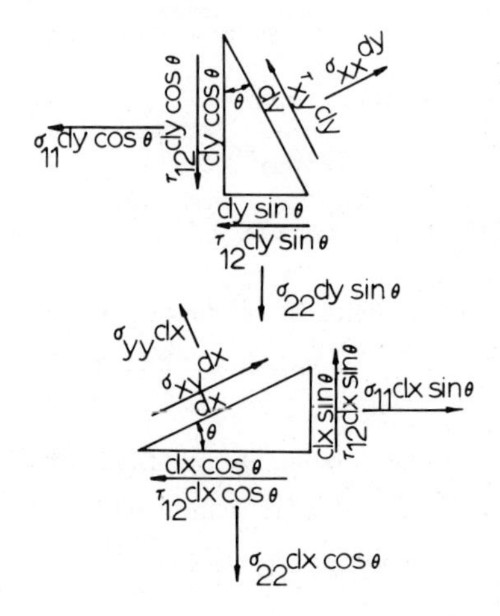

Figure 15.2.3

$$\begin{Bmatrix} \sigma_{xx} \\ \sigma_{yy} \\ \sigma_{xy} \end{Bmatrix} = [T_1] \begin{Bmatrix} \sigma_{11} \\ \sigma_{22} \\ \sigma_{12} \end{Bmatrix} \tag{15.2.23}$$

where

$$[T_1] = \begin{bmatrix} \cos^2\theta & \sin^2\theta & 2\sin\theta\cos\theta \\ \sin^2\theta & \cos^2\theta & -2\sin\theta\cos\theta \\ -\sin\theta\cos\theta & \sin\theta\cos\theta & \cos^2\theta - \sin^2\theta \end{bmatrix} \tag{15.2.24}$$

Similarly, we obtain for strain

$$\begin{Bmatrix} \varepsilon_{xx} \\ \varepsilon_{yy} \\ \varepsilon_{xy} \end{Bmatrix} = [T_2] \begin{Bmatrix} \varepsilon_{11} \\ \varepsilon_{22} \\ \varepsilon_{12} \end{Bmatrix} \tag{15.2.25}$$

where

$$[T_2] = \begin{bmatrix} \cos^2\theta & \sin^2\theta & \sin\theta\cos\theta \\ \sin^2\theta & \cos^2\theta & -\sin\theta\cos\theta \\ -2\sin\theta\cos\theta & 2\sin\theta\cos\theta & \cos^2\theta - \sin^2\theta \end{bmatrix} \tag{15.2.26}$$

Substituting gives

$$\begin{Bmatrix} \sigma_{11} \\ \sigma_{22} \\ \sigma_{12} \end{Bmatrix} = [\bar{Q}] \begin{Bmatrix} \varepsilon_{11} \\ \varepsilon_{22} \\ \varepsilon_{12} \end{Bmatrix} \tag{15.2.27}$$

where

$$[\bar{Q}] = [T_1]^{-1}[Q][T_2] \tag{15.2.28}$$

The coefficients $\bar{Q}_{ij}$ of this matrix are

$$\bar{Q}_{11} = U_1 + U_2 \cos 2\theta + U_3 \cos 4\theta \tag{15.2.29}$$

$$\bar{Q}_{22} = U_1 - U_2 \cos 2\theta + U_3 \cos 4\theta \tag{15.2.30}$$

$$\bar{Q}_{12} = U_4 - U_3 \cos 4\theta = \bar{Q}_{21} \tag{15.2.31}$$

$$\bar{Q}_{33} = U_5 - U_3 \cos 4\theta \tag{15.2.32}$$

$$\bar{Q}_{13} = -\frac{1}{2} U_2 \sin 2\theta - U_3 \sin 4\theta = \bar{Q}_{31} \tag{15.2.33}$$

$$\bar{Q}_{23} = -\frac{1}{2} U_2 \sin 2\theta + U_3 \sin 4\theta = \bar{Q}_{32} \tag{15.2.34}$$

where

$$U_1 = \frac{1}{8}(3Q_{11} + 3Q_{22} + 2Q_{12} + 4Q_{33}) \tag{15.2.35}$$

$$U_2 = \frac{1}{2}(Q_{11} - Q_{22}) \tag{15.2.36}$$

$$U_3 = \frac{1}{8}(Q_{11} + Q_{22} - 2Q_{12} - 4Q_{33}) \tag{15.2.37}$$

$$U_4 = \frac{1}{8}(Q_{11} + Q_{22} + 6Q_{12} - 4Q_{33}) \tag{15.2.38}$$

$$U_5 = \frac{1}{8}(Q_{11} + Q_{22} - 2Q_{12} + 4Q_{33}) \tag{15.2.39}$$

15.3 LAMINATED COMPOSITE

Let us assume again that the shell is thin, even if it is composed of n laminations. Furthermore, we will assume again that displacements vary linearly through the shell thickness. This implies that all relationships of Sec. 2.4 hold. Thus,

$$\begin{Bmatrix} \varepsilon_{11} \\ \varepsilon_{22} \\ \varepsilon_{12} \end{Bmatrix} = \begin{Bmatrix} \varepsilon_{11}^{\circ} \\ \varepsilon_{22}^{\circ} \\ \varepsilon_{12}^{\circ} \end{Bmatrix} + \alpha_3 \begin{Bmatrix} k_{11} \\ k_{22} \\ k_{12} \end{Bmatrix} \tag{15.3.1}$$

Introducing the subscript k to denote the kth lamina, we may express the stress in the kth lamina by combining Eqs. 15.3.1 and 15.2.27.

$$\begin{Bmatrix} \sigma_{11} \\ \sigma_{22} \\ \sigma_{12} \end{Bmatrix}_k = [\bar{Q}]_k \begin{Bmatrix} \varepsilon_{11}^{\circ} \\ \varepsilon_{22}^{\circ} \\ \varepsilon_{12}^{\circ} \end{Bmatrix} + \alpha_3 [\bar{Q}]_k \begin{Bmatrix} k_{11} \\ k_{22} \\ k_{12} \end{Bmatrix} \tag{15.3.2}$$

The stress resultants are

$$\begin{Bmatrix} N_{11} \\ N_{22} \\ N_{12} \end{Bmatrix} = \int_{\alpha_3} \begin{Bmatrix} \sigma_{11} \\ \sigma_{22} \\ \sigma_{12} \end{Bmatrix} d\alpha_3 \tag{15.3.3}$$

or, since we have n laminae as shown in Fig. 15.3.1,

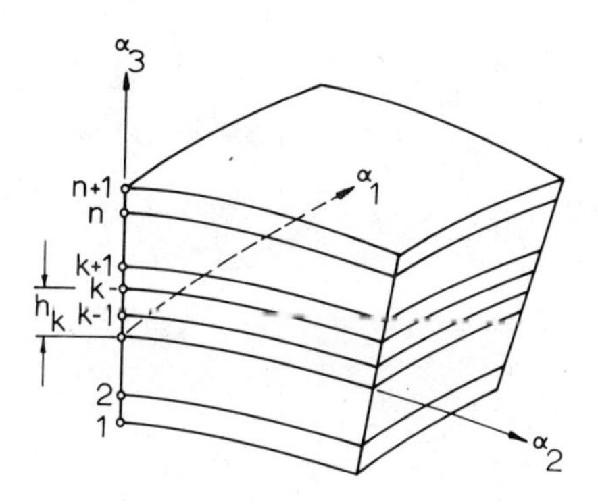

Figure 15.3.1

$$\begin{Bmatrix} N_{11} \\ N_{22} \\ N_{12} \end{Bmatrix} = \sum_{k=1}^{n} \int_{h_k}^{h_{k+1}} \begin{Bmatrix} \sigma_{11} \\ \sigma_{22} \\ \sigma_{12} \end{Bmatrix} d\alpha_3 \tag{15.3.4}$$

The subscript k is used here such that h_k defines the distance from the reference surface to the bottom surface of the k-th lamina. Substituting Eq. (15.3.2) gives

$$\begin{Bmatrix} N_{11} \\ N_{22} \\ N_{12} \end{Bmatrix} = \sum_{k=1}^{n} \left\{ [\bar{Q}]_k \begin{Bmatrix} \varepsilon^\circ_{11} \\ \varepsilon^\circ_{22} \\ \varepsilon^\circ_{12} \end{Bmatrix} \int_{h_k}^{h_{k+1}} d\alpha_3 + [\bar{Q}]_k \begin{Bmatrix} k_{11} \\ k_{22} \\ k_{12} \end{Bmatrix} \int_{h_k}^{h_{k+1}} \alpha_3 \, d\alpha_3 \right\} \tag{15.3.5}$$

This allows us to write

$$\begin{Bmatrix} N_{11} \\ N_{22} \\ N_{12} \end{Bmatrix} = [A] \begin{Bmatrix} \varepsilon^\circ_{11} \\ \varepsilon^\circ_{22} \\ \varepsilon^\circ_{12} \end{Bmatrix} + [B] \begin{Bmatrix} k_{11} \\ k_{22} \\ k_{12} \end{Bmatrix} \tag{15.3.6}$$

where

$$[A] = \sum_{k=1}^{n} [\bar{Q}]_k (h_{k+1} - h_k) \tag{15.3.7}$$

$$[B] = \frac{1}{2} \sum_{k=1}^{n} [\bar{Q}]_k (h_{k+1}^2 - h_k^2) \tag{15.3.8}$$

Each term is therefore given by

$$A_{ij} = \sum_{k=1}^{n} (\bar{Q}_{ij})_k (h_{k+1} - h_k) \tag{15.3.9}$$

$$B_{ij} = \frac{1}{2} \sum_{k=1}^{n} (\bar{Q}_{ij})_k (h_{k+1}^2 - h_k^2) \tag{15.3.10}$$

An interesting result, different from that for the isotropic material equations of Sec. 2.5, is that the stress resultants are in general also a function of the bending strains. Only if it is possible to select the reference plane such that

$$\sum_{k=1}^{n} (\bar{Q}_{ij})_k (h_{k+1}^2 - h_k^2) = 0 \tag{15.3.11}$$

do we have uncoupling. For a single lamina homogeneous and isotropic material, our classical shell case, $n = 1$, and thus we have to satisfy only

$$\bar{Q}_{ij} (h_2^2 - h_1^2) = 0 \tag{15.3.12}$$

This is done by selecting $h_2 = -h_1$, which means that the reference surface is halfway between the inner and outer surface, which implies that all B_{ij} are zero.

In general, any composite material whose laminae are homogeneous and isotropic can be made to have a zero [B] matrix. Also, composite materials that are arranged such that each lamina's orthotropic principal directions coincide with the composite material principal directions can be made to have a zero [B] matrix. In most other cases, we will find that it is impossible to find a location for the reference surface that satisfies this condition. In other words, a neutral surface does not exist for many composites.

The moment resultants are

$$\begin{Bmatrix} M_{11} \\ M_{22} \\ M_{12} \end{Bmatrix} = \int_{\alpha_3} \begin{Bmatrix} \sigma_{11} \\ \sigma_{22} \\ \sigma_{12} \end{Bmatrix} \alpha_3 \, d\alpha_3 = \sum_{k=1}^{n} \int_{h_k}^{h_{k+1}} \begin{Bmatrix} \sigma_{11} \\ \sigma_{22} \\ \sigma_{12} \end{Bmatrix}_k d\alpha_3 \tag{15.3.13}$$

Substituting Eq. 15.3.2 give

$$\begin{Bmatrix} M_{11} \\ M_{22} \\ M_{12} \end{Bmatrix} = [B] \begin{Bmatrix} \varepsilon^\circ_{11} \\ \varepsilon^\circ_{22} \\ \varepsilon^\circ_{12} \end{Bmatrix} + [D] \begin{Bmatrix} k_{11} \\ k_{22} \\ k_{12} \end{Bmatrix} \tag{15.3.14}$$

where each term in the two matrices is given by

$$B_{ij} = \frac{1}{2} \sum_{k=1}^{n} (\bar{Q}_{ij})_k (h^2_{k+1} - h^2_k) \tag{15.3.15}$$

$$D_{ij} = \frac{1}{3} \sum_{k=1}^{n} (\bar{Q}_{ij})_k (h^3_{k+1} - h^3_k) \tag{15.3.16}$$

As expected, the coupling matrix [B] is the same as before and all comments concerning its vanishing apply as before.

It is customary, to combine the expressions for force and moment resultants

$$\begin{Bmatrix} N_{11} \\ N_{22} \\ N_{12} \\ M_{11} \\ M_{22} \\ M_{12} \end{Bmatrix} = \left[\begin{array}{ccc|ccc} A_{11} & A_{12} & A_{13} & B_{11} & B_{12} & B_{13} \\ A_{21} & A_{22} & A_{23} & B_{21} & B_{22} & B_{23} \\ A_{31} & A_{32} & A_{33} & B_{31} & B_{32} & B_{33} \\ \hline B_{11} & B_{12} & B_{13} & D_{11} & D_{12} & D_{13} \\ B_{21} & B_{22} & B_{23} & D_{21} & D_{22} & D_{23} \\ B_{31} & B_{32} & B_{33} & D_{31} & D_{32} & D_{33} \end{array}\right] \begin{Bmatrix} \varepsilon^\circ_{11} \\ \varepsilon^\circ_{22} \\ \varepsilon^\circ_{12} \\ k_{11} \\ k_{22} \\ k_{12} \end{Bmatrix} \tag{15.3.17}$$

Because of the symmetry of the $\bar{Q}_{ij}$ terms, we have also that

$$\begin{aligned} A_{ij} &= A_{ji} \\ B_{ij} &= B_{ji} \\ D_{ij} &= D_{ji} \end{aligned} \tag{15.3.18}$$

15.4 EQUATION OF MOTION

If we examine the development described in Chap. 2, we note that it is not at all influenced by the fact that we have now a much more complicated relationship between strains and the force and moment resultants. Thus, Love's equations [(2.7.20) to (2.7.24)] and boundary condition expressions in force and moment resultants form are still valid. Mass densities are averaged over the thickness of the shell.

However, as soon as the equations are expressed in terms of displacements, the added complexity becomes apparent. So far, only a few special cases of composite material plate or shell eigenvalues have been obtained analytically. They are almost invariably orthotropic material structures.

15.5 THE ORTHOTROPIC PLATE

Let us, for instance, find the eigenvalues of a rectangular simply supported orthotropic plate. In this case $\alpha_1 = x$, $\alpha_2 = y$, $A_1 = 1$, $A_2 = 1$, $1/R_1 = 0$, $1/R_2 = 0$. This gives, for transverse deflection, the equations

$$-\frac{\partial Q_{x3}}{\partial x} - \frac{\partial Q_{y3}}{\partial y} + \rho h \ddot{u}_3 = q_3 \tag{15.5.1}$$

where

$$Q_{x3} = \frac{\partial M_{xx}}{\partial x} + \frac{\partial M_{xy}}{\partial y} \tag{15.5.2}$$

$$Q_{y3} = \frac{\partial M_{xy}}{\partial x} + \frac{\partial M_{yy}}{\partial y} \tag{15.5.3}$$

Setting $q_3 = 0$ and substituting gives

$$-\frac{\partial^2 M_{xx}}{\partial x^2} - 2\frac{\partial^2 M_{xy}}{\partial x \partial y} - \frac{\partial^2 M_{yy}}{\partial y^2} + \rho h \ddot{u}_3 = q_3 \tag{15.5.4}$$

For orthotropic material, with the reference plane coinciding with the neutral plane, we obtain from Eq. (15.3.17)

$$\begin{Bmatrix} N_{xx} \\ N_{yy} \\ N_{xy} \\ M_{xx} \\ M_{yy} \\ M_{xy} \end{Bmatrix} = \left[\begin{array}{ccc|ccc} A_{11} & A_{12} & 0 & 0 & 0 & 0 \\ A_{12} & A_{22} & 0 & 0 & 0 & 0 \\ 0 & 0 & A_{33} & 0 & 0 & 0 \\ \hline 0 & 0 & 0 & D_{11} & D_{12} & 0 \\ 0 & 0 & 0 & D_{12} & D_{22} & 0 \\ 0 & 0 & 0 & 0 & 0 & D_{33} \end{array}\right] \begin{Bmatrix} \varepsilon^\circ_{xx} \\ \varepsilon^\circ_{yy} \\ \varepsilon^\circ_{xy} \\ k_{xx} \\ k_{yy} \\ k_{xy} \end{Bmatrix} \tag{15.5.5}$$

Thus, for our purpose here,

$$M_{xx} = D_{11}k_{xx} + D_{12}k_{yy} \tag{15.5.6}$$

$$M_{yy} = D_{12}k_{xx} + D_{22}k_{yy} \tag{15.5.7}$$

$$M_{xy} = D_{33}k_{xy} \tag{15.5.8}$$

Substituting this in Eq. (15.5.4) gives

$$-\left(D_{11}\frac{\partial^2 k_{xx}}{\partial x^2} + D_{12}\frac{\partial^2 k_{yy}}{\partial x^2}\right) - 2D_{33}\frac{\partial^2 k_{xy}}{\partial x \partial y} - \left(D_{12}\frac{\partial^2 k_{xx}}{\partial y^2} + D_{22}\frac{\partial^2 k_{yy}}{\partial y^2}\right) + \rho h \ddot{u}_3 = q_3 \tag{15.5.9}$$

Since

$$k_{xx} = -\frac{\partial^2 u_3}{\partial x^2} \tag{15.5.10}$$

$$k_{yy} = -\frac{\partial^2 u_3}{\partial y^2} \tag{15.5.11}$$

$$k_{xy} = -2\frac{\partial^2 u_3}{\partial x \partial y} \tag{15.5.12}$$

we obtain

$$D_{11}\frac{\partial^4 u_3}{\partial x^4} + 2(D_{12} + 2D_{33})\frac{\partial^4 u_3}{\partial x^2 \partial y^2} + D_{22}\frac{\partial^4 u_3}{\partial y^4} + \rho h\ddot{u}_3 = q_3 \tag{15.5.13}$$

For a simply supported plate, the boundary conditions are

$$u_3(0,y,t) = u_3(a,y,t) = u_3(x,0,t) = u_3(x,b,t) = 0 \tag{15.5.14}$$

$$M_{xx}(0,y,t) = M_{xx}(a,y,t) = 0 \tag{15.5.15}$$

$$M_{yy}(x,0,t) = M_{yy}(x,b,t) = 0 \tag{15.5.16}$$

To solve for the eigenvalues, we set $q_3 = 0$, and

$$u_3(x,y,t) = U_3(x,y)e^{j\omega t} \tag{15.5.17}$$

where the mode shape $U_3(x,y)$ is assumed to be

$$U_3(x,y) = \sin\frac{m\pi x}{a}\sin\frac{n\pi y}{b} \tag{15.5.18}$$

This satisfies the partial differential equation and the boundary conditions. The natural frequencies turn out to be [15.2]

$$\omega_{mn} = \pi^2\sqrt{D_{11}\left(\frac{m}{a}\right)^4 + 2(D_{12} + 2D_{33})\left(\frac{m}{a}\right)^2\left(\frac{n}{b}\right)^2 + D_{22}\left(\frac{n}{b}\right)^4}\sqrt{\frac{1}{\rho h}} \tag{15.5.19}$$

Let us reduce this formula to that for a homogeneous and isotropic plate. In this case $D_{11} = D_{22} = D$, $D_{12} = \mu D$, $D_{33} = (1 - \mu)D/2$, and the result agrees with that of Sec. 5.4.2.

For a discussion of anisotropic plate vibration, see, for instance, ref. 15.3. Analytical solutions in this area are mainly of an iterative nature.

15.6 CIRCULAR CYLINDRICAL SHELL

Let us utilize the Donnell-Mushtari-Vlasov approximations. In this case we follow the procedure outlined in Sec. 6.7. After neglecting the influence of inertia in the in-plane direction and the shear term $Q_{3\theta}/a$, we obtain ($A_1 = 1$, $A_1 d\alpha_1 = dx$, $1/R_1 = 0$, $A_2 = a$, $d\alpha_2 = d\theta$, $R_2 = a$)

$$a \frac{\partial N_{xx}}{\partial x} + \frac{\partial N_{x\theta}}{\partial \theta} = 0 \tag{15.6.1}$$

$$a \frac{\partial N_{x\theta}}{\partial x} + \frac{\partial N_{\theta\theta}}{\partial \theta} = 0 \tag{15.6.2}$$

$$- a \frac{\partial Q_{x3}}{\partial x} - \frac{\partial Q_{\theta 3}}{\partial \theta} + N_{\theta\theta} + a\rho h\ddot{u}_3 = aq_3 \tag{15.6.3}$$

where

$$Q_{x3} = \frac{\partial M_{xx}}{\partial x} + \frac{1}{a} \frac{\partial M_{x\theta}}{\partial \theta} \tag{15.6.4}$$

$$Q_{\theta 3} = \frac{\partial M_{x\theta}}{\partial x} + \frac{1}{a} \frac{\partial M_{\theta\theta}}{\partial \theta} \tag{15.6.5}$$

Equations (15.6.1) and (15.6.2) are satisfied by introducing the same stress function as in Sec. 6.7

$$N_{xx} = \frac{1}{a^2} \frac{\partial^2 \phi}{\partial \theta^2} \tag{15.6.6}$$

$$N_{\theta\theta} = \frac{\partial^2 \phi}{\partial x^2} \tag{15.6.7}$$

$$N_{x\theta} = - \frac{1}{a} \frac{\partial^2 \phi}{\partial x \partial \theta} \tag{15.6.8}$$

Substituting Eqs. (15.6.4), (15.6.5), and (15.6.7) in Eq. (15.6.3) gives

$$-a \frac{\partial^2 M_{xx}}{\partial x^2} - 2 \frac{\partial^2 M_{x}}{\partial x \partial \theta} - \frac{1}{a} \frac{\partial^2 M_{\theta\theta}}{\partial \theta^2} + \frac{\partial^2 \phi}{\partial x^2} + a\rho h\ddot{u}_3 = aq_3 \tag{15.6.9}$$

Substituting

$$M_{xx} = D_{11}k_{xx} + D_{12}k_{\theta\theta} \tag{15.6.10}$$

$$M_{\theta\theta} = D_{22}k_{\theta\theta} + D_{12}k_{xx} \tag{15.6.11}$$

$$M_{x\theta} = D_{33}k_{x\theta} \tag{15.6.12}$$

give

$$-\left(D_{11}\frac{\partial^2 k_{xx}}{\partial x^2} + D_{12}\frac{\partial^2 k_{\theta\theta}}{\partial x^2}\right) - \frac{2D_{33}}{a}\frac{\partial^2 k_{xy}}{\partial x \partial\theta}$$

$$-\frac{1}{a^2}\left(D_{12}\frac{\partial^2 k_{xx}}{\partial\theta^2} + D_{22}\frac{\partial^2 k_{\theta\theta}}{\partial\theta^2}\right) + \frac{1}{a}\frac{\partial^2\phi}{\partial x^2} + \rho h\ddot{u}_3 = q_3 \tag{15.6.13}$$

Substituting

$$k_{xx} = -\frac{\partial^2 u_3}{\partial x^2} \tag{15.6.14}$$

$$k_{\theta\theta} = -\frac{1}{a^2}\frac{\partial^2 u_3}{\partial\theta^2} \tag{15.6.15}$$

$$k_{x\theta} = -\frac{2}{a}\frac{\partial^2 u_3}{\partial x\partial\theta} \tag{15.6.16}$$

results in the final equation

$$D_{11}\frac{\partial^4 u_3}{\partial x^4} + 2(D_{12} + 2D_{33})\frac{1}{a^2}\frac{\partial^4 u_3}{\partial x^2\partial\theta^2} + D_{22}\frac{1}{a^4}\frac{\partial^4 u_3}{\partial\theta^4}$$

$$+\frac{1}{a}\frac{\partial^2\phi}{\partial x^2} + \rho h\ddot{u}_3 = q_3 \tag{15.6.17}$$

Examining as a check the homogeneous and isotropic case, where

$$D_{11} = D_{22} = D \tag{15.6.18}$$

$$D_{12} = \mu D \tag{15.6.19}$$

$$D_{33} = (1 - \mu)D/2 \tag{15.6.20}$$

and where $D = Eh^3/12(1 - \mu^2)$, we obtain the first Donnell-Mushtari-Vlasov equation

$$D\nabla^4 u_3 + \frac{1}{a^2} \frac{\partial^2 \phi}{\partial x^2} + \rho h \ddot{u}_3 = q_3 \tag{15.6.21}$$

Next, we start with the compatibility equation (6.7.12).

$$\frac{k_{xx}}{a} + \frac{\partial^2 \varepsilon^\circ_{\theta\theta}}{\partial x^2} - \frac{1}{a} \frac{\partial^2 \varepsilon^\circ_{x\theta}}{\partial\theta\partial x} + \frac{1}{a^2} \frac{\partial^2 \varepsilon^\circ_{xx}}{\partial\theta^2} = 0 \tag{15.6.22}$$

Since

$$N_{xx} = A_{11}\varepsilon^\circ_{xx} + A_{12}\varepsilon^\circ_{\theta\theta} \tag{15.6.23}$$

$$N_{\theta\theta} = A_{12}\varepsilon^\circ_{xx} + A_{22}\varepsilon^\circ_{\theta\theta} \tag{15.6.24}$$

$$N_{x\theta} = A_{33}\varepsilon^\circ_{x\theta} \tag{15.6.25}$$

We obtain

$$\varepsilon^\circ_{xx} = P_{11}N_{xx} - P_{12}N_{\theta\theta} \tag{15.6.26}$$

$$\varepsilon^\circ_{\theta\theta} = P_{22}N_{\theta\theta} - P_{12}N_{xx} \tag{15.6.27}$$

$$\varepsilon^\circ_{x\theta} = P_{33}N_{x\theta} \tag{15.6.28}$$

where

$$P_{11} = \frac{A_{22}}{\alpha} \tag{15.6.29}$$

$$P_{22} = \frac{A_{11}}{\alpha} \tag{15.6.30}$$

$$P_{12} = \frac{A_{12}}{\alpha} \tag{15.6.31}$$

$$\alpha = A_{11}A_{22} - A_{12}^2 \tag{15.6.32}$$

$$P_{33} = \frac{1}{A_{33}} \tag{15.6.33}$$

Substitution gives

$$\frac{k_{xx}}{a} + P_{22}\frac{\partial^2 N_{\theta\theta}}{\partial x^2} - P_{12}\frac{\partial^2 N_{xx}}{\partial x^2} - \frac{1}{a}P_{33}\frac{\partial^2 N_{x\theta}}{\partial\theta\partial x}$$

$$+\frac{1}{a^2}P_{11}\frac{\partial^2 N_{xx}}{\partial\theta^2} - \frac{1}{a^2}P_{12}\frac{\partial^2 N_{\theta\theta}}{\partial\theta^2} = 0 \qquad (15.6.34)$$

Substituting next Eqs. (15.6.6) to (15.6.8) and Eq. (15.6.14) gives

$$-\frac{1}{a}\frac{\partial^2 u_3}{\partial x^2} + P_{22}\frac{\partial^4\phi}{\partial x^4} + P_{11}\frac{1}{a^4}\frac{\partial^4\phi}{\partial\theta^4} + \frac{1}{a^2}(P_{33} - 2P_{12})\frac{\partial^4\phi}{\partial x^2\partial\theta^2} = 0$$

$$(15.6.35)$$

This may also be written as

$$\frac{A_{12}^2 - A_{11}A_{22}}{a}\frac{\partial^2 u_3}{\partial x^2} + A_{11}\frac{\partial^4\phi}{\partial x^4} + \frac{A_{22}}{a^4}\frac{\partial^4\phi}{\partial\theta^4}$$

$$+\frac{A_{11}A_{22} - A_{12}^2 - 2A_{12}A_{33}}{A_{33}a^2}\frac{\partial^4\phi}{\partial x^2\partial\theta^2} = 0 \qquad (15.6.36)$$

In order to check this equation, we examine again the isotropic case:

$$A_{11} = A_{22} = K \qquad (15.6.37)$$

$$A_{12} = \mu K \qquad (15.6.38)$$

$$A_{33} = (1 - \mu)K/2 \qquad (15.6.39)$$

where $K = Eh/(1 - \mu^2)$. This gives

$$\frac{Eh}{a}\frac{\partial^2 u_3}{\partial x^2} - \nabla^4\phi = 0 \qquad (15.6.40)$$

As expected, this is the second Donnell-Mushtari-Vlasov equation.

Let us now find the natural frequencies and modes of a closed circular shell, simply supported at both ends. The boundary conditions are

$$u_3(0,\theta,t) = u_3(L,\theta,t) = 0 \qquad (15.6.41)$$

$$M_{xx}(0,\theta,t) = M_{xx}(L,\theta,t) = 0 \tag{15.6.42}$$

We are able to satisfy both of these boundary conditions and the two governing equations, Eqs. (15.6.17) and (15.6.36), by

$$u_3(x,\theta,t) = U_3(x,\theta)e^{j\omega t} \tag{15.6.43}$$

$$\phi(x,\theta,t) = \Phi(x,\theta)e^{j\omega t} \tag{15.6.44}$$

where

$$U_3(x,\theta) = U_{mn} \sin\frac{m\pi x}{L} \cos n(\theta - \phi) \tag{15.6.45}$$

$$\Phi(x,\theta) = \Phi_{mn} \sin\frac{m\pi x}{L} \operatorname{cox} n(\theta - \phi) \tag{15.6.46}$$

Equation (15.6.17) becomes

$$\left[D_{11}\left(\frac{m\pi}{L}\right)^4 + 2(D_{12} + 2D_{33})\left(\frac{n}{a}\right)^2\left(\frac{m\pi}{L}\right)^2 + D_{22}\left(\frac{n}{a}\right)^4 - \rho h\omega^2\right]U_{mn} - \frac{1}{a}\left(\frac{m\pi}{L}\right)^2 \Phi_{mn} = 0 \tag{15.6.47}$$

and Eq. (15.6.36) becomes

$$\frac{A_{11}A_{22} - A_{12}^2}{a}\left(\frac{m\pi}{L}\right)^2 U_{mn} + \left[A_{11}\left(\frac{m\pi}{L}\right)^4 + A_{22}\left(\frac{n}{a}\right)^4 + \frac{A_{11}A_{22}-A_{12}^2-2A_{12}A_{33}}{A_{33}}\left(\frac{n}{a}\right)^2\left(\frac{m\pi}{L}\right)^2\right]\Phi_{mn} = 0 \tag{15.6.48}$$

For these two equations to be meaningfully satisfied, the determinant has to be equal to zero. This gives us the natural frequencies of the orthotropic shell for those modes where transverse deflection components dominate:

$$\omega^2 = \omega_{mn}^2 = \frac{1}{\rho h}\left(\left[D_{11}\left(\frac{m\pi}{L}\right)^4 + 2(D_{12} + 2D_{33})\left(\frac{n}{a}\right)^2\left(\frac{m\pi}{L}\right)^2 + D_{22}\left(\frac{n}{a}\right)^4\right] + \frac{(A_{11}A_{22}-A_{12}^2)\left(\frac{m\pi}{L}\right)^4}{a^2\left\{A_{11}\left(\frac{m\pi}{L}\right)^4 + A_{22}\left(\frac{n}{a}\right)^4 + \frac{A_{11}A_{22} - A_{12}^2 - 2A_{12}A_{33}}{A_{33}}\left(\frac{n}{a}\right)^2\left(\frac{m\pi}{L}\right)^2\right\}}\right) \tag{15.6.49}$$

The result shows that the circumferential bending stiffness component D_{22} gains influence with increasing values of n while the axial bending stiffness component D_{11} increases its influence as m increases. A similar influence division exists for the membrane stiffness terms A_{11} and A_{22}. The structure of the formula shows clearly that if it is desired to raise the natural frequencies of the shell in general, both stringer and ring stiffeners have to be employed, since it is permissible to think of stiffeners as making an isotropic shell orthotropic. Neither stringers nor rings alone can raise the natural frequencies for all m,n combinations.

Let us check Eq. (15.6.49) against the isotropic case treated in Sec. 6.12. Substituting Eqs. (15.6.17) to (15.6.19) and (15.6.37) to (15.6.39) gives, as expected,

$$\omega_{mn}^2 = \frac{E}{\rho a^2}\left\{\frac{\left(\frac{m\pi a}{L}\right)^4}{\left[\left(\frac{m\pi a}{L}\right)^2 + n^2\right]^2} + \frac{\left(\frac{h}{a}\right)^2}{12(1-\mu^2)}\left[\left(\frac{m\pi a}{L}\right)^2 + n^2\right]^2\right\} \tag{15.6.50}$$

A treatment of an orthotropic circular cylindrical shell that does not utilize the above simplifications is given in Ref. 15.4. A literature review of orthotropic cylindrical and conical shell eigenvalue solutions can be found in Ref. 15.5.

REFERENCES

15.1 T. E. Ashton, T. C. Halpin, and P. H. Petit, *Primer on Composite Materials Analysis,* Technomic, Stanford, Conn., 1969.

15.2 R. F. S. Hearmon, "The frequency of flexural vibration of rectangular orthotropic plates with clamped or supported edges," *J. Appl. Mec., vol. 26,* no. 3-4, 1959, pp. 537-540.

15.3 R. M. Jones, *Mechanics of Composite Materials,* McGraw-Hill, New York, 1975.

15.4 S. B. Dong, "Free vibration of laminated orthotropic cylindrical shells," *J. Acoustical Soc. Amer., vol. 44,* no. 6, 1968, pp. 1628-1635.

15.5 A. W. Leissa, *Vibration of Shells,* NASA SP-288, U. S. Government Printing Office, Washington, D.C., 1973.

INDEX